Lecture Notes in Computer Science 4621

Commenced Publication in 1973
Founding and Former Series Editors:
Gerhard Goos, Juris Hartmanis, and Jan van Leeuwen

Dorothea Wagner   Roger Wattenhofer (Eds.)

Algorithms for Sensor and Ad Hoc Networks

Advanced Lectures

 Springer

Volume Editors

Dorothea Wagner
University of Karlsruhe
Faculty of Informatics, ITI Wagner
Box 6980, 76128 Karlsruhe, Germany
E-mail: wagner@iti.uni-karlsruhe.de

Roger Wattenhofer
ETH Zurich
Distributed Computing Group
Gloriastrasse 35, 8092 Zurich, Switzerland
E-mail: wattenhofer@tik.ee.ethz.ch

Library of Congress Control Number: 2007935200

CR Subject Classification (1998): C.2, F.2, D.2, G.3-4, I.2.11, C.4, I.6, K.6.5, D.4.6

LNCS Sublibrary: SL 1 – Theoretical Computer Science and General Issues

ISSN   0302-9743

ISBN   978-3-540-74990-5 Springer Berlin Heidelberg New York

Springer is a part of Springer Science+Business Media

springer.com

Typesetting: Camera-ready by author, data conversion by Markus Richter, Heidelberg
Printed on acid-free paper   SPIN: 12124572   06/3180   5 4 3 2 1 0

Preface

Thousands of mini computers (comparable to a stick of chewing gum in size), equipped with sensors, are deployed in some terrain or other. After activation the sensors form a self-organized network and provide data, for example about a forthcoming earthquake.

The trend towards wireless communication increasingly affects electronic devices in almost every sphere of life. Conventional wireless networks rely on infrastructure such as base stations; mobile devices interact with these base stations in a client/server fashion. In contrast, current research is focusing on networks that are completely unstructured, but are nevertheless able to communicate (via several hops) with each other, despite the low coverage of their antennas. Such systems are called *sensor* or *ad hoc networks*, depending on the point of view and the application.

Wireless ad hoc and sensor networks have gained an incredible research momentum. Computer scientists and engineers of all flavors are embracing the area. Sensor networks have been adopted by researchers in many fields: from hardware technology to operating systems, from antenna design to databases, from information theory to networking, from graph theory to computational geometry.

Both the tremendous growth of the subject and the broad interdisciplinary community make research progress in wireless ad hoc and sensor networking incredibly difficult to follow and apprehend. What are the key concepts of wireless multi-hop networks? Which of the fundamentals that will still be valid in 10 or 20 years' time? What are the main techniques, and why do they work?

This book is a naïve snapshot of the current research on wireless ad hoc and sensor networks. Whenever possible, we focus on algorithmic results, that is, algorithms and protocols that allow for an analysis of their efficiency and efficacy. Often these algorithmic results are complemented by lower bounds, showing that some problems cannot be solved in a better way. Many of our chapters deal with distributed algorithms, in particular local and localized algorithms. Many problems are of an inherently distributed nature, as nodes locally sense data, or locally need to decide on a media access scheme.

Our survey is by no means complete; some topics presented in the book may be identified as wrong paths in a few years' time, and other important aspects might be missing. Most likely, some topics have made it into the book because of their *algorithmic* beauty rather than their *practical* importance. Not surprisingly, one might add. After all, these topics were either proposed or selected (from a much larger list of topics) by the participants of a *GI-Dagstuhl Seminar* held in Dagstuhl on November 23-25, 2005.

The idea of the GI-Dagstuhl Seminars is to provide young researchers with the opportunity to become actively involved in new relevant and interesting areas of computer science. Based on a list of topics and references offered by the organizers, the participants prepared overview lectures that were presented and discussed at the research seminar in Dagstuhl. Each chapter was then elaborated and carefully cross-reviewed by the participants. Although we are aware that further progress has been made since this book was written, we hope to provide at least a first overview of algorithmic results in the field, making the book a suitable basis for an advanced course.

It is our pleasure to thank the young researchers who put a huge amount of work into this book, not only as authors of the chapters but also as reviewers of other chapters. Special thanks go to Steffen Mecke and Frank Schulz who invested a lot of administrative work in the preparation of the seminar. Moreover, Steffen Mecke handled most of the technical parts of the editing process. Finally, we would like to thank the *Gesellschaft für Informatik e.V.* (GI) and *IBFI Schloss Dagstuhl* for supporting this book and the GI-Dagstuhl Seminar.

April 2007   Dorothea Wagner
                                                             Roger Wattenhofer

List of Contributors

Zinaida Benenson
Department of Computer Science
University of Mannheim
A 5,6
68159 Mannheim, Germany
zina@uni-mannheim.de

Erik-Oliver Blaß
Institute of Telematics
University of Karlsruhe
Zirkel 2
76128 Karlsruhe, Germany
blass@tm.uka.de

Kevin Buchin
Institut für Informatik
Freie Universität Berlin
Takustraße 9
14195 Berlin, Germany
buchin@inf.fu-berlin.de

Maike Buchin
Institut für Informatik
Freie Universität Berlin
Takustraße 9
14195 Berlin, Germany
mbuchin@inf.fu-berlin.de

Erik Buchmann
IPD Böhm
Universität Karlsruhe
Am Fasanengarten 5
76128 Karlsruhe, Germany
buchmann@ipd.uni-karlsruhe.de

Marcel Busse
Universität Mannheim
A5, 6
68159 Mannheim, Germany
busse@informatik.uni-mannheim.de

Michael Dom
Institut für Informatik
Friedrich-Schiller-Universität Jena
Ernst-Abbe-Platz 2
07743 Jena, Germany
dom@minet.uni-jena.de

Benjamin Fabian
Institute of Information Systems
Humboldt-Universität Berlin
Spandauer Straße 1
10178 Berlin, Germany
bfabian@wiwi.hu-berlin.de

Matthias Fischmann
Institute of Information Systems
Humboldt-Universität Berlin
Spandauer Straße 1
10178 Berlin, Germany
fis@wiwi.hu-berlin.de

Seda F. Gürses
Institute of Information Systems
Humboldt-Universität Berlin
Spandauer Straße 1
10178 Berlin, Germany
seda@wiwi.hu-berlin.de

Daniel Fleischer
Fachbereich Informatik/Inf.-Wiss.
Universität Konstanz
Fach D 67
78457 Konstanz, Germany
Daniel.Fleischer@uni-konstanz.de

Christian Frank
ETH Zurich
Clausiusstr. 59, IFW
8092 Zurich, Switzerland
chfrank@inf.ethz.ch

Christian Gunia
Institute of Computer Science
University of Freiburg
Georges-Köhler-Allee 79
79110 Freiburg, Germany
gunia@informatik.uni-freiburg.de

Hans-Joachim Hof
Institute of Telematics
University of Karlsruhe
Zirkel 2
76128 Karlsruhe, Germany
hof@tm.uka.de

Alexander Kröller
Institute for
Mathematical Optimization
TU Braunschweig
Pockelsstraße 14
38106 Braunschweig, Germany
a.kroeller@tu-bs.de

Olaf Landsiedel
Distributed Systems Group
RWTH Aachen
Ahornstraße 55
52074 Aachen, Germany

Steffen Mecke
ITI Wagner
Universität Karlsruhe
Am Fasanengarten 5
76128 Karlsruhe, Germany
mecke@ira.uka.de

Thomas Moscibroda
Microsoft Research
One Microsoft Way
Redmond, WA 98052, USA
moscitho@microsoft.com

Christian Pich
Universität Konstanz
Fachbereich Informatik/Inf.-Wiss.
Fach D 67
78457 Konstanz, Germany
Christian.Pich@uni-konstanz.de

Leonid Scharf
Institut für Informatik
Freie Universität Berlin
Takustraße 9
14195 Berlin, Germany
leo@llscharf.de

Ludmila Scharf
Institut für Informatik
Freie Universität Berlin
Takustraße 9
14195 Berlin, Germany
scharf@inf.fu-berlin.de

Thilo Streichert
Dept. of Computer Science 12
Friedrich-Alexander-University
Erlangen-Nuremberg
Am Weichselgarten 3
91058 Erlangen, Germany
streichert@cs.fau.de

Frank Schulz
ITI Wagner
Universität Karlsruhe
Am Fasanengarten 5
76128 Karlsruhe, Germany
frankschulz@gmx.de

Dorothea Wagner
Universität Karlsruhe
ITI Wagner
Am Fasanengarten 5
76128 Karlsruhe, Germany
wagner@iti.uni-karlsruhe.de

Roger Wattenhofer
ETH Zürich
Distributed Computing Group
Gloriastrasse 35
8092 Zurich, Switzerland
wattenhofer@tik.ee.ethz.ch

Aaron Zollinger
ETH Zürich / UC Berkeley
zollinger@tik.ee.ethz.ch

Contents

1

Applications of Sensor Networks

Hans-Joachim Hof

1.1 Introduction

Networks of small sensor nodes, so-called sensor networks, allow to monitor and analyze complex phenomena over a large region and for a long period of time. Recent advances in sensor network research allow for small and cheap sensor nodes which can obtain a lot of data about physical values, e.g. temperature, humidity, lightning condition, pressure, noise level, carbon dioxide level, oxygen level, soil makeup, soil moisture, magnetic field, current characteristics of objects such as speed, direction, and size, the presence or absence of certain kinds of objects, and all kinds of values about machinery, e.g. mechanical stress level or movement. This huge choice of options allow to use sensor network applications in a number of scenarios, e.g. habitat and environment monitoring, health care, military surveillance, industrial machinery surveillance, home automation, als well as smart and interactive places. The application of a sensor network usually determines the design of the sensor nodes and the design of the network itself. No general architecture for sensor networks exists at the moment. This chapter gives an overview of existing sensor network applications, shows some currently available sensor network hardware platforms and gives an outlook on upcoming applications for sensor networks.

A sensor network can be of great benefit when used in areas where dense monitoring and analysis of complex phenomena over a large region is required for a long period of time. The design criteria and requirements of sensor networks differ from application to application. Some typical requirements are:

- Failure resistance: Sensor networks are prone to node failure. Thus, it is crucial that algorithms for sensor networks can deal with node failures in an efficient way.
- Scalability: As sensor nodes get cheaper and cheaper, it is highly likely that the sensor networks consist of a huge number of nodes. Algorithms

D. Wagner and R. Wattenhofer (Eds.): Algorithms for Sensor and Ad Hoc Networks, LNCS 4621, pp. 1–20, 2007.

for sensor networks must therefore scale well with thousands and tens of thousands of sensor nodes.

- Simplicity: Due to the design of sensor nodes, the available resources for algorithms of the application layer are severely restricted. Algorithms for sensor networks must be very efficient considering computation cycles and memory usage.
- Unattended operation: Sensor networks are usually deployed in unattended areas, or administration tasks should be kept at a minimum. It is therefore necessary for an algorithm for sensor networks to work unattended after the deployment of the sensor node. However, it is usually no problem to configure sensor nodes before or during deployment.
- Dynamic: Sensor nodes must adapt to changes of the environment, e.g. changed connectivity or changing environmental stimuli.

This paper is structured as follows: First, some applications of sensor networks are presented. Second, some common sensor network hardware platforms are discussed. The paper ends with an outlook on upcoming applications.

1.2 Applications of Sensor Networks

In this chapter, current sensor network applications are shown. The chapter covers habitat monitoring, environment monitoring, health care, and industrial applications.

1.2.1 Habitat Monitoring

The main objective of habitat monitoring is to track and observe wild life. In the past, habitat monitoring has been done by researchers hiding and observing the wild life, or cameras were used for observations. However, these techniques are intrusive and uncomfortable, and long term observations are difficult and expensive. Usually, live data is not available. Sensor networks offer a better way for habitat monitoring. The sensor network technology is less intrusive than any other technique. Thus, the wild life is less affected, resulting in better research results. Also, long-term observations are possible and sensor networks can be designed in a way that live data is available on the Internet. Human interaction is usually needed only for setup of the sensor network and for removal of the sensors after the end of the observation. Hence, sensor networks help to reduce the costs of habitat monitoring research projects. Two examples of habitat monitoring, the Great Duck Island project and ZebraNet, are presented in the following.

Great Duck Island. The *Great Duck Island project* [273] was one of the first applications of sensor networks in habitat monitoring research. The main objective of the research project was to monitor the micro climates (e.g. temperature and humidity) in and around nesting burrows used by the Leach's

Storm Petrel. The great advantage of this sensor network compared to standard habitat monitoring was that it is non-intrusive and non-disruptive. The project is named after a small island at the coast of Maine, Great Duck Island, where the research took place. At first, a network of 32 sensor nodes was deployed (see Figure 1.1). The sensor network platform consists of processor radio boards commonly referred to as motes. MICA Motes (see Chapter 1.3.1) were used as sensor nodes. They were manually placed in the nesting burrows by researchers. The sensor nodes periodically sent their sensor readings to a base station and got back to sleep mode. The base station used a satellite link to offer access to real-time data over the Internet. To get information about the micro climate in the nesting burrows, the sensor nodes collected data about temperature, humidity, barometric pressure and mid-range infrared. Between spring 2002 and November 2002, over 1 million of sensor readings were logged from the sensor network. In June 2003, a larger sensor network, consisting of 56 nodes, was deployed, and it was extended in July 2003 by 49 additional sensor nodes and again augmented by 60 more sensor nodes and 25 weather station nodes in August 2003. Hence, the network consisted of more than 100 sensor nodes at the end of 2003. The network used multi-hop routing from the nodes to the base station. The software of the sensor nodes was based on the sensor network operating system TinyOS [364]. The sensor network of the Great Duck Island project was preconfigured and did not self configure, e.g. each sensor node got assigned a unique network layer address during compilation of the code prior to deployment.

Fig. 1.1. Placement of sensor nodes in the Great Duck Island project: sensor nodes were placed in the nesting burrows of storm petrels (1) and outside of the burrow (2). Sensor readings are relayed to a base station (3), which transmits them to a laptop in the research station (4), that sends it via satellite (5) to a lab in California. Image source: http://www.wired.com/wired/archive/11.12/network.html

ZebraNet. Another habitat monitoring project that uses a sensor network is *ZebraNet* [403]. The goal of the ZebraNet research project is to understand the migration patterns of zebras and how these patterns are affected by changes in environmental conditions like the weather or plant life. While the Great Duck Island project observed the nestings of birds and not the birds, in ZebraNet the sensor nodes are attached to the animals themselves. Those sensor nodes are integrated into a collar. For deployment of the sensor nodes, zebras are caught using a tranquilizer gun. The sensor collar is then mounted at the necks of the unconscious zebras. The design of the sensor nodes is limited by the maximum weight that a zebra can carry on its neck without being severely hindered. Figure 1.2 shows a zebra equipped with such a sensor collar. The design of the sensor nodes and its deployments make the ZebraNet project different from the Great Duck Island project: while in the first project the sensor nodes are mobile, they are fixed in the second project. This also has implications for the design of the base station. As zebras have a quite large territory and the sensor nodes have a limited communication range, it is not guaranteed that a sensor node has always a communication path to the base station. Therefore, the base station of ZebraNet is also mobile. It can be realized by a jeep which regularly drives through the zebras territory, or by a plane that regularly flies over the territory.

The sensor nodes of ZebraNet collect biometric data of the zebras, e.g. heart rate, body temperature, and the frequency of feeding. The main focus of the research project lies on the migration pattern of zebras. Therefore, the sensors acquire a GPS position regularly. Latency is not an issue in ZebraNet,

Fig. 1.2. Zebra with mounted sensor collar. Image source: `http://web.princeton.edu/sites/pumpala/pumpala/princeton%20research/princeton_research.htm`

but reliable delivery of data is important. ZebraNet's sensor nodes send data to peer sensor nodes, and each sensor node stores its own sensor data and those data of other sensors that were in range in the past. They upload data to a mobile base station whenever it is available. However, as data is swapped from node to node, it is not necessary to encounter all animals to get full data coverage. Sensor nodes use two communication modules: one for short-range transmission and one for long-range transmission. Of course, the short-range transmitter is much more energy-efficient than the long-range transmitter. The sensor collars are equipped with a solar panel to recharge the battery of the sensor nodes. However, the system is designed in such a way that a sensor node can run without any recharge for 5 days. The sensor nodes of ZebraNet use a complex duty cycle to allow a network lifetime of one year:

- 30 GPS position samples are taken per hour for 24 hours a day.
- A detailed zebra activity log is taken for 3 minutes every hour.
- The sensor nodes search for other sensor nodes in their direct neighborhood for a total of 6 hours per day. In this time, data is transfered between neighbor nodes using low-power short-range radio.
- The sensor nodes search for the mobile base station for a total of 3 hours a day. However, the time interval a sensor node searches for the base station overlaps with the time interval the sensor node searches for other sensor nodes in the direct neighborhood. To search for the base station, long-range radio communication is used.
- During a 5 day period, a total of 640 kilobytes of data is sent to the mobile base station using long-range radio communication.

Data storage is limited, therefore there is a prioritisation when to delete data: First, a node deletes data obtained from other nodes. Then, it deletes its own data, starting with the data with the oldest timestamps. If data is transmitted to the base station, it is registered in a delete list. The delete list is also communicated to peer sensor nodes so they know which data they can erase. All incoming data is compared to the delete list to avoid storage of data which is already known to the base station.

1.2.2 Environment Monitoring

While habitat monitoring deals with observation of animals and their surroundings, the task of environmental monitoring is to sense the state of the environment. Sensor networks in environment monitoring are extremely helpful in biology research and they allow exploration of areas that were not accessible up to date. In the past, a small number of wired sensors was used in biology. With a sensor network consisting of a large number of nodes, it is possible to explore the macrocosmos in a much better way because it is easier to obtain data of natural phenomena at many different places. Hence, better prediction models can be build. Other applications of environmental monitoring include structural monitoring, which ensures the integrity of bridges,

Fig. 1.3. Berkeley motes in a redwood tree. Image source: `http://www.cs.berkeley.edu/~culler/talks/mobihoc.ppt`

buildings, and other man-made structures. In the following, four environment monitoring research projects are presented.

Redwood Ecophysiology. A lot of research has been done to research the micro climates of forests. However, past research work obtained data either at the ground or above the canopy of the trees. The goal of the *Redwood Ecophysiology* project [92] is to learn about the ecophysiology (ecophysiology studies the interaction of physiological traits with the abiotic environment) of redwood trees, especially about the micro climate of the redwood trees between the ground and the canopy. They chose to use a sensor network for this research project. The sensor network was deployed by manually placing sensor nodes on different elevations of a number of redwood trees. Each tree was equipped with 40 to 50 sensor nodes. Figure 1.3 shows some sensor nodes in a redwood tree. The sensors measured sapflow, temperature, and humidity. The sensor nodes used a simple sleep schedule: each single sensor reported its sensor data every five minutes to a data logger and it was inactive the rest of the time. The experiment lasted for 25 days. Figure 1.4 shows a schematic view of the node placement.

Meteorology and Hydrology in Yosemite National Park. In [269], a sensor network was deployed to monitor the water system across and within the Sierra Nevada, especially regarding natural climate fluctuation, global warming, and the growing needs of water consumers. Research of the water system in the Sierra Nevada is difficult, because the monitoring takes place in severe terrain with limited access. Therefore, the sensor network was designed

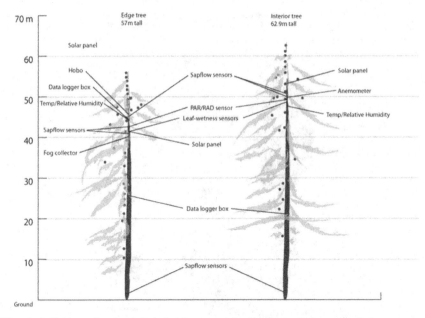

Fig. 1.4. Schematic view of node placement in the Redwood ecophysiology project. Image source: `http://www.cs.berkeley.edu/~culler/talks/mobihoc.ppt`)

to operate with little or no human interaction. A prototype network of meteorological and hydrological sensors was deployed in Yosemite National Park in different elevation zones ranging from 1200 to 3700 meters. The network used different communication technologies to fit in the area: radio, cell-phone, landline, and satellite transmissions. Twenty sensors were installed in basins. They hourly reported water level and temperature of the water. Also, 25 meteorological stations were installed, which measured air temperature and relative humidity. It turned out that communication was the major problem in areas like the Yosemite National Park because the park has a high relief. For the project's success it was important to monitor the environment conditions in river valleys where the communication conditions are especially challenging. Also, battery life and long-term data logger backup were major design issues. Low power, low cost data loggers were important parts of the sensor network. These loggers were equipped with 32 MB of memory. Energy was the limiting factor, and solar panels were out of scope, because their large, reflecting surface conflicts with wildlife protection.

GlacsWeb. Glaciers are very important in climate research because climate changes can be identified by observing size and movement of the glaciers. *Glacs Web* [280] is a sensor network for monitoring glaciers. A sample network was installed at Briksdalsbreen, Norway, in 2004. The aim of GlacsWeb is to understand glacier dynamics in response to climate change. Sensor nodes (see

Fig. 1.5. Basestation of GlacsWeb. Image source: `http://envisense.org/` `glacsweb/norway04/index.html`

Figure 1.6) were deployed into the ice and on the till. A base station (see Figure 1.5) was installed on the surface of the glacier. The sensor network was designed to acquire data about weather conditions and the GPS position at the base station from the surface of the glacier. The sensor nodes used a simple duty cycle. Every sensor node sampled every 4 hours temperature, strain (due to stress from the ice), pressure (if immersed in water), orientation (in 3 dimensions), resistivity (to determine if the sensor is sitting in sediment till, water, or ice) and battery voltage. Thus, there were 6 sets of readings for each sensor node every day. The sensor nodes directly communicated with the base station once a day at a fixed time. All sensor nodes were queried during a 5 minutes interval each day. The base station recorded its location once a week using GPS. Obtaining the GPS position took 10 minutes. During those 10 minutes, remote administration of the base station was also possible. After this, the base station sent all its readings to a reference station PC via long range radio. The radio range of the sensor nodes was significantly reduced in ice to less than 40 m while it is 500 meters in air. The sensor nodes had a clock drift of 2 seconds a day. The base station synchronized them daily. The base station was equipped with a solar panel which failed when the panels

Fig. 1.6. A sensor node of GlacsWeb. Image source: http://radio.weblogs. com/0105910/2004/05/31.html

were covered with snow. Future versions of the base station will also have a wind turbine.

The Four Seasons Project. Structure health monitoring of human-made structures is especially important in areas with frequent earthquakes. California is such an area. One of the buildings damaged during the 1994 earthquake in California is the Four Seasons building in Sherman Oaks (see Figure 1.7). This building had been used in the *Four Seasons* project [394] for structural health monitoring research. A sensor network had been deployed in the building to find out what the reasons for the observed damages were. The building is scheduled for demolition and was used for a series of force-vibration tests. MICA Motes (see Chapter 1.3.1) were used for the sensor nodes. They acquired vibration data at different locations in the building. Figure 1.8 shows one of the vibration sensors. The sensor nodes organized themselves in a large sensor network and sent their data to a base station. Special effort was made to have reliable data transport. Another problem was data compression. As the Four Seasons project is a data acquisition system which needs high accuracy, a lossy compression scheme could not be used. Thus, an algorithm was used that encodes sequences of samples whose values lie within a small range (a so-called silence period) in "run-length code", meaning that one sample value and the number of samples is given. This compression scheme is easy to compute and in the cases where there is a lot of activity, data is given totally accurate. Another problem was synchronization of vibration data collected by different nodes. Achieving global time synchronization would have involved too much effort. Thus, the total time of the packages in the network is calculated. The algorithm for time synchronization takes as granted that the time a node needs to handle a package is multiple attitudes higher than the time delay of transmission. Hence, each node added the time it needed for proceeding a package to a header field.

Fig. 1.7. Four Seasons Building, which was equipped with sensor nodes for structural health monitoring. Image source: `http://research.cens.ucla.edu/pls/portal/docs/1/12805.JPG`

Fig. 1.8. One of the vibration sensors used in the Four Seasons Building. Image source: `http://research.cens.ucla.edu/pls/portal/docs/1/12821.JPG`

1.2.3 Health Care

In health care, especially in hospitals, a lot of sensors are used today to monitor the vital signs and states of patients 24 hours a day. Most of those sensors are wired, so patients are often severely restricted in their mobility. Hence, the use of wireless sensors would result in a great gain in life quality for the patients and it even would give them more security, because an automated system may react fast on emergency situations. Another issue in health care is the seamless transfer of patients between first aid personal, emergency rescuers and doctors at a hospital, so that a seamless monitoring and data transfer can take place. Sensor networks can solve a lot of problems in health care scenarios. The most important design criteria for applications in health care are security and reliability.

CodeBlue. *CodeBlue* [263] is a system to enhance emergency medical care with the goal to have a seamless patient transfer between a disaster area and a hospital. Wearable vital sign sensors (body sensors) are used to track a patient's status and location. They also operated as active tags, which enhance first responders ability to assess patients on the scene, ensure seamless transfer of data among emergency personal, and enable an efficient allocation of hospital resources. Current body sensors are able to obtain heart rate, oxygen saturation, end-tidal CO_2, and serum chemistries measurements. The body sensors of the CodeBlue project are based on MICA2 Motes (see Chapter 1.3.1). Figure 1.9 shows a typical body sensor node. The network of Code-Blue does not rely on any infrastructure but is formed ad hoc. It is intended to span a large disaster area or a whole building, e.g. a hospital. The system is designed to scale with a very dense network consisting of lots of nodes. The nodes of the network are heterogeneous and range from simple body sensors and PDAs to PCs. CodeBlue addresses data discovery and data naming, robust routing, prioritisation of critical data, security, and tracking of device locations. For data discovery, a publish/subscribe-model is used. Body sensors publish its data and devices used by nurses or doctors, e.g. a PDA, can subscribe to that data stream. CodeBlue also uses data filtering and data aggregation. For example, a doctor may be interested in the full data stream of one patient, and wants only to be notified if some other patients are in an unusual or critical state. The CodeBlue project uses a best-effort security model: if an external authority can be reached, strong guarantees are needed to access data. However, if no external authorities are available because of poor connectivity or infrastructure loss, weaker guarantees may be used. This situation may arise for example in a disaster area where it is not possible to spend time setting up an infrastructure, keying in passwords or exchanging keys.

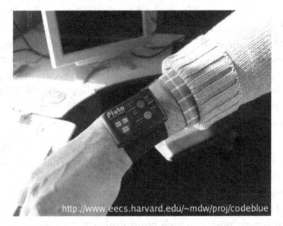

Fig. 1.9. A body sensor node of the CodeBlue Project which is build into a wristband. Image source: `http://www.eecs.harvard.edu/~mdw/proj/codeblue/`

1.2.4 Military Application

Sensor networks in military applications are often used for surveillance missions. The focus of surveillance missions is to collect or verify as much information as possible about the enemy's capabilites and about the positions of hostile targets. Sensor networks are used to replace soldiers because surveillance missions often involve high risks and require a high degree of stealthiness. Hence, the ability to deploy unmanned surveillance missions, by using wireless sensor networks, is of great practical importance for the military.

EnviroTrack. *EnviroTrack* is a middleware for tracking of objects. In [179] the design and implementation of a system for energy-efficient surveillance based on EnviroTrack is described. The main objective of the system is to track the positions of moving vehicles in an energy-efficient and stealthy manner. For the prototype, a network consisting of 70 sensor nodes was used. The sensor nodes are based on MICA2 Motes (see Chapter 1.3.1). They are equipped with dual-axis magnetometers. Issues in the design of the network were long network lifetime, adjustable sensitivity, stealthiness, and effectiveness.

To prolong the network lifetime, it is important to use as little energy as possible. The system allows for a trade-off between energy consumption of the sensor nodes and the accuracy of the tracking. To save energy, only a subset of nodes, the so-called sentries, are active at a given time. The sentries monitor events. When an event occurs, the sentry nodes awake the other sensor nodes of the network and form groups for collaborate tracking. Hence, in the absence of an event, most sensor nodes are in a sleep mode, using only very little energy. The sensor nodes can adjust the sensitivity of their sensors to adapt to different terrains and security requirements. To achieve stealthiness, zero communication exposure is desired in the absence of significant events. The sensor network can only be used in challenging military scenarios if the tracking is effective. Especially, the estimated location of a moving object must be precise enough and the event must be communicated as fast and reliable as possible to the base station. As the collaborative detection and tracking process relies on the spatio-temporal correlation between the tracking reports sent by multiple sensor motes, it is important that the sensor nodes are time-synchronized and that the positions of the sensor nodes are known.

Counter Sniper System. Snipers are a danger to military operations, especially because they are usually difficult to locate. The aim of the counter sniper system *PinPtr* [349] is to detect and accurately locate shooters based on acoustic signals. PinPtr allows to locate a sniper with one meter accuracy in average and the tracking of a shot needs 2 seconds. The system consists of a large number of sensor nodes, which communicate through an ad hoc wireless network. The large number of sensors is needed to overcome multi-path effects of the acoustic signals. All sensor nodes report to a base station, which is also used by the user to access the system. The sensors measure acoustic signals, in detail muzzle blasts and shock waves of guns. The sensor nodes of PinPtr are based on MICA2 Motes (see Figure 1.10). A special sensor board was implemented and connected to the MICA2 Motes. On this sensor board, an FPGA is responsible for fast signal processing. Precise time synchronization is crucial for the application because the application uses the fact that an acoustic signal travels with a certain velocity. PinPtr uses the Flooding Time Synchronization Protocol [277] for time synchronization. Communication problems do not arise from multi-hop communication but from the situation that in the case of an event (a shot), a lot of sensors want to send their sensor data at the same time. This problem is solved by data aggregation at the sensor nodes. To allow fast response of the system after an event, the base station must receive as many sensor data as possible in the first second after the event. PinPtr uses a fast, gradient-based "best-effort" converge-cast protocol with built-in data aggregation. Data needs to include the position of a sensor node so that it can be used to calculate the position of a sniper. PinPtr uses a self-localization algorithm based on acoustic range estimation between pairs of sensor nodes. This method can be used to locate sensors in a range of 9 meters and the accuracy is 10 centimeters.

Fig. 1.10. A sensor node of PinPtr. Image source: `http://www.isis.vanderbilt.edu/projects/nest/applications.html`

1.2.5 Industrial Application

Industrial production relies on more and more complex machinery. Breakdown of a machine can severely interfere the production, resulting in a loss of money. Thus, a close observation of industrial machinery is of advantage.

Condition Based Monitoring. Maintenance of machinery usually prevents unplanned breakdowns. There are different ways for machinery maintenance: Run-to-fail maintenance does not include any maintenance at all. A machine is repaired or changed when it fails. Scheduled maintenance has a schedule when to do what kind of maintenance. The schedule may for example state that the oil of a machinery must be changed every 3 month. While prolonging the lifetime of a machine, scheduled maintenance is based on expert knowledge and not on the actual state of the machine. Thus, it is possible that a machine is maintained even if it does not need maintenance, resulting in unnecessary costs and outage time. Condition-based maintenance deals with this disadvantage. When using condition-based maintenance, a machine is maintained regarding its condition. However, prior to condition-based maintenance, it is necessary to obtain the current condition of the machine. Sensor networks may be used for this. In [82] the company Rockwell Scientific ported existing diagnostic routines and expert systems to wireless sensor network hardware and modified the software for autonomous data collection and analysis. The main task in developing diagnostic software for a sensor network was to abstract and design the algorithms independent from the actual machine so the sensors did not need to know on which machine they were running.

1.2.6 Home Automation and Smart and Interactive Places

The goal of home automation and smart and interactive places is to offer people context-aware assistance with the tasks at their hands. Easy installation and wireless communication are important to allow a user to augment his space in a convenient way.

Smart Kindergarten. In the *Smart Kindergarten* project [351], a classroom was equipped with a sensor network to learn about behaviour patterns of children in a group setting. The goal of the research project was to learn from the Smart Kindergarten classroom how to create more interactive learning environments and how to improve the quality of teaching. The sensor nodes were integrated in tags which the children wore (see Figure 1.11), and in everyday objects like textbooks or tabletops. The sensors obtained three data sets: The physical location, the orientation, and the speech of a child. All sensor nodes sent their data to a central database which was used later for evaluation of data.

1.3 Current Hardware Platforms

Sensor network design is highly application specific. To achieve the goal of the application, it is necessary to choose the right sensor network hardware platform, because the limitations of the hardware greatly influence the design of the sensor network, its protocols, and its algorithms. Some popular platforms

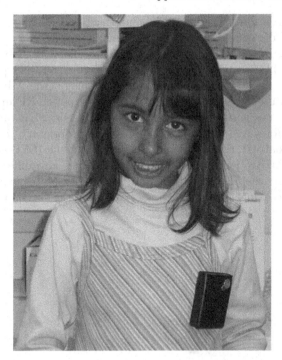

Fig. 1.11. A child equipped with a Smart Kindergarten Badge. Image source: http://nesl.ee.ucla.edu/projects/smartkg/photos.htm

will be introduced in this chapter to give the reader a basic understanding of hardware limitations. In the following, the MICA hardware platform (MICA2, MICAz, MICA2DOT), the Intel hardware platform (MOTE2), the Telos platform (TelosB), the BTnode platform and the ScatterWeb platform (Electronic Sensor Board) will be described in brief.

1.3.1 The MICA Sensor Network Hardware Platform

The MICA sensor network hardware platform was one of the first available hardware platforms, hence it is widely used. The platform was developed by the University of Berkeley and is now maintained by Crossbow Technology. Nodes of the MICA platform, the so called MICA Motes, use an Atmel ATmega128L micro controller [21]. The ATmega128L has an 8-bit RISC Harvard architecture. Hence, it uses different address spaces for code and data. A 128 KByte Flash memory is used to store code. 4 KBytes of SRAM are available for data. In addition, the Atmel ATmega128L has 4 KBytes of EEPROM for nonvolatile storage of data. The clock frequency of the micro controller can be regulated in a range from 0 to 8 MHz by software. The Atmel ATmega128L has two 8-bit timers, two 16-bit timers, and a 10 bit A/D converter. The ATmega has a JTAG interface, an SPI bus interface, and UARTs. The CPU can be in active mode, or in one of six energy saving modes.

Different types of MICA Motes are available: MICA2 (see Figure 1.12) [285], MICAz [287], and MICA2DOT (see Figure 1.13)[286]. The motes differ for example in the communication technology: the MICA2 and the MICA2DOT Motes use the Chipcon CC1000 low power RF module, which can be used at frequencies of 916 MHZ, 868 MHZ, 433 MHz, and 315 MHz. The MICA2 Motes and MICA2DOT Motes use proprietary communication protocols. In contrast, the MICAz Motes implement the IEEE 802.15.4 standard and use ZigBee for communication. The motes also differ in their size and shape: while the MICA2 Motes, and the MICAz Motes look quite alike, the MICA2DOT Motes are much smaller and have the size of a US quarter.

Fig. 1.12. MICA2 Mote. Image source: `http://computer.howstuffworks.com/mote.htm`

Fig. 1.13. MICADOT Mote. Image source: `http://sensorweb.geoict.net/news.htm`

1.3.2 The Intel Sensor Network Hardware Platform Mote2

The Intel sensor network platform Intel Mote2 [195] (see Figure 1.14)is based on the Intel PXA27x XScale CPU [399]. The PXA27x is a 32 bit CPU with a clock frequency of 104 to 624 MHz. However, only three clock frequencies are used for the Mote2 platform: 320 MHz, 416 MHz, and 520 MHz. The PXA27x uses 265 KByte of SRAM and has 32 KByte instruction cache and 32 KByte data cache. In addition to the internal CPU memory, the Mote2 platform uses 32 MB of external Flash and 32MB of external SDRAM. In summary, the PXA27x is one of the most powerful CPUs used in sensor networks. In addition to a relatively large amount of memory, the PXA27x also has a wide array of interfaces, including an FIR/SIR infrared interface, an AC97 interface, an I^2C bus interface, an USB host controller, an UART interfaces, a JTAG interface and an LCD driver. The CPU can be in active mode or in one of four energy saving modes. The Intel Mote2 platform uses the Chipcon CC2420 Chip for 802.15.4/ZigBee communication. The Intel Mote2 can be battery powered or powered by the USB-Port.

Fig. 1.14. Intel Mote2. Image source: http://www.ixbt.com/short/images/ news121206_007.jpg

1.3.3 TelosB Sensor Network Hardware Platform

The TelosB sensor network platform [359] is the current research version of the MICA Motes. Different from the MICA Motes, the TelosB motes use the

TI MSP430 micro processor [299], which was designed especially for ultra low power operation. The MSP430 is a 16-bit RISC micro controller with a von Neuman architecture. It has a clock speed of 8 MHz. The MSP430 of the TelosB motes has 10 KBytes of RAM and up to 1 MB of flash ROM. It has a 16 KBytes EEPROM for non-volatile data storage, e.g. to hold a configuration. The MSP430 has two 16 bit timers, two 12 bit D/A converter, one 12 bit A/D converter and up to 2 UART interfaces, an SPI interface, an I^2C interface, and 48 I/O pins. The MSP430 can be in active mode, or in one of 5 energy saving modes. Changing from one of the sleep modes to active mode is very fast. The MSP430 is very energy efficient. The TelosB motes use the Chipcon CC2420 chip for 802.15.4/ZigBee communication. It achieves a data rate of up to 250 kbps. Figure 1.15 shows a TelosB mote.

Fig. 1.15. TelosB mote. Image source: http://www.xbow.jp/tpr2400.jpg

1.3.4 Electronic Sensor Board

The Electronic Sensor Board hardware platform was developed in the Scatter-Web project [342] at Freie Universitaet Berlin. The Electronic Sensor Board uses the TI MSP430 Micro controller (see Chapter 1.3.3). For communication, the TR1001 low power RF module is used. It operates on 868 MHz. The Electronic Sensor Board platform uses a proprietary communication protocol. Figure 1.16 shows an Electronic Sensor Board.

1.3.5 BTnode

The BTnode rev3 sensor network hardware platform [53] was developed at ETH Zürich. The BTnode rev3 uses the Atmel ATmega128L micro processor (see chapter 1.3.1). Two different communication chips are used: the Zeevo ZV4002 communication chip is used for Bluetooth communication and the Chipcon CC1000 chip is used for proprietary communication protocols. The BTnode ref3 has 244 KByte of external RAM. Figure 1.17 shows a BTnode.

Fig. 1.16. Electronic Sensor Board. Image source: `http://www.inf.fu-berlin.de/`
`inst/ag-tech/scatterweb_net/hardware/esb/overview.htm`

1.4 Upcoming Applications

Sensor networks allow applications that were not possible in the time before
sensor networks. This is especially true for applications that require dense
monitoring and analysis of complex phenomena covering a large region and
lasting for a long time. Meteorologic research is one of the research fields that
deals with complex phenomena. Even today, we still do not know a lot about
the climate of earth. Hence, more sensor network applications in meteorology
research can be expected for the near future.

Sensor networks are also good to advance into areas that have limited
accessibility. The oceans are one example. A better knowledge of the oceans
will result in a better understanding of the climate because the oceans are
an important factor, e.g. for the emergence of hurricanes. However, to explore
the oceans with sensor networks, research into underwater communication and
sensor design for underwater missions is necessary.

In the long run, sensor networks may be of good use in space-related
research. For example instead of sending one single sensor system like the
Mars Rover to a distant planet, it would perhaps be possible in the future
to deploy a sensor network consisting of thousands of nodes. As the system

Fig. 1.17. BTnode rev3. Image source: `http://www.btnode.ethz.ch/`

would consist of a lot of nodes, a total failure of the overall system would be not very likely. Thus, the risk of a failure of a mission would be smaller.

1.5 Chapter Notes

There are few textbooks about applications of sensor networks yet. However, the paper [335] and the sensor network reading list [227] give a good overview of current applications.

2

Modeling Sensor and Ad Hoc Networks

Frank Schulz

This chapter introduces four major aspects of modeling ad hoc and sensor networks: distributed algorithms, communication, energy, and mobility. Commonly used formal models will be introduced as basis for the design and evaluation of algorithms solving problems in this domain. We assume basic knowledge about graph theory and algorithms.

In the introduction the terms ad hoc and sensor networks are defined and the OSI layer model is introduced, which provides several levels of abstraction for systems dealing with communication. Algorithms in the context of sensor networks are usually distributed, and thus basic concepts of the theory of distributed systems are summarized in Section 2.2. A big difference to classical distributed systems comes into play through radio communication: the problem of modeling the communication topology, which is assumed to be given in the classical theory of distributed systems, is discussed in Section 2.3. It is even possible that the communication network may change with time. Especially for sensor networks energy is the crucial resource, which is the topic of Section 2.4. Finally, in Section 2.5 mobility models are outlined, and we conclude with Section 2.6 giving references to the sources that have been used in this chapter and pointers for further reading.

2.1 Introduction

We define an *ad hoc network* to consist of a set of nodes (also called processors) that are able to communicate with each other if a communication link is available—we focus on a scenario where nodes communicate over radio (e.g., Bluetooth or WLAN). This means that, in general, the availability of a communication link between two nodes can change with time. Nodes may be able to move, which further contributes to the instability of communication links.

Sensor networks are ad hoc networks with special characteristics: the most important ones are the limited available energy, which results for example in

D. Wagner and R. Wattenhofer (Eds.): Algorithms for Sensor and Ad Hoc Networks, LNCS 4621, pp. 21–36, 2007.

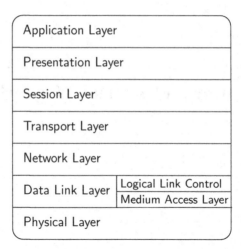

Fig. 2.1. The Open Systems Interconnection Reference Model (OSI Model) standardized by ISO.

limited communication power and range, and the large number of sensors. For sensor networks, it is sometimes assumed that the positions of the nodes do not change, at least during the execution of an algorithm, for example if sensor nodes are once deployed in some area and are not able to move.

Different tasks within sensor and ad hoc networks are usually divided according to the layers of the OSI model (cf. Figure 2.1). In the course of this book the Physical layer (radio communication), the Medium Access Control (MAC) Layer, the Network Layer, and the Application Layer occur frequently. The remaining layers (Presentation Layer, Session Layer and Transport Layer) are often not needed.

The task of designing algorithms requires an abstraction from the real complex system. All relevant features have to be included in the model, but all not really important technical details shall be hidden. Sometimes, such an abstraction does not fit into the layer model, and it is not possible to consider the layers isolated from each other. This is also done in cross-layer optimization, where the motivation is to reduce the overhead that is caused by vertical communication between the layers.

2.2 Distributed Algorithms

In this section, we give a short overview on the design and analysis of distributed algorithms, as a foundation for the modeling of ad hoc and sensor networks.

Definition 2.2.1. *A* distributed system *is a set of autonomous computing units, which are able to communicate.*

Examples of distributed systems are multiprocessor computers, clusters of workstations, peer-to-peer networks like SETI@home [368], and sensor networks.

Typical problems that have to be solved include asynchronous communication, local knowledge, and disruptions of communication links or node failures. We distinguish different kinds of distributed systems, for example depending on the communication paradigm (synchronous or asynchronous), or whether computing units are distinguishable or not, and we are going to define complexity measures of a distributed algorithm.

2.2.1 Formal Model

We will use the following notations: The system consists of n processors (computing units) $p_1, \ldots, p_n$. Two processors can be connected by a bidirectional communication link. These communication links are represented by the *communication graph* $G = (V, E)$, consisting of nodes V and edges E. There is one node per processor, and an edge $e_{ij} = \{p_i, p_j\}$ exists if and only if there is a communication link between p_i and p_j. We use the terms processor, node, and computing unit interchangeably in the following. Unless stated otherwise we assume communication links to be bidirectional, which leads to an undirected communication graph.

As a first example we consider the following *Broadcast Problem:* From a dedicated root node $p_r \in V$ we want to send a message M, which is known to that node, to all other nodes in the network. We are given further a spanning tree T of the communication graph G, rooted at p_r. Each computing unit $p_i \in V$ knows its parent and its children in T. Now, the message M shall be transmitted to all other computing units in G.

Each processor p_i has a set of states Q_i, including special initial and final states. From a final state only final states are reachable. There are communication buffers $\text{outbuf}_i[l]$ and $\text{inbuf}_i[l]$ $(1 \leq l \leq d_i)$, each of which consists of a sequence of messages. At the beginning of the algorithm, all $\text{inbuf}_i[l]$ are empty. Note that the communication buffers can also be modeled through the states Q_i.

Let us consider our Broadcast Problem. A state of a processor p_i includes

- $\text{parent}(p_i) = \text{parent}_i$, either another processor index or nil;
- $\text{children}(p_i) = \text{children}_i$, a set of processor indices; and
- terminated_i, a **boolean** variable.

In the initial state, we have parent_i and children_i according to the given spanning tree, $\text{terminated}_i = \textbf{false}$, $\text{outbuf}_r[j] = M$ for all $j \in \text{children}_r$. All other outbuf_i are empty. A state with $\text{terminated}_i = \textbf{true}$ is a final state.

2.2.2 Distributed Algorithm

A distributed algorithm is given by a sequence of computation and communication steps:

computation step comp(i) at p_i
- p_i is in state $q_i \in Q_i$
- depending on q_i and inbuf$_i$ a state change is performed
- at the end of the step inbuf$_i$ is empty
- for each incident edge a message can be added to outbuf$_i[l]$

communication step del(i, j, M): Send M through e_{ij}
- a message M is transferred from outbuf$_i[l_i]$ to inbuf$_j[l_j]$

The state of the distributed system at each step of the algorithm is given by a configuration $C = (c_1, \ldots, c_n)$: c_i contains the state q_i of processor p_i as well as inbuf$_i$ and outbuf$_i$. A configuration describes a snapshot of the complete distributed system at some time during the algorithm. We define the *execution* of a distributed algorithm as an infinite sequence of configurations and steps:

$$C_1, \phi_1, C_2, \phi_2, \ldots$$

Consider again the Broadcast Problem. A computation step at processor p_i can be implemented as follows: Now we turn to the validity of an execution.

Algorithm 1: Broadcast Algorithm

---

if $M \in \textit{inbuf}_i[k]$ *for some* k then
    add M to outbuf$_i[j]$, for all $j \in$ children$_i$
    set terminated$_i$ =true
if $i = r$ *and* $\textit{terminated}_r = \textit{false}$ then
    set terminated$_r$ = true
else
    do nothing

A *valid* execution has to fulfill certain conditions, which depend on the type of communication—asynchronous or synchronous.

With *asynchronous communication*, there is no guarantee with respect to the duration of a message transmission, and the processors are not coordinated. Consider an execution $C_1, \phi_1, C_2, \phi_2, \ldots$. It is valid if (i) C_{k+1} is the result of step ϕ_k applied to the state C_k, (ii) each processor has infinitely many computation steps, and (iii) every message is eventually delivered. The initial states in the initial configuration C_1 have empty inbuf$_i$.

Synchronous communication is based on *rounds*. One round consists of three parts: (i) each processor can send a message to each neighbor; (ii) these messages are immediately delivered; and (iii) each processor executes one computation step. We can consider synchronous communication as a special case of asynchronous communication: One round consists of communication steps until all outbuf$_i$ are empty, followed by one computation step per processor.

2.2.3 Correctness

We defined valid executions of a distributed algorithm to be infinite. So we need now to define when such an algorithm is terminated. Recall that each processor has a set of final states, and that from a final state only final states are reachable.

Definition 2.2.2. *A valid execution is* terminated *if all processors are in a final state. A distributed algorithm is* correct *if all valid executions terminate and after termination the states of the processors constitute a correct solution of the problem.*

We consider now the correctness of the Broadcast Algorithm.

Lemma 2.2.3. *The Broadcast Algorithm is correct: Each execution terminates after a finite number of steps, and each processor has received message M.*

Proof. By definition of the algorithm, terminated$_i$ is set to true if and only if M has been received. Hence, we have to show that all processors are in a final state after a finite number of steps. We assume the contrary, and consider the processor p_i that does not terminate and has the minimal height in the spanning tree among all processors that do not terminate. We have $p_i \neq r$ since r terminates in the first computation step. Further, since we chose p_i to have minimal height in the tree, the parent of p_i in the tree is terminated. This implies that the parent of p_i has sent message M, and since all messages are eventually delivered and each processor is assigned infinitely many computation steps, p_i receives M after a finite number of steps and by definition of the algorithm p_i terminates after receipt of M, in contradiction to the above assumption.

2.2.4 Complexity

We are going to introduce two different ways to measure the complexity of a distributed algorithm: on the one hand the total number of messages are considered, and on the other the time until termination of the algorithm is measured.

Definition 2.2.4. *The* message complexity *of a distributed algorithm is the maximum number of messages sent before termination over all valid executions of the algorithm.*

Obviously, it can be applied to both asynchronous and synchronous communication. Let us consider the message complexity of the Broadcast Algorithm:

Lemma 2.2.5. *The Broadcast Algorithm has message complexity of $n - 1$.*

Proof. The message M is sent exactly one time for every edge in the spanning tree, since each processor sends the message to its children and terminates then. The spanning tree has $n - 1$ edges.

Now we turn to the complexity measure dealing with the time until termination of the algorithm. With synchronous communication we can give a very easy definition of time complexity.

Definition 2.2.6. *The* time complexity of a synchronous distributed algorithm *is the maximum number of rounds of an execution of the algorithm.*

For our sample Broadcast Algorithm we now consider the time complexity in the synchronous communication model.

Lemma 2.2.7. *For each valid synchronous execution a processor with distance t from the root r in the spanning tree receives message M in round t.*

Proof. Induction by t. Each child of r receives M in the first round, so the statement is true for $t = 1$. Assume each processor with distance $t - 1$ has received M in round $t - 1$. Consider a processor p_i with distance t. The parent p_j has distance $t - 1$, and by assumption p_j has received M in round $t - 1$. By definition of the algorithm, p_j sends M in the following round to p_i, so p_i receives M in round t, which concludes the proof.

Lemma 2.2.8. *The Broadcast Algorithm with synchronous communication has time complexity of h, with h being the height of the spanning tree.*

Proof. The maximum distance from r in the spanning tree is the height h of the spanning tree. Hence, the time complexity of h follows by Lemma 2.2.7.

In the asynchronous communication model, things are more complicated because there is no guarantee on the duration of the transmission of a message. To overcome this problem, we assume a maximum transmission time of 1. We define a *timed execution* of a distributed algorithm and assign each step of an execution a non-negative time value. Time starts at 0 and the assigned time values are non-decreasing in the order of execution. Further, the assigned time values of a single processor are strictly increasing in the order of execution. We define the *duration* of a message transfer to be the time between the addition of a message to $outbuf_i[l_i]$ and the removal from the corresponding $inbuf_j[l_j]$. Time values are normalized such that the maximum duration is 1.

Definition 2.2.9. *The* time complexity of an asynchronous distributed algorithm *is the maximum time until termination over all valid timed executions of the algorithm.*

2.2.5 Symmetry Breaking

Consider the following Leader Election Problem: We assume the communication graph to be a ring consisting of n processors, where each processor knows its left and right neighbor. The problem to be solved is to assign one of the processors to be the leader.

Definition 2.2.10. *A distributed algorithm is called* uniform, *if the number n of processors is not used in the algorithm.*

Definition 2.2.11. *A distributed algorithm is called* anonymous, *if the processors are all equal, i.e., they have no ID values to distinguish processors, and all processors execute identical programs.*

Theorem 2.2.12. *There is no anonymous distributed algorithm for the Leader Election Problem in rings.*

Proof. Assume the existence of such an algorithm, and assume further without loss of generality that the algorithm is synchronous. The execution of the algorithm is determined by the algorithm up to the order of steps within one round. Since the algorithm is anonymous, the state transition functions for all processors are identical, and because of the symmetry of the communication graph, after each round all processors are in the same state (induction by the number of rounds). Hence, if one processor is assigned to be leader, all other processors are leaders as well, and the algorithm can never terminate in a state where only one processor is elected as leader.

We can overcome this negative result by breaking the symmetry through maintaining IDs of processors, i.e., by dismissing the anonymity. The following non-anonymous (but still uniform) algorithm solves the Leader Election Problem: Each processor sends its ID to the left neighbor. If a processor receives a larger ID from its right neighbor, it sends this larger ID to its left neighbor. If the received ID is equal to the own ID, it sends a message "terminate" to its left neighbor, elects itself as leader and terminates. Otherwise, if the received ID is smaller than the own ID, the processor does nothing. If a processor receives the message "terminate" it does not elect itself and terminates.

The message complexity of this algorithm is $\mathcal{O}(n^2)$. It can be improved to $\Theta(n \log n)$. There are also other possibilities to break the symmetry, for example by randomization.

2.3 Communication

In the previous section we assumed the communication graph to be given. However, for ad hoc and sensor networks it is a major problem how this topology of the nodes can be modeled in a realistic way. Models for the communication graphs are essential to the (theoretical) analysis of the algorithms

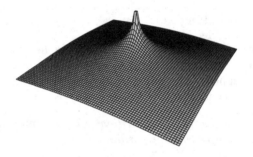

Fig. 2.2. Path loss in the ideal case.

and their simulation. The main difficulty is the realistic modeling of radio communication.

In this section, we focus on the question which nodes *are able* to communicate. A related topic is topology control, which deals with the problem which nodes *shall* communicate over which edges in order to save energy, to reduce interference, or to obtain a planar subgraph for routing.

Another related question is that the communication graph may change during the execution of an algorithm, caused either by moving nodes or by unstable communication links. Most theoretical analyses assume a stable communication graph.

2.3.1 The Ideal Case

For a start we assume the ideal case of a radio communication without any disruptions.

Path Loss. A node that wants to communicate with another node transmits a radio signal with some transmit power P_t. The signal strength gets weaker with the Euclidean distance to the transmitter: the path loss $PL(d)$ is proportional to d^x, where x is the so-called *path-loss exponent x*, which depends on the medium through which the signal is transmitted and usually is assumed to be in the range from 2 to 4, sometimes maybe up to 6 inside buildings. More detailed, the path loss at distance d can be modeled using the path loss $PL(d_0)$ at a certain initial distance d_0; these two values are constants depending, e.g., on the antenna characteristics and the wavelength. Now, the path loss at distance d in dB (given some power p in Watt, the corresponding value in dB is $10\log_{10}(p/1W)$):

$$PL(d)[dB] = PL(d_0)[dB] + 10x\log(d/d_0).$$

The power in dB at some receiver at distance d from the transmitter is then given through the power in dB at the transmitter minus the path loss in dB:

$$P_r(d)[dB] = P_t[dB] - PL(d)[dB].$$

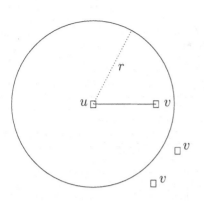
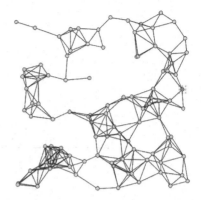

Fig. 2.3. The left figure illustrates the definition of a unit disk graph, and on the right a sample unit disk graph with 100 nodes is shown.

Measuring the power in Watt, this equation amounts for $d \geq d_0$ to

$$P_r(d) = \quad \cdot \frac{P}{d},$$

with some constant   depending on the antenna characteristics, wavelength, etc. Figure 2.2 illustrates the path loss around a transmitting node in the ideal case. Assuming that communication is possible if and only if the received power is larger than some constant, two nodes can communicate with each other if and only if the Euclidean distance between the two nodes is smaller or equal to a given constant transmission range. Based on this observation we can model the topology of the processor nodes as a so-called unit disk graph.

Unit Disk Graph. From the node positions and a given communication range r we define the edges of the communication graph as follows in the unit disk graph model: There is an edge between two nodes if the Euclidean distance is less than the communication range. Figure 2.3 shows a sample unit disk graph. More generally, we can define the unit disk graph as a property of a given graph G:

Definition 2.3.1. *A graph $G = (V, E)$ is a* unit disk graph (UDG) *if and only if there is an embedding of the nodes in the plane such that $\{u, v\} \in E$ if and only if the Euclidean distance between u and v is less than or equal to 1.*

This definition is equivalent to the following statement: A graph $G = (V, E)$ is a unit disk graph if and only if there exists an embedding of the nodes in the plane such that $\{u, v\} \in E$ if and only if the unit disks around u and v intersect.

On the one hand, there are several problems that are easier to solve in unit disk graphs than in general graphs, while on the other hand there are problems that remain $\mathcal{NP}$-complete. To get an idea of proof techniques used with unit disk graphs we consider the following lemma:

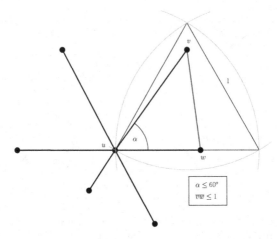

Fig. 2.4. Star Lemma: a unit disk graph does not contain $K_{1,6}$ as node-induced subgraph.

Lemma 2.3.2 ([276, Lemma 3.2]). *Let $G = (V, E)$ be a unit disk graph. Then G does not contain $K_{1,6}$ as node induced subgraph.*

Proof. Figure 2.4 illustrates the proof of the lemma. A node induced star $K_{1,6}$ always has two edges whose angle is less than or equal to 60 degrees, which implies that there must also be an edge between two of the tentacles of the star, and also this edge would be induced by the same 7 nodes. Hence, it is not possible to find 7 nodes inducing a star $K_{1,6}$.

An application of Lemma 2.3.2 is the following factor 5 approximation of the dominating set problem in unit disk graphs. See Chapter 3 (pages 37 et seqq.) for more details on such problems.

Theorem 2.3.3 ([276, Theorem 4.8]). *Let $G = (V, E)$ be a unit disk graph, D_T a minimum dominating set and I a maximal independent set. Then, I is a dominating set with $|I| \leq 5| D_T|$.*

Note that a maximal independent set can be computed easily in linear time (as long as there are nodes in $V \setminus I$ whose neighbors are all not in I add one of these nodes to I), and the above is a linear-time factor 5 approximation.

Proof. We first show that a node v in D_T dominates at most 5 nodes in I. This is true because (i) if $v \in I$, then v dominates none of the nodes in I, and (ii) otherwise assume v dominates $v_1, \ldots, v_6 \in I$; then, v and the v_i induce a $K_{1,6}$ (since the nodes in I are not connected), in contradiction to Lemma 2.3.2. Since D_T dominates V and thus also I, there can only be 5 times more nodes in I than in D_T.

2.3.2 Disruption

The major drawback of the geometric unit disk graphs introduced above is the use of communication ranges to define whether communication between two nodes is possible or not. In reality, the radio signal is disrupted by various effects, and the receiving area defined by a certain signal strength necessary for reception is much more complex than a simple circle that we got in the ideal case in the previous section.

On the one hand, large scale fading of the radio signal is caused by reflection (e.g., on the ground), diffraction (e.g., by obstacles like buildings), and by scattering. Experiments have shown that these effects can be modeled by a log-normally distributed random variable X with a certain standard deviation σ in addition to the path loss in the ideal case, as introduced in the previous section:

$$\overline{P_r}(d) = P_t - PL(d) - X_\sigma.$$

Figure 2.5 shows a sample of such a distribution.

On the other hand, the radio signal is further disturbed by small-scale fading effects caused for example by interference, multipath or Doppler effects. These disruptions cause small-case effects in the sense that a movement by a very small distance can result in a very different signal strength.

Statistical models as the log-normal shadowing described above are used by ad hoc network simulators like ns-2 [119]. When a communication step is simulated, according to the position of the nodes, signal strength, and assumed disruptions (obstacles, reflection, interference, small and large scale fading) an incoming signal strength is computed according to a probability distribution, and it is decided whether the message could be received or not.

Quasi Unit Disk Graph. Now, we want to discuss how the unit disk graph model can be relaxed to cope with the just described disruptions of the radio signal leading to strangely shaped transmission areas that are definitely not disks. Therefore, we introduce—in addition to a given range r where for distances smaller or equal to r a communication is guaranteed—a second range $R > r$. For distances larger than R it is now guaranteed that definitely no communication is possible anymore. Between r and R communication may or may not be possible.

For a given set of points a graph on the point set is an r-R-quasi unit disk graph if for each pair of nodes with distance smaller or equal to r there is an edge in the graph, and for each pair of nodes with distance larger than R there is no edge in the graph. Note that there may be many different r-R-quasi unit disk graphs for a given set of points, since edges in the range $(r, R]$ are optional. Figure 2.5 illustrates this definition of quasi unit disk graphs.

From a more graph-theoretic point of view we introduce a parameter $0 \leq d \leq 1$ and define for a given graph $G = (V, E)$ whether G is a d-quasi unit disk graph or not as follows:

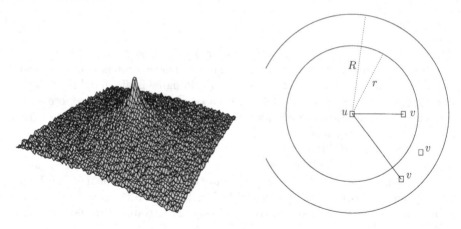

Fig. 2.5. On the left a random sample distribution of log-normal shadowing around a transmitter is shown, and on the right an illustration of the quasi unit disk graph model.

Definition 2.3.4. *Let $d \in [0,1]$ be a parameter. A graph $G = (V,E)$ is a d-quasi unit disk graph (d-QUDG) if and only if there is an embedding of the nodes in the plane such that $dist(u,v) \leq d \Rightarrow \{u,v\} \in E$ and $dist(u,v) > 1 \Rightarrow \{u,v\} \notin E$.*

2.3.3 Variations of Unit Disk Graphs

It is common to use subgraphs of unit disk graphs as communication graphs. Some routing algorithms, for example, require planar communication graphs (cf. Chapter 9). The *Gabriel graph* of a set of nodes with positions in the plane is a planar graph, which can be computed by an easy distributed algorithm. Related to the Gabriel graph is the *relative neighborhood graph*, which is a subgraph of the Gabriel graph that is still connected, provided the unit disk graph is connected. Such graphs are important for topology control and discussed in more detail in chapter 5.

For restricted versions of unit disk graphs some algorithms can be proven to be more efficient than for general unit disk graphs. If one assumes that two nodes are never closer than some constant, one can show that the unit disk graph is of bounded degree. This assumption is referred to as the (1)-*Model*. For example, a routing algorithm [241] (cf. Chapter 9) can be proven to be more efficient for unit disk graphs with bounded degree than for general unit disk graphs.

2.4 Energy

So far we have evaluated the complexity of the algorithms in the number of messages or the time until termination of the algorithm. However, especially for sensor networks, the energy consumption of an algorithm is crucial—usually scenarios are considered where sensors are equipped with non-rechargeable batteries, and once the battery is finished the sensor is useless.

Energy consumption of sensor and ad hoc nodes can be distinguished into energy needed for transmission on the one hand and listening on the other hand. The classical analysis of distributed algorithms focuses on communication, either counting the number of rounds of communication or the number of messages sent. This can be extended in a straightforward way to an *energy complexity* measure counting the energy needed for the transmission of the messages sent during the algorithm. However, measurements have shown that most of the energy is spent while listening to incoming signals. To cope with this fact, *power saving* techniques set nodes to sleep mode, according to some schedule, in order to save energy.

2.4.1 Energy Complexity

In Section 2.3.1 we have seen that the path loss of a radio signal is proportional to d^x, where d is the Euclidean distance to the transmitter, and x is a path loss exponent. The signal can be received if the remaining power at the receiver is above an antenna-specific constant. If we assume now that the transmitter wants to transmit its signal over a given distance d, the power needed to transmit the signal is again proportional to the distance d, with the same exponent x.

$$P_r = a \cdot \frac{P_t}{d^x} \Rightarrow P_t = \frac{P_r}{a} \cdot d^x.$$

Now, one can argue that the transmitter can adjust its transmit power for each message it sends, and the energy consumption of a distributed algorithm can be defined through assigning each edge in the communication graph an estimate for the energy needed for the corresponding radio communication—according to the above discussion, a function in $\Theta(d^x)$ would be appropriate. Now, the energy complexity of an algorithm can be defined to be the sum of these communication costs over all transferred messages.

We consider again the routing example discussed in Section 2.3.3 taken from [241]. The authors consider the above described energy complexity as a function of the energy consumption of an (energy) optimal path p^* in the communication graph. Let $c(d) = d^x$ be the edge cost function with some exponent x. An interesting result is that the energy complexity of the routing algorithm is in $O(c(p^*)^2)$ for linear bounded cost functions (e.g., for $x = 0$ the message complexity and for $x = 1$ the Euclidean distance). On the other hand, the authors argue that it is not possible to design a local geometric algorithm whose cost can be bound by a polynomial in $c(p^*)$ for cost functions that are not linear bounded (e.g., with exponents $x \in [2,4]$). The reason for this difference between message and Euclidean distance complexity on the one hand and energy complexity on the other hand is that with the energy costs, a "geometrically" long path can have an arbitrary small energy cost, provided it consists of small enough single edges.

Recall, however, that assigning edge costs proportional to $\Theta(d^x)$ bears many problems—the most severe may be the fact that the ideal model of unit communication ranges is far from reality—and in practice power saving techniques outlined in the following section are more established.

2.4.2 Power Saving

Power saving techniques rely on the fact that processor nodes have the ability to be set into a sleep state with a minimum amount of energy consumption. Usually, the communication device is completely powered off while sleeping.

One possibility to define sleep schedules is usually defined by topology control on the network layer. Only a subset of the nodes is active at a time, and the decision which nodes shall be active is based on the communication graph, maintaining a connected subgraph. For example, small connected dominating sets (cf. Chapter 3) of the communication graph could be chosen as active nodes. Sleep schedules can also be defined by the MAC layer. For example, with time division multiplexing some nodes may save energy in empty time slots.

Generally, power saving is a tradeoff between efficiency of the network— measured for example as throughput, connectivity or latency—and the amount of energy consumed by the network. Another objective, especially for sensor networks, is to develop power saving techniques that maximize the lifetime of the whole system.

2.5 Mobility

We present some common mobility models for sensor and ad hoc networks. The main idea is to randomly simulate the movement of nodes. We distinguish two types of models: Entity models, where the single nodes move independently of each other, and group mobility models, where some of the nodes are forming groups. Such models are mainly used in simulators, rather not in theoretical analyses.

2.5.1 Entity Mobility Models

The first and most straightforward, but also least realistic model is the Random Walk Model, essentially resembling a Brownian motion. The second model, the Random Waypoint model, tries to incorporate the fact that the nodes travel strictly and straight-line goal-directed through the network.

Random Walk. Given a time unit t, each node repeats the following algorithm: select independently at random a direction and speed; move in that direction with the selected speed for a time interval t. The positions of the nodes are updated after each time interval t. A variation of this model uses the same constant speed value for all nodes.

Random Waypoint. In the Random Waypoint Model, the algorithm repeated for each node is as follows: Choose randomly a destination and speed; travel towards destination; wait a randomly selected time at destination. The positions of the nodes are updated in discrete time intervals.

2.5.2 Reference Point Group Mobility

Concerning group mobility models we describe the Reference Point Group Mobility Model. It is based on entity mobility models and includes most of the known models as special cases (also the entity mobility models).

Nodes are partitioned into groups. The group center moves according to some entity mobility model (e.g., the Random Waypoint Model). Inside each group, for each node a reference point relative to the group center is maintained. In addition to the movement of the group center, the single nodes move, in smaller scale, according to some entity mobility model, relative to the reference point. To calculate the position of each node, the group center movement and the individual movement around the reference point are superposed.

2.6 Chapter Notes

Examples of textbooks introducing the basics on algorithms, graph theory and complexity theory are "Introduction to Algorithms" by Cormen, Leiserson, Rivest and Stein [84], "Graph Theory" by Diestel [101] and "Computers and Intractability: A Guide to the Theory of $\mathcal{NP}$-Completeness" by Garey and Johnson [150]. The section on distributed algorithms is based on the textbook "Distributed Computing" by Attiya and Welch [22], which is also a good resource for further reading in this area. A survey by Schmid and Wattenhofer [343] gives a good overview on recent algorithmic models for sensor networks, putting models for connectivity, interference and algorithms into perspective.

The facts on modeling radio communication are taken from the textbook "Wireless Communication: Principles and Practice" by Theodore Rappaport [329], which provides a deep insight in the foundations of wireless radio communication. The first discussion of problems related to unit disk graphs is provided by Clark, Colbourn and Johnson in [75], and simple heuristics for unit disk graphs are shown by Marathe et. al. in [276]. Relations of geometric graphs defined by forbidden regions, like Gabriel graphs and relative neighborhood graphs are nicely shown in [62].

A surveying discussion of the topic energy consumption is presented by Feeney [122], which has also been used as the basis of Section 2.4. The mobility models shown in Section 2.5 are taken from [59]. The described entity mobility models, random walk and random waypoint, and variations of these—for example, reflection or wrapping at the border of a given region—can be generalized to the Random Trip Model introduced by Le Boudec and Vojnović [46].

3

Clustering

Thomas Moscibroda

3.1 Introduction

Wireless ad hoc and sensor networks consist of autonomous devices communicating via radio. There is no central server or stationary infrastructure which could be used for organizing the network. Hence, the coordination necessary for efficient communication has to be carried out by the nodes themselves. That is, the nodes have to build up a communication service by applying a distributed algorithm.

Consider for instance the task of routing a message between two distant, non-neighboring nodes. In absence of a stationary infrastructure, intermediate network nodes have to serve as routers. Since the topology of the network may be constantly changing, routing algorithms for ad hoc networks differ significantly from standard routing schemes used in wired networks. One effective way of improving the performance of routing algorithms is to group nodes into *clusters*. The routing is then done between *clusterheads* who act as routers. All other nodes communicate via a neighboring clusterhead.

In literature, numerous proposals to structure wireless networks have been presented. Besides facilitating the communication between distant nodes (routing protocols, e.g., [382, 398, 354, 388]), the main purposes of such structures are to enable efficient communication between adjacent nodes (MAC layer protocols) [324, 181], to improve the energy efficiency of the network [99], or to enable a proper initialization of the network [231, 295]. As diverse as these different purposes may be, there are a few key primitives on which most of the applied structures are built. The most prominent such primitive is clustering based on classic graph-theoretic objects such as dominating sets or independent sets. It is therefore interesting to investigate the possibilities and limitations of (distributed) clustering algorithms for such problems. Finally, other distributed algorithms for clustering in ad hoc and sensor networks include [200, 238, 237] (general graphs) and [147, 310, 388] (restricted graphs) . An empirical comparison of some of the above mentioned clustering algorithms has been conducted in [33].

D. Wagner and R. Wattenhofer (Eds.): Algorithms for Sensor and Ad Hoc Networks, LNCS 4621, pp. 37–61, 2007.

What is a good clustering? Depending on the specific network problem at hand, the answer to this question may be varying. But in light of the wireless and multi-hop nature of ad hoc and sensor networks, a good clustering should satisfy some desirable properties. In order to allow efficient communication between each pair of nodes, every node should have at least one clusterhead in its neighborhood. From a graph theory point of view, this first property demands a *dominating set* (and preferably a *minimum dominating set*).

Definition 3.1.1 (Minimum Dominating Set). *In a graph* $G = (V, E)$, *a dominating set is a subset* $S \subseteq V$ *such that every node is either in* S *or has at least one neighbor in* S. *The* minimum dominating set problem *asks for a dominating set* S *of minimum cardinality.*

For the purpose of dominating set based routing, a mere dominating set structure may not be sufficient since it is not guaranteed that there exists a connected path between two nodes consisting entirely of dominators. For routing, we are therefore interested in a dominating set which is connected.

Definition 3.1.2 (Minimum Connected Dominating Set). *In a graph* $G = (V, E)$, *a connected dominating set is a subset* $S \subseteq V$ *such that the subgraph induced by* S *is connected and* S *forms a dominating set in* G. *The* minimum connected dominating set problem *asks for a dominating set* S *of minimum cardinality.*

Both the general and the connected versions of the minimum dominating set problem are $\mathcal{NP}$-hard, even in the unit disk graph model.

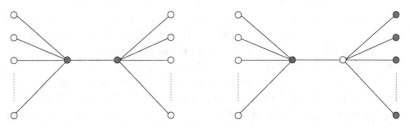

Fig. 3.1. An example graph with minimum (connected) dominating set of size 2 (left) and a maximal independent set of size $n/2$ (right).

The definition of dominating sets leaves open the possibility of neighboring nodes being clusterheads. In fact, a good solution to the minimum dominating set problem may even *require* neighboring nodes to be become clusterheads as exemplified in Figure 3.1. However, in light of the wireless nature of communication in ad hoc and sensor networks, having neighboring and thus directly interfering clusterheads may be undesirable. The task of setting up an efficient cluster-based MAC layer, for instance, is greatly facilitated if no two cluster-heads are in mutual transmission range. This additional requirement leads to the graph theoretic notion of a *maximal independent set*.

Definition 3.1.3 (Maximal Independent Set). *In a graph $G = (V, E)$, an independent set $S \subseteq V$ of G is a subset of nodes such that no two nodes $\forall u, v \in S$ are neighbors in G. S is a* maximal independent set *(MIS) if any node not in S has a neighbor in S, i.e., if S forms a dominating set.*

Notice that a *maximal independent set* (MIS) should not be confused with a *maximum independent set*. Specifically, the maximum independent set problem asks for an independent set S in a graph with *maximum cardinality*. In a MIS, however, we merely seek a set of mutually independent (i.e., non-neighboring) nodes S, such that every node has at least one neighbor in S. The two problems differ in the sense that computing a maximum independent set is $\mathcal{NP}$-complete, whereas computing an MIS in the centralized way is trivial: simply pick one independent node one-by-one and remove all its neighboring nodes.

In this chapter, we give an overview over *distributed MIS and dominating set algorithms*. Since algorithms that are intended to be employed on network devices must inherently be distributed, it clearly makes sense to study such clustering problems in a distributed context. However, it is important to note that not all distributed algorithms are equally suitable to be applied in ad hoc and sensor networks. Particularly, distributed algorithms that are based on maintaining a *global view* of the network are very time (and energy) consuming and thus hardly feasible. In other words, algorithms that have a running time that can grow linearly in the number of nodes are typically unemployable in large-scale ad hoc or sensor systems.

Instead, we are primarily interested in so-called *local (or localized) algorithms* featuring a (very) *low time complexity*, even at the cost of somewhat suboptimal solutions [237]. When studying such *local algorithms*—i.e., algorithms with a running time that is guaranteed to be less than linear in the number of nodes—it is of theoretical and practical interest to investigate the achievable trade-off between the amount of communication (time complexity) and the quality of the obtained global solution (approximation ratio). In other words, we seek to find out how much communication and time is needed in an ad hoc and sensor network, such that its nodes can achieve a good global solution.

The remainder of this chapter is organized as follows. In Section 3.2, we discuss different models in the context of clustering algorithms. Section 3.3 is concerned with several algorithmic approaches to clustering in *unit disk graphs*. The distributed complexity of computing a minimum dominating set in *general graphs* is analyzed in Section 3.4. Finally, we present upper and lower bounds on the achievable trade-off between time complexity and the resulting distributed approximation. The chapter is concluded with a catalogue of open problems in Section 3.5.

3.2 Models

The complexity of clustering significantly depends on the underlying network model. In this section, we give a brief classification of the major models that have been used for the analysis of clustering algorithms. Our list is by no means comprehensive, but it attempts to capture the most fundamental models.

Computational models in the context of ad hoc and sensor networks can be classified along numerous (sometimes orthogonal) coordinates. Probably the most striking distinction is whether we consider *centralized* or *distributed algorithms*. As already argued in the introduction, efficient solutions for ad hoc and sensor networks are based on *local distributed algorithms*. The centralized case is nonetheless of importance because insight gained in the centralized case can often lead to the design of efficient local solutions.

As also discussed in Chapter 2, an alternative way of classifying models is how the wireless communication nature is modeled. Typically, ad hoc and sensor networks are modeled as a graph $G = (V, E)$. In the literature, the two most frequently adopted graph models are *general graphs* and *unit disk graphs* (UDG). In both cases, vertices represent network nodes and an edge between two nodes represents a communication link between them. In the general graph model, the graph G can be arbitrary. In a unit disk graph, nodes are located in the Euclidean plane and there is a link between two nodes if and only if their mutual distance is at most 1. Evidently, the UDG model is a simplification in many ways, because in reality, there is no sharp threshold for the transmission and interference range as implied by it. On the other hand, the UDG model does manage to capture the inherently local nature of wireless communication - something that is not properly modeled in general graphs. Thus, modeling wireless multi-hop networks as unit disk graphs and general graphs can be considered as being overly optimistic and pessimistic, respectively. In Sections 3.3 and 3.4, we deal with algorithms for unit disk graphs and general graphs, respectively.

UDG models can further be classified according to the hardware capabilities nodes are assumed to have. For instance, nodes may be capable of sensing the *distance* to neighboring nodes or they may even know their *coordinates*. In the harshest model, nodes know merely about their neighbors, but they can neither measure distances nor angles. In Section 3.3, we show that the specific set of assumptions has a tremendous impact on the distributed complexity of clustering.

In many cases, the underlying communication model may need to be adjusted depending on the specific *purpose* for which the network structure is to be employed. In the *message passing model*, a node can send a message to each of its neighbors without collision. The underlying assumption is that the network has already been properly initialized. Particularly, it is assumed that there exists an efficient MAC layer protocol which provides reliable point-to-point connections between neighboring nodes. This MAC layer deals with crucial aspects such as collisions and interference and thus provides a high level abstraction to higher level protocols and applications. If, however, the clustering is used for the purpose of *setting up* such a MAC layer (or an initial structure in general), it is not reasonable to study clustering algorithms in message passing models. In fact, this highlights a "chicken-and-egg" problem in the sense that clustering is often a key component in establishing an initial structure, but many existing clustering algorithms rely on the existence of such an initial structure. Based on a model initiated in [231], we briefly discuss clustering in harsh radio network models (as opposed to the message passing model) in Section 3.3.5.

Finally, it is frequently assumed that the physical distribution of nodes follows a specific distribution. Particularly, nodes are often assumed to be uniformly and independently distributed in the plane. In this chapter, we take a more worst-case oriented view and do not follow this approach. Throughout the chapter, we assume that nodes can be arbitrarily deployed and located, i.e., all upper bounds hold even in worst case scenarios.

3.3 Clustering Algorithms for Unit Disk Graphs

The problem of finding a minimum dominating set in *general graphs* is $\mathcal{NP}$-hard. Furthermore, it has been shown in [123] that the best possible approximation ratio for this problem is $\ln \Delta$, where Δ is the highest degree in the graph, unless $\mathcal{NP}$ has deterministic $n^{O(\log \log n)}$-time algorithms. In *unit disk graphs*, the problem (as well as its *connected* counterpart) remains $\mathcal{NP}$-hard, but constant approximations become possible. In unit disk graphs and some of its generalizations, the problem even allows for a *polynomial time approximation scheme* (PTAS) [192, 70, 116], even without geometric representation [303].

In the distributed case, the complexity of computing a good dominating set depends on the model assumptions. In the easiest model - when all nodes know their coordinates - a constant approximation to MDS can be computed with only two rounds of communication! To see this, partition the plane by a grid into squares (cells) of side length $1/\sqrt{2}$. Each cell defines a cluster. By checking its coordinates, each node can now decide to which cluster it belongs. Since all nodes within one cell are at most at distance 1 to each other, nodes in each cluster can simply elect one "clusterleader" per cell (for instance the node with the lowest ID). The union of all these leaders covers the entire graph and it is easy to see that it is a constant approximation to the optimal solution: For every cell, the optimal solution places at least one dominator in the cell or in one of the neighboring cells.

If nodes do not know their coordinates, the problem becomes more involved. In this section, we first study different solutions for computing (connected) dominating sets in unit disk graphs in case nodes have neither coordinate nor distance or angle information. In Section 3.3.4, we discuss how distance information can be exploited to devise significantly faster algorithms.

3.3.1 Simple Solutions

In absence of any distance, angle, or coordinate information, a straightforward and promising approach to solving the minimum dominating set problem appears to be the following:

Algorithm 2: One-round MDS Algorithm

1: Each node v randomly selects an ID $\nu(v)$ from an interval $[1, \ldots, n^2]$.
2: Nodes exchange $\nu(v)$ with all neighbors.
3: Each node v elects the node $w \in \nu(v)$ in its neighborhood with the largest $\nu(w)$.
4: All nodes that have been elected at least once become dominator.

We first show that the initial randomization step is necessary. Without it, (i.e., when real node IDs are used) the algorithm's performance can be $n/2$ times worse than the optimum. If nodes are ordered in a line with descending IDs, as in Figure 3.2, all nodes on the left half will be elected from a node on the right half. Hence, algorithm's solution consists of $n/2$ nodes, while all nodes can be covered by picking the single node in the middle.

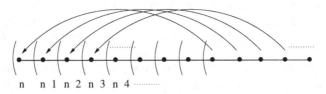

$$n \quad n\,1\;n\,2\;n\,3\;n\,4 \quad \cdots$$

Fig. 3.2. A worst case example of Algorithm 2 without randomization. All nodes on the left will join the dominating set.

Intuitively, a bad example as in Figure 3.2 seems unlikely in case the node IDs (v) are selected randomly. However, as first analyzed in [147], even in this case, the algorithm's performance can be disastrous in the worst-case.

Theorem 3.3.1 ([147]). *Algorithm 2 computes a correct dominating set in two communication rounds. In worst case node distributions, the selected dominating set is $(\sqrt{n})$ times larger than the optimal solution in expectation.*

Proof. Consider the configuration in Figure 3.3. Each point in set represents a set of $\sqrt{n} - 1$ nodes that are located at the same location. Each of the $\sqrt{n}$ points in B is a single node which can see a distinct set of $\sqrt{n}-1$ nodes in . In total, there are $\sqrt{n}$ nodes in set B and $\sqrt{n}(\sqrt{n}-1)$ nodes in set . Each node in B sees $\sqrt{n}$ nodes in B (including itself) and $\sqrt{n}-1$ nodes in set . It follows that each node in B nominates a point in with probability $\frac{\sqrt{n}-1}{2\sqrt{n}-1} \quad 1/2$.

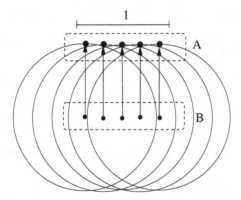

Fig. 3.3. A worst case example of Algorithm 2. Each point in set A consists of $\sqrt{n}-1$ nodes, while each point in set B represents a single node. If node identifiers are selected randomly, at least $(\sqrt{n}-1)/2$ nodes join the dominating set in expectation.

Thus, the expected number of dominators in   is $(\sqrt{n})$. On the other hand, the optimal solution could cover all nodes with only 2 dominators, one in set   and one in B.   □

Theorem 3.3.1 shows that the low time-complexity of Algorithm 2 comes at the cost of a bad performance guarantee in the worst case. Interestingly, it can be shown that many other clustering algorithms proposed in the literature on ad hoc and sensor networks can be reduced to Algorithm 2 when it comes to its worst-case behavior. That is, protocols such as [398, 94] inherit the bad worst case behavior of the simple Algorithm 2.

3.3.2 CDS Via Connecting A MIS

The downside of the constant-time algorithm presented in the previous section is its bad behavior in the worst-case. In order to come up with better solutions, we now study a different algorithmic approach, namely to use a MIS as an approximation for a minimum dominating set in the unit disk graph. As we have seen in Figure 3.1, the cardinality of a MIS in a general graph may be arbitrarily larger than the cardinality of a minimum dominating set. It follows that there is, in general, no approximation algorithm for the minimum dominating set problem that could always result in a solution constituting a MIS. On the other hand, it is intuitively clear that such a severe example as in Figure 3.1 may not exist in the case of unit disk graphs. Specifically, in unit disk graphs, it is impossible to pack nodes as dense without some of them becoming neighbors. This intuition is formalized by the following lemmas which show that a MIS is a good (constant) approximation to the minimum dominating set problem in unit disk graphs [276, 382].

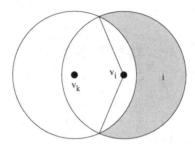

Fig. 3.4. All nodes in S_i lie in a sector of at most 240 degree.

Lemma 3.3.2. *Consider a unit disk graph G and let OPT be the optimal minimum dominating set in G. The size of any independent set S in G is at most $5 \cdot |OPT|$.*

Proof. Since S is an independent set, no node in OPT can dominate more than 5 nodes in S. Hence, $OPT \geq |S|/5$. □

Lemma 3.3.3. *Consider a unit disk graph G and let OPT be the optimal minimum connected dominating set in G. The size of any independent set S in G is at most $4 \cdot |OPT| + 1$.*

Proof. Let T be a spanning tree of the optimal connected dominating set OPT. Let $v_1, v_2, \ldots, v_{|T|}$ be a preorder traversal of T. Further, for $i = 1, \ldots, |OPT|$, let S_i be the set of MIS nodes that are adjacent to node v_i, but not to any node $v_1, \ldots, v_{i-1}$, formally $S_i = \{v \in S \mid v \in N_i \; v \notin \bigcup_{j=1}^{i-1} N_j\}$, where N_i denotes the set of neighbors of v_i. By the unit disk graph property, v_1 can be adjacent to at most 5 independent nodes, i.e., $|S_1| \leq 5$. Because OPT denotes a connected dominating set, it holds that for $i = 2, \ldots, |OPT|$, at least one node, say v, in $v_1, \ldots, v_{i-1}$ is connected to v_i. By the definition of S_i, all nodes in S_i must be located in a sector U_i of at most 240 degrees within the coverage range of node v_i (See Figure 3.4), implying that $|S_i| \leq 4$. Hence, $|S| = \sum_{i=1}^{|T|} |S_i| \leq 4|OPT| + 1$. □

Given Lemmas 3.3.2 and 3.3.3, we can devise a constant approximation algorithm for the minimum dominating set problem in the unit disk graph by computing a MIS. Assuming that each node has a unique ID (or in absence thereof, is capable of choosing a random number from a large enough range), the simplest way of selecting a MIS is the following (distributed) procedure.

Each node $v \in V$ having ID $id(v)$ waits with its decision until one of the two following conditions hold:

a) v has a neighbor who joins the MIS $\Longrightarrow v$ becomes covered.
b) v has the largest ID among all uncovered neighbors $\Longrightarrow v$ joins the MIS.

This procedure yields a correct MIS. However, it is easy to observe that the worst-case time and message complexities are (n) and $(|E|)$, respectively, because there might be a single point of activity at any time. Consider for instance n nodes located on a line having monotonously decreasing IDs. In this case, the node on the very right will have to wait $n - 1$ time-slots before it can finally decide whether to join the MIS.

3.3.3 Efficient MIS Algorithm

Algorithm 3 presents an elegant and much faster solution to the problem of computing a MIS. This classic algorithm by Luby is randomized and achieves a MIS in time $O(\log n)$ (instead of $\Omega(n)$ as above) with high probability [266]. The algorithm operates in synchronous rounds, grouped into phases. The three steps of each phase can be implemented efficiently even in a distributed setting. At the beginning of each phase, every node informs its neighbors about its being active or passive. Similarly, the second and third step can be implemented each using a single round of communication. Clearly, the algorithm produces a correct MIS because two neighboring nodes can never join the MIS. Furthermore, if a node does not have a neighboring node that has joined the MIS, it will sooner or later mark itself and consequently join the MIS. Throughout the algorithm, a node's *current degree* $d(v)$ is the number of its active neighbors, and N_v^+ denotes node v's neighborhood (including v itself).

Algorithm 3: Fast distributed MIS Algorithm

Code executed by node v, which is initially active.
while v is active **do**
  1: v *marks* itself with probability $\frac{1}{2d(v)}$, where $d(v)$ is the current degree of v.
  2: If no higher degree neighbor of v is also marked, v joins the MIS.
    If two neighbors have the same degree, ties are broken by identifier.
  3: If there is a node $u \in N_v^+$ that has joined the MIS, become passive.
end while

We now prove the algorithm's expected running time of $\mathcal{O}(\log n)$. The first lemma bounds from below a node's probability of joining the MIS in a given phase.

Lemma 3.3.4. *A node v joins the MIS in step 2 of an arbitrary phase with probability at least $\frac{1}{4d(v)}$.*

Proof. Let M denote the set of marked nodes after step 1 of the phase. $H(v)$ is the set of neighbors of v having a higher current degree (or same degree with higher identifier). The probability that v does not join the MIS in spite of marking itself in step 1 is

$$P[v \notin MIS | v \in M] \quad = \quad P[\exists w \in H(v), w \in M | v \in M]$$

$$\leq \quad \sum_{w \in H(v)} P[w \in M] = \sum_{w \in H(v)} \frac{1}{2d(w)}$$

$$\underset{d(v) \leq d(w)}{\leq} \sum_{w \in H(v)} \frac{1}{2d(v)} \leq \frac{d(v)}{2d(v)} = \frac{1}{2}.$$

The probability that v joins the MIS in step 2 is therefore lower bounded by

$$P[v \in MIS] = P[v \in MIS | v \in M] \cdot P[v \in M] \geq \frac{1}{2} \frac{1}{2d(v)}.$$

$\square$

We call a node *good* if the sum of the marking probabilities in its neighborhood is larger than a certain constant. Formally, a node v is called good if $\sum_{w \in N_v} \frac{1}{2d(w)} \geq \frac{1}{6}$. Intuitively, good nodes have many low-degree neighbors and hence, the probability of one of them joining the MIS is high. This idea is formalized in the next lemma.

Lemma 3.3.5. *A good node is removed in step 3 with probability at least $\frac{1}{36}$.*

Proof. We distinguish two cases, starting with the simple one.

Case 1: There is a neighbor $w \in N_v$ with degree $d(w) \leq 2$. By Lemma 3.3.4, w joins the MIS in step 2 (and v is removed in step 3) with probability at least $\frac{1}{8}$.

Case 2: All neighbors of v have degree 3 or more. For any neighbor $w \in N_v$, we have $\frac{1}{2d(w)} \leq \frac{1}{6}$. Since by definition of a good node $\sum_{w \in N_v} \frac{1}{2d(w)} \geq \frac{1}{6}$, there is a non-empty subset of neighbors $X \subseteq N_v$ such that,

$$\frac{1}{6} \leq \sum_{w \in X} \frac{1}{2d(w)} \leq \frac{1}{3}.$$

We can now bound the probability that v will be removed. Again, if one of v's neighbors joins the MIS in step 2, v is removed in step 3. Thus, we have

$$P[v \text{ is removed}] \geq P[\exists u \in X, u \in MIS]$$

$$\geq \sum_{u \in X} P[u \in MIS] - \sum_{\substack{u,w \in X \\ u \neq w}} P[u \in MIS \wedge w \in MIS].$$

For the last inequality, we used the inclusion-exclusion principle truncated after the second order terms. Recall that M is the set of marked nodes after step 1, then

$$P[v \text{ is removed}] \geq \sum_{u \in X} P[u \in MIS] - \sum_{\substack{u,w \in X \\ u \neq w}} P[u \in MIS \wedge w \in MIS]$$

$$\geq \sum_{u \in X} P[u \in MIS] - \sum_{u \in X} \sum_{w \in X} P[u \in M] \cdot P[w \in M]$$

$$\geq \sum_{u \in X} \frac{1}{4d(u)} - \sum_{u \in X} \sum_{w \in X} \frac{1}{2d(u)} \frac{1}{2d(w)}$$

$$\geq \sum_{u \in X} \frac{1}{2d(u)} \left(\frac{1}{2} - \sum_{w \in X} \frac{1}{2d(w)} \right) \geq \frac{1}{6} \left(\frac{1}{2} - \frac{1}{3} \right) = \frac{1}{36}.$$

$\square$

It would be nice if there were many good nodes in each phase. Particularly, if a constant fraction of the nodes in each phase were good, the logarithmic running time would follow immediately. Unfortunately, this is not the case. In a star-graph, for instance, only the center node is good. The situation looks better, however, when we consider edges. In particular, let an edge be *good* if exactly one of its endpoints is good.

Lemma 3.3.6. *At any time, at least half of the edges are good.*

Proof. Direct each edge towards the higher degree node (or towards the node with higher identifier in case of a tie). We first show that for a *bad* node v (a node that is not good), the number of outgoing edges is at least twice the ingoing edges. If this was not the case, then at least one third of v's neighbors (denoted by set S) have a degree at most $d(v)$. Then,

$$\sum_{w \in N_v} \frac{1}{2d(w)} \geq \sum_{w \in S} \frac{1}{2d(v)} \geq \frac{d(v)}{3} \frac{1}{2d(v)} = \frac{1}{6},$$

which contradicts the assumption that v is a bad node.

Hence, the number of edges directed into bad nodes is at most half the number of edges directed out of bad nodes. This implies that at least half of the edges are directed into good nodes, rendering them good edges. $\square$

Finally, we are ready to prove the theorem.

Theorem 3.3.7. *Algorithm 3 terminates in time $O(\log n)$ with constant probability.*

Proof. By Lemma 3.3.5, a good node (and therefore a good edge!) will be deleted with constant probability. Due to Lemma 3.3.6, we know that at least half of the edges are good. Consequently, a constant fraction of the edges is deleted in each phase. After $O(\log |E|)$ phases, all edges are deleted. Because $|E| \leq n^2$ and hence, $\log |E| \leq 2 \log n$, it follows that the algorithm requires $O(\log |E|) = O(\log n)$ phases. Finally, each phase consists of a constant number of communication rounds. $\square$

It is important to notice that nowhere in the analysis of Algorithm 3, we made use of the fact that the underlying graph forms a unit disk graph. And in fact, Algorithm 3 works even in *general graphs*. Moreover, we will show in Section 3.4.4 and specifically in Theorem 3.4.5 that in general graphs, the running time of Algorithm 3 is close to optimal [233]. It is therefore natural to ask the question whether there exist faster algorithms for special graphs such as unit disk graphs. In fact, the Luby MIS algorithm presented in this section has long been the fastest known distributed MIS algorithm even in the case of unit disk graphs. In the next section, however, we prove that it is indeed possible to come up with a much faster solution if nodes can sense distances to their neighbors.

3.3.4 MIS with Known Distances

In the previous section, nodes did not have any additional knowledge such as coordinates, angles, or distances to neighboring nodes. As we show in this section, being able to measure distances to neighbors significantly helps in reducing the time-complexity required for computing a MIS. Specifically, if distances are known and if we assume message-size to be unbounded, it is possible to (deterministically) compute a MIS in time $O(\log^* n)$ [234], which is asymptotically optimal [260]. The function $\log^* n$ denotes the number of times the logarithm has to be taken before - starting from n - the value decreases below the constant 2. Note that $\log^* n$ is a function that grows extremely slowly in n. For instance, for $a = 10^{500}$, it holds that $\log^* a < 4$.

Before discussing this algorithm, however, it must be noted that the first paper to show how knowing distances helps in clustering was [147]. In this paper, Gao et al. present an algorithm that computes a constant approximation to the minimum dominating set problem in unit disk graphs in time $O(\log \log n)$, thus outperforming the MIS procedure of Section 3.3.3. The algorithm in [147] has recently been generalized in [237].

In this section, we present a simplified version of the $O(\log^* n)$ MIS algorithm. In particular, we present an algorithm that computes a *2-hop MIS*. In a 2-hop MIS, no two MIS nodes can be neighbors, and every non-MIS node must have a MIS node within its two-hop neighborhood. The motivation behind our focusing on 2-hop MIS is the following claim, which states that a 2-hop MIS can be transformed into a proper MIS in a unit disk graph in time $O(\log^* n)$ using techniques presented in [260, 81, 234].

Claim. Let $\mathcal{M}$ be a 2-hop MIS in a unit disk graph G. Given $\mathcal{M}$, there is a deterministic distributed algorithm which computes a MIS in G in time $O(\log^* n)$.

We first give a potentially slow deterministic distributed 2-hop MIS algorithm, which we then show to be implementable in time $O(\log^* n)$ [234]. For the slow algorithm (Algorithm 4), we make the additional assumption that all nodes

---

Algorithm 4: Network Decomposition: Clustering

1: $r := \min\{2^{-\lambda} \mid \lambda \in \mathbb{N} \wedge 2^{-\lambda} \geq d_{\min}\}$;
2: $\mathcal{V} := V$; $\mathcal{M} := \emptyset$;
3: **while** $r \leq 1$ **do**
4:    $\mathcal{G} := (\mathcal{V}, \mathcal{E})$ with $\mathcal{E} = \{\{u, v\} \mid d(u, v) < r\}$;
5:    compute MIS on $\mathcal{G}$ using algorithms in [81, 260];
6:    $\mathcal{V} := \{v \in \mathcal{V} \mid v \text{ in MIS}\}$;
7:    $r := r \cdot 2$
8: **end while**;

---

know the minimum distance $d_{\min}$ between any two nodes. This assumption is not required in the fast implementation anymore.

The idea of Algorithm 4 is to repeatedly decrease the number of potential MIS nodes in each iteration. Initially, the set $\mathcal{V}$ of possible MIS nodes contains all nodes. Starting from a small value, the radius r is increased by a factor 2 in every iteration of the while loop. In each iteration, the algorithm chooses a subset of the nodes in $\mathcal{V}$ such that the selected nodes form a maximal independent set on the graph consisting of all edges of length at most r. Only selected nodes remain in $\mathcal{V}$ for the next iteration. Initially, this r is set to the smallest value $2^{-\lambda}$, $\lambda \in \mathbb{N}$, which is larger than $d_{\min}$.

Lemma 3.3.8. *Algorithm 4 computes a correct 2-hop MIS.*

Proof. The algorithm maintains a set $\mathcal{V}$ of nodes which are candidates for becoming MIS nodes. In each iteration, some nodes are removed from $\mathcal{V}$. We prove that for all nodes u which are removed, there is a node v with $d(u, v) \leq 2$ which remains in $\mathcal{V}$ until the end. Let r_u be the radius at which u is removed from $\mathcal{V}$. For every node that is removed from $\mathcal{V}$, there is a node at distance at most r_u which stays in $\mathcal{V}$, because the independent set computed in line 5 is maximal. Hence, after removing u, there exists a node $u_0 \in \mathcal{V}$ with $d(u, u_0) \leq r_u$. If u_0 is removed in the subsequent iteration, there is a node u_1 with $d(u_0, u_1) \leq 2r_u$ that remains in $\mathcal{V}$. Hence, we get a sequence $u_0, u_1, \ldots, u_i, \ldots$ of nodes where $d(u_{i-1}, u_i) \leq 2^i r_u$, such that u_i remains in $\mathcal{V}$, i iterations after the removal of u. Summing up these distances results in a geometric series; for the distance between u and u_i, it follows

$$d(u, u_i) \leq \sum_{j=0}^{i} 2^j r_u < 2^{i+1} r_u = 2 r_{u_i},$$

where r_{u_i} is the radius of the iteration in which node u_i remains in $\mathcal{V}$ and in which u_{i-1} is removed from $\mathcal{V}$. Let v be the last node in the sequence, that is, v is a MIS node. Because the radius of the last iteration of Algorithm 4 is 1, it follows that $d(u, v) < 2$. Thus, every node is at most at distance 2 from its closest MIS node. Finally, the proof is concluded by noticing that in the last iteration, a MIS on the graph with edges of length at most 1 is computed, i.e., no two nodes in $\mathcal{V}$ at the end of the algorithm's execution are neighbors.   □

We now have a closer look at the complexity of a single iteration of the while loop of Algorithm 4. Apart from line 5, everything can easily be computed in a constant number of communication rounds. The time complexity of computing the MIS in line 5 depends on the maximum degree Δ of the graph. For small Δ the fastest known algorithms are based on coloring algorithms. A coloring with K colors can be turned into a MIS in K rounds of communication. First all nodes with color 1 join the MIS. Then, nodes with color 2 join the MIS if they do not yet have a neighbor in the MIS, and so forth. Thus, if the graph can be colored using K colors in t rounds, a MIS can be computed in a distributed manner in $t + K$ rounds.

In [81], an extremely elegant algorithm which colors a graph with 3^{Δ} colors in $O(\log^* n)$ rounds is described. By the above reasoning, this yields a MIS algorithm with time complexity $O(\log^* n + 3^{\Delta})$. The algorithm in [81] was later improved in [163] where an algorithm for computing a MIS in time $O(\log \Delta(\Delta^2 + \log^* n))$ is given. Finally, in [260], it has been shown that it is even possible come up with an algorithm that has time complexity $O(\Delta^2 + \log^* n)$ for computing a MIS. The important thing to note is that the time complexity of computing the MIS in line 5 depends on the maximum degree Δ of the graph. Particularly, for constant Δ, all three algorithms compute a MIS in $O(\log^* n)$ rounds. Fortunately, in a unit disk graph, the maximum degree of the graph $\mathcal{G}$ in each iteration is bounded by a constant, as shown in the subsequent lemma.

Lemma 3.3.9. *In each iteration of Algorithm 4, the maximum degree of $\mathcal{G}$ is constant.*

Proof. Consider an arbitrary iteration of the algorithm and let ℓ be the minimum distance between any two nodes of $\mathcal{V}$. As the algorithm computes an independent set in each iteration, we have $\ell > r/2$. Therefore, disks of radius $r/4$ around each node in $\mathcal{V}$ do not overlap. The number of such disks that can maximally be packed to cover nodes in a disk with radius r is bounded by a constant.   □

In combination with one of the above MIS algorithms, Lemma 3.3.9 implies the following corollary.

Corollary 3.3.10. *The distributed time complexity of a single iteration of the while loop of Algorithm 4 is $O(\log^* n)$.*

The final problem we face is that the number of iterations executed by Algorithm 4 is bounded only by $\log(1/d_{\min})$, which can be arbitrarily large. Hence, the overall time complexity of Algorithm 4 is in $O(\log(1/d_{\min}) \cdot \log^* n)$. In order to reduce this to the claimed $O(\log^* n)$, we need to add some general facts concerning the synchronous message passing model of distributed computing. If nodes communicate for k rounds, they can only gather information which is at most k hops away. With unbounded message size, every distributed k-round algorithm can principally be formulated as follows (for a detailed explanation of this transformation, we refer the reader to the chapter on lower bounds, specifically to Section 7.2.2):

1. Collect complete k-neighborhood in graph in k communication rounds.
2. Compute the output by locally simulating the relevant part of the distributed algorithm (no communication needed).

Collecting the complete k-neighborhood can be achieved if all nodes send their complete states to all their neighbors in every round. After round i, all nodes know their i-neighborhood. Learning the i-neighborhoods of all neighbors in round $i + 1$ suffices to know the $i + 1$-neighborhood.

Keeping this observation in mind, we now return again to a single iteration of the while loop in Algorithm 4 in which the radius is r_h. The graph $\mathcal{G}$ contains all information needed to compute a single iteration. Hence, all messages are sent on edges having length at most r_h. Thus, if nodes communicate for k rounds and all messages are sent over edges of length at most r_h, then all collected information is derived from at most distance $k \cdot r_h$. That is, in order to compute everything locally, nodes can collect the complete neighborhood up to distance kr_h, which can be achieved with at most $2kr_h$ communication rounds. Applying this reasoning to Algorithm 4, we obtain Theorem 3.3.11.

Theorem 3.3.11. *Algorithm 4 can be implemented with $O(\log^* n)$ communication rounds.*

Proof. By Corollary 3.3.10, the number of rounds per iteration is $O(\log^* n)$. Therefore, nodes need to collect information from distance at most $O(r \log^* n)$. In order to obtain the total distance from which nodes need information, nodes can sum up the distances for all iterations. Even though nodes do not know the number of iterations in advance, the sum can be upper bounded by taking two times the maximum summand, because r grows exponentially by a factor of 2 in each iteration. It follows that the whole while loop can be computed in $O(\log^* n)$ rounds, which completes the proof.   □

Note that when collecting the whole neighborhood, it is not necessary that nodes know the minimum distance $d_{\min}$ between nodes. Because the radius grows exponentially, the locality of the problem is independent of the starting radius. Each node can for example use the smallest distance in its neighborhood in order to locally simulate the distributed Algorithm 4.

The fast implementation of Algorithm 4 requires unbounded message size. This disadvantage renders the algorithm rather impracticable. On the other hand, the algorithm demonstrates that theoretically, the distributed time complexity for computing a MIS can be significantly below $O(\log n)$ if distances can be measured. It should further be noted that the algorithm discussed in this section can be generalized to work not only in unit disk graphs, but in a much more general class of graphs. As shown in [234], the algorithm achieves the same asymptotic results in all network models in which nodes are located in a metric space with constant doubling dimension.

Finally, it is important to note that Algorithm 4 is *asymptotically optimal.* This follows from the seminal $\Omega(\log^* n)$ lower bound given by Linial in [260] (cf Section 7.2). This lower bound proves that computing a MIS in a ring requires at least time $\Omega(\log^* n)$. Since unit disk graphs (even with distance information) can form a ring, it is clear that this lower bound applies to our setting. It is intriguing that from the point of view of locality and distributed complexity, a simple toy-network such as the ring is as hard as the vast family of unit disk graphs and even more general classes of graphs!

3.3.5 Clustering in More Realistic Models

In the previous sections, we have considered distributed algorithms in the classic message passing model, i.e., nodes were able to send messages to all their neighbors without fearing collisions or interference. This implies that there exists an underlying medium access control (MAC) layer which provides a corresponding abstraction to higher-layer protocols and applications.

When being deployed, however, wireless ad hoc and sensor networks are *unstructured* and chaotic [231], lacking any established MAC layer. Before any reasonable communication can be carried out, and before the network can start performing its intended task, the nodes must set up some kind of structure or MAC layer that allows an efficient communication scheme. Once this *initial structure* is achieved, more sophisticated algorithms and protocols may be used on top of it. This self-organized transition from a chaotic, unstructured network into a network containing an initial structure – *initialization phase* – is of practical importance. In systems such as Bluetooth, for example, the initialization tends to be slow even for a small number of devices.

In this section, we focus on the construction of an *initial clustering* as a structure for subsequent network organization tasks. In view of our goal of setting up an initial clustering in a newly deployed network, clustering algorithms for the initialization phase must not rely on any previously established MAC layer. Instead, we need algorithms that quickly come up with a clustering entirely from *scratch.* Clearly, this precludes algorithms working under any sort of *message passing model* in which nodes know their neighbors a priori, and in which messages can reliably be sent to neighbors. Studying classic network coordination problems such as clustering in absence of an established MAC layer highlights the *chicken-and-egg* problem of the initialization phase [232]. A structure ("chicken") helps achieving a clustering ("egg"), and vice versa. The problem is that in a newly deployed ad hoc/sensor network, there is typically no built-in structure, i.e. there are neither "chickens" nor "eggs."

An important aspect when studying clustering during the initialization phase of ad hoc/sensor networks is to use an appropriate model. On the one hand, the model should be realistic enough to actually capture the particularly harsh characteristics of the deployment phase. But on the other hand, it ought to be concise enough to allow for stringent reasoning and proofs. Recently, the *unstructured radio network model* has been proposed as a model that attempts to combine both of these contradictory aims [232, 231]. It makes the following assumptions.

1. Networks are multi-hop, that is, there exist nodes that are not within mutual transmission range. Therefore, it may occur that some neighbors of a sending node receive a message, while others experience interference from other senders and do not receive the message. The wireless nature of the communication graph is modeled as a unit disk graph.
2. Collisions can occur, i.e., a message is received only if exactly one neighbor transmits in the corresponding time-slot. Furthermore, nodes do not feature a reliable *collision detection* mechanism. This assumption is often realistic, considering that nodes may be tiny sensors with equipment restricted to the minimum due to limitations in energy consumption, weight, or cost. Also, a sender does not know how many (if any at all!) neighbors have received its transmission correctly.
3. Nodes can wake up *asynchronously*. In a wireless, multi-hop environment, it is realistic to assume that some nodes wake up (e.g. become deployed, or switched on) later than others. Thus, nodes do not have access to a global clock. Contrary to work on the so-called *wake-up problem* [151, 204], nodes are *not* woken up by incoming messages, that is, sleeping nodes neither send nor receive any messages. Finally, the node's wake-up pattern can be completely arbitrary.
4. At the time of their waking-up, nodes have only limited knowledge about the total number of nodes in the network and no knowledge about the nodes' distribution or wake-up pattern. Particularly, they have no a priori information about the number of neighbors and when waking up, they do not know how many neighbors have already started executing the algorithm.

Naturally, algorithms for such unstructured and chaotic networks have a different flavor compared to "traditional" algorithms that operate on a given network graph that is static and well-known to all nodes. Nonetheless, it is interesting that efficient solutions for problems such as MIS can be found even in this harsh model. In the following, we review some of the recent results on clustering in radio networks.

In [231], a distributed algorithm has been proposed which computes a constant approximation to the minimum dominating set (roughly) in time $O(\log^2 n)$ in unit disk graphs. The paper then generalizes the result to more elaborate network models beyond unit disk graphs. On the downside, the algorithm requires three independent communication channels which may be realized using an FDMA scheme. The algorithm in [231] was subsequently improved in [295] in which the authors present an algorithm that computes a MIS in the same model in time $O(\log^2 n)$ with high probability. Moreover, the algorithm in [295] requires only a single communication channel and is therefore more useful from a practical point of view. In a somewhat easier radio network model, [310] presents an algorithm that computes a connected dominating set in time $O(\log^2 n)$. Finally, it should be noted that in view of a $\Omega(\log^2 n/\log\log n)$ lower bound given in [204], the running time of $O(\log^2 n)$ for computing a MIS or any (non-trivial) dominating set is close to being asymptotically optimal in the unstructured radio network model.

3.4 Clustering Algorithms for General Graphs

We now return to the *synchronous message passing model* of distributed computing in which every node can send a message to every neighbor in each round of communication. As we have seen in Section 3.3.3, there is a distributed algorithm for computing a maximal independent set in arbitrary graphs in logarithmic time. While in unit disk graphs, a MIS constitutes a constant approximation to the MDS problem, the same does not hold in arbitrary graphs. In this section, we attempt to give an overview over distributed approximation algorithms for the minimum dominating set problem in arbitrary graphs.

In the centralized case, the simple greedy algorithm yields an $O(\log \Delta)$ approximation, where Δ denotes the maximum degree in the graph [74]: As long as there are uncovered nodes, the greedy algorithm adds a node to the dominating set which covers the biggest number of uncovered nodes. More formally, at any time during the execution of the algorithm, let the *span* $d(v)$ of a node v be the number of uncovered nodes in v's neighborhood. The greedy algorithm repeatedly adds the node with the maximum span to the dominating set, until all nodes are covered. Furthermore, it has been shown that under reasonable complexity assumptions, the greedy algorithm is asymptotically optimal [123].

While the greedy algorithm does not automatically yield a connected dominating set, it is straightforward to obtain a connected dominating set from a given dominating set without deteriorating its approximation ratio too much. The algorithm's idea is to simply *connect* the dominating set obtained by the greedy algorithm. It iteratively merges two connected components by adding at most 2 nodes to the dominating set [172]. This idea of connecting a given dominating set by *bridges* of length at most 2 has been reused - albeit in a more subtle way - in the *distributed* algorithms studied for instance in [105, 382].

3.4.1 Distributed Greedy Algorithm

Can the sequential greedy algorithm presented in the previous section be parallelized in an efficient way? In the sequential greedy algorithm, the algorithm picks a single node with maximum span. A simple distributed version of this process can be implemented based on the following observation: The span of a node can only be reduced if any of the nodes at distance at most 2 are included in the dominating set. Therefore, if the span of node v is greater than the span of any other node at distance at most 2 from v, the greedy algorithm chooses v before any of the nodes at distance at most 2. This leads to a very simple distributed version of the greedy algorithm: At the beginning of each step, every node computes its span. A node v joins the dominating set if it has the largest span in its 2-hop neighborhood. It then informs its neighbors about their being covered. Note that every such step can be implemented in 2 communication rounds in a message passing model using a local flooding.

Theorem 3.4.1. *The Distributed Greedy Algorithm computes in at most n communication rounds an $\ln \Delta$-approximation for the minimum dominating set problem.*

Proof. The approximation quality follows directly from the above observation and the analysis of the sequential greedy algorithm. The time complexity is at most linear because in every other round, at least one node is added to the dominating set.   □

The approximation ratio of the above distributed algorithm is identical to the sequential algorithm and thus asymptotically optimal. As for the distributed time complexity, however, there are graphs on which in each iteration of the while loop, only one node is added to the dominating set. As an example, consider the graph in Figure 3.5. An optimal dominating set consists of all nodes on the center axis. The *distributed greedy algorithm* computes an optimal dominating set, however, the nodes are selected one-by-one from left to right. Hence, while the running time of the algorithm cannot exceed n, it can be as bad as $\Omega(\sqrt{n})$.

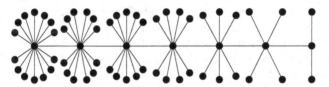

Fig. 3.5. Distributed greedy algorithm: Bad example

In the sequel, we give a brief overview over some of the most recent results on the distributed approximation of dominating sets in arbitrary graphs.

3.4.2 Efficient Distributed Dominating Set Algorithm

Jia, Rajaraman, and Suel [200] have proposed an elegant distributed algorithm which manages to circumvent the problem mentioned in the previous section. In particular, their *Local Randomized Greedy* (LRG) algorithm selects nodes not only based on their span, but also based on how the nodes covered by an individual node are also covered by other nodes.

Let N_v^2 be the set of all nodes that are at most two hops away from v. The LRG algorithm proceeds in rounds, and in each round, every node v executes the computation shown in Algorithm 5 until every node is covered.

Algorithm 5: Local Randomized Greedy Algorithm (LRG)

1: **Span Calculation:** Calculate span $d(v)$. Let the rounded span $\hat{d}(v)$ be the smallest power of 2 which is at least $d(v)$.

2: **Candidate Selection:** A node $v \in V$ becomes a *candidate* if $\hat{d}(v)$ is at least $\hat{d}(w)$ for all $w \in N_v^2$. For each candidate v, *cover* $C(v)$ denotes the set of uncovered nodes that v covers, i.e., $|C(v)| = d(v)$.

3: **Support Calculation:** Each uncovered node u calculates its support $s(u)$, which is the number of candidates that cover u:
$s(u) = |\{v \in V \,|\, u \in C(v)\}|$

4: **Dominator Selection:** Each candidate joins the dominating set D with probability $P_D(v) = 1/med(v)$, where $med(v)$ is the median support of all nodes $u \in C(v)$.

Intuitively, the LRG algorithm tries in each round to select a small number of dominators while covering a large fraction of uncovered nodes. Let the number of candidates that cover a node u be its *support* $s(u)$. Ideally, if a node u is covered by $s(u)$ candidates, exactly one (or only few) of these candidates should be selected as a dominator. An easy approach would therefore be to select each of these candidates with probability $1/s(u)$. The outcome of this random choice would cover u with a constant probability, and the expected number of dominators would be 1. The inherent problem with this idea is that there exists no assignment of probabilities to candidates which would guarantee the above property for *all* nodes $u \in V$.

The LRG algorithm addresses this challenge by assigning each candidate a probability that is the reciprocal of the median support $s(u)$ of all nodes it covers. One effect of this choice of $P_D(v)$ is that a node u becomes covered with constant probability if its support $s(u) \geq 1/P_D(v)$ for more than half the candidates v that cover u. Additionally, this choice of $P_D(v)$ does not yield too many dominators in each round and thus prevents a deterioration of the resulting approximation ratio. Specifically, as shown in [200], the LRG algorithm achieves the following performance guarantees.

Theorem 3.4.2. *The* LRG *Algorithm computes a dominating set of expected size* $O(\log \Delta)$ *in time* $O(\log n \log \Delta)$ *with high probability, where* Δ *is the maximum degree in the network.*

In their paper [200], the authors give an example showing that the analysis of both the algorithm's running time and approximation ratio is tight. Furthermore, they generalize the algorithm to solve dominating set problems in which nodes need to be covered by multiple dominators. Such multiple coverage dominating sets may be particularly useful for the purpose of providing fault-tolerance to failure-prone or dynamic ad hoc and sensor networks.

3.4.3 Constant Time Dominating Set Algorithm

The LRG algorithm as presented in the previous section has only polylogarithmic running time. While this is certainly efficient, the dominating set problem appears to be inherently *local* by nature and it is not clear whether a logarithmic or polylogarithmic number of communication rounds are actually required in order to obtain a non-trivial approximation result. Intuitively, it seems plausible that after communicating with its neighbors for a few times, every node should be capable of deciding whether or not to join the dominating set. Moreover, it seems unlikely that very distant nodes can influence a node's decision whether to join the dominating set.

The question whether the minimum dominating set problem can be nontrivially approximated with a constant number of communication rounds has been studied in [238] and subsequently - as part of a more general solution for arbitrary covering and packing linear programs in [237].

The approach taken in these papers differs from the LRG algorithm in several ways. Most strikingly, the algorithms in [238] and [237] do not *directly* compute a dominating set. Instead, these algorithms make the detour of first distributedly approximating the *fractional version* of the dominating set problem which is captured by the following linear program in which for each node $v \in V$, the variable x_v is set to 1 if and only if v joins the dominating set, and to 0 otherwise:

$$
\begin{aligned}
\min \quad & \sum_{v \in V} x_v \\
\text{s.t.} \quad & \sum_{w \in N_v} x_w \geq 1 \qquad \forall v \in V \\
& x_v \geq 0 \qquad \forall v \in V.
\end{aligned}
$$

For any parameter k (which can be a constant), the algorithms run the distributed LP approximation for $O(k)$ communication rounds. In a second step, the solution to the minimum dominating set problem is obtained by rounding the fractional x_v values using a distributed randomized rounding scheme. This scheme requires merely a constant number of additional communication rounds.

Assume that in $O(k)$ rounds, the distributed algorithm achieves an α approximation to the fractional dominating set problem. Then, the final rounding step typically incurs an additional factor of $O(\log \Delta)$ in the approximation factor [238], yielding a final approximation guarantee of $O(\alpha \log \Delta)$ in time $O(k)$. So, the question is how well can the linear program be approximated in $O(k)$ rounds.

The algorithm presented in [238] achieves an approximation ratio of $O(\sqrt{k}\Delta^{1/\sqrt{k}})$ for the linear program, yielding a final approximation guarantee of $O(\sqrt{k}\Delta^{1/\sqrt{k}} \log \Delta)$ for the minimum dominating set problem. When choosing $k = O(\log^2 \Delta)$, the approximation ratio becomes $O(\log^2 \Delta)$. The improved algorithm of [237] achieves an approximation ratio $O(\Delta^{1/\sqrt{k}} \log \Delta)$ in time $O(k)$. Invoking this algorithm with a running time parameter of $k = O(\log^2 \Delta)$, the approximation ratio becomes $O(\log \Delta)$, roughly the same trade-off as achieved by the LRG algorithm presented in Section 3.4.2. Unlike the LRG algorithm, however, the running time of algorithm [237] can be parametrized, i.e. it yields non-trivial results even with a constant number of communication rounds.

3.4.4 Lower Bound for MIS and Dominating Set

The results in the previous section show that even in a constant number of communication rounds, the minimum dominating set problem can be approximated to within a non-trivial factor. This naturally raises the question how well the problem can be approximated in a given number of communication rounds. In fact, studying the distributed approximability of the minimum dominating set problem sheds light into the achievable trade-off between the amount of communication (time-complexity) and the quality of the resulting solution (approximation ratio). Moreover, it turns out that insight into this fundamental trade-off also yields time lower bounds for the distributed computation of maximal independent sets.

In [233], Kuhn, Moscibroda, and Wattenhofer present general lower bounds for the distributed approximability of the minimum dominating set problem, for an arbitrary number of communication rounds k. In particular, the results are summarized in the following theorems:

Theorem 3.4.3. *In k communication rounds, every (possibly randomized) distributed algorithm for the minimum dominating set problem has an approximation ratio of at least*

$$\Omega\left(\frac{n^{c/k^2}}{k}\right) \quad or \quad \Omega\left(\frac{\Delta^{1/k}}{k}\right)$$

for some constant $c \geq 1/4$, even if messages are unbounded and nodes have unique IDs.

Theorem 3.4.4. *In order to achieve a constant or polylogarithmic approximation ratio, every (possibly randomized) distributed algorithm requires at least*

$$\Omega\left(\sqrt{\frac{\log n}{\log \log n}}\right) \quad or \quad \Omega\left(\frac{\log \Delta}{\log \log \Delta}\right)$$

communication rounds, even if messages are unbounded and nodes have unique IDs.

Theorem 3.4.5. *In order to compute a maximal independent set (MIS), every (possibly randomized) distributed algorithm requires at least*

$$\Omega\left(\sqrt{\frac{\log n}{\log \log n}}\right) \quad or \quad \Omega\left(\frac{\log \Delta}{\log \log \Delta}\right)$$

communication rounds, even if messages are unbounded and nodes have unique IDs.

It is interesting to compare these lower bounds with the upper bounds that we have studied earlier in this chapter. The LRG algorithm of Section 3.4.2 as well as the algorithm for general covering and packing problems given in [237] achieve a logarithmic approximation in $O(\log n \log \Delta)$ communication rounds. The above lower bound shows that at least $\Omega(\sqrt{\log n / \log \log n})$ or $\Omega(\log \Delta / \log \log \Delta)$ rounds are required by any algorithm. Hence, while these upper and lower bounds are not tight, they are not too far apart either. Particularly, they indicate that the trade-off between the amount of communication and the achieved approximation ratio is close to optimal in both algorithms.

Moreover, in Section 3.3.3, we have seen Luby's randomized MIS algorithm [266] which computes a maximal independent set in time $O(\log n)$. The lower bound proves that $\Omega(\sqrt{\log n / \log \log n})$ rounds are required, again indicating that this MIS algorithm is close to being asymptotically optimal.

Finally, it is interesting to note that the lower bound does not hold in the unit disk graph model and hence, more efficient solutions are possible there. Specifically, the algorithm discussed in Section 3.3.4 achieves a MIS (and therefore a constant approximation to the dominating set problem) in unit disk graphs in time $O(\log^* n)$ [234], which is much faster than the $\Omega(\sqrt{\log n / \log \log n})$. The comparison of these two results offers an intriguing insight into the respective complexities of the two most basic and frequently studied computational models in ad hoc and sensor networks.

3.5 Conclusions and Open Problems

Clustering ad hoc and sensor networks poses numerous challenging algorithmic questions. In this chapter, we have presented different algorithms and techniques for achieving good quality clusterings that require only a limited number of communication rounds. On the other hand, we have also seen how overly simple solutions can lead to poor results in worst case scenarios, both from the point of view of time complexity and approximation guarantee. It has also become apparent that depending on the specific underlying model assumptions, more or less efficient clustering solutions become possible or remain impossible, respectively. While in general graphs every distributed algorithm requires at least time $\Omega(\sqrt{\log n / \log \log n})$ for computing a MIS, for instance, there exist $O(\log^* n)$ time algorithms for unit disk graphs if distances can be measured.

Many interesting and important questions remain open. In the unit disk graph model without distances, the best known algorithms are not faster than algorithms in general graphs. Intuitively, one would expect that faster algorithms should be possible, particularly in view of the $O(\log^* n)$ algorithm in unit disk graphs if distances are known. Besides this, it would be interesting to study variants of clusterings which guarantee *fault-tolerance* or maintain *energy efficient* structures. Also, studying aspects such as *mobility* from an algorithmic perspective is an important challenge. Finally, there is the question of how to maintain a clustering over long period of time without unfairly using up all energy at clusterheads. This lifetime-maximization problem can be mapped to the so-called (fractional) maximum domatic partition problem and has been studied, for instance in [296, 316]

3.6 Chapter Notes

The usefulness of dominating set or MIS-based clustering in the context of wireless ad hoc and sensor networks has been documented in a number of studies. For instance, clustering has been shown to be a useful tool for routing (e.g. [382, 398, 354, 388]), for setting up and maintaining MAC layer protocols (e.g. [324, 181]), or to improve the network's energy-efficiency during both initialization and the operational phase (e.g. [99, 291]).

The $\ln \Delta$-approximability lower bound for the minimum dominating set problem was proven in [123] and this bound is achieved by the standard greedy algorithm [74]. In the unit disk graphs and generalizations thereof, the problem remains $\mathcal{NP}$-hard, but *polynomial time approximation schemes* (PTAS) become possible [192, 70, 116]. In [303], Nieberg and Hurink show that in unit disk graphs, a PTAS is possible even in the absence of the graph's geometric representation in the plane. Most recently, Ambühl et al prove that the weighted version of the minimum dominating set problem can be approximated within a constant factor in unit disk graphs [14].

In [246], Kutten and Peleg present the first *distributed* minimum dominating set algorithm that computes a dominating set of size at most $n/2$ in time $O(\log^* n)$ in general graphs. The algorithm proposed in [398] and its recent adaptation in [94] computes a connected dominating set in constant time, but does not have good worst-case guarantees. The distributed algorithm presented in Section 3.4.2 was proposed in [200]. Constant-time distributed algorithms achieving non-trivial approximation guarantees have been devised in [238, 237], as well as in [236] for the fault-tolerant version of the problem. Finally, the time and approximability lower-bounds for distributed dominating set and MIS algorithms mentioned in Section 3.4.4 were derived in [233].

In unit disk graphs with known distances, the randomized algorithm of [147] computes an $O(1)$-approximation in expectation and features a running time of $O(\log \log n)$. This result was recently generalized to the fault-tolerant version in [236]. The $O(\log^* n)$-time algorithm presented in Section 3.3.4 was proposed in [234]. In the absence of any distance information, the algorithm of [382] computes an MIS (and hence, a constant approximation to the minimum dominating set problem) in linear time. This can be improved using Luby's MIS algorithm presented in Section 3.3.3 [260]. A deterministic, distributed MIS algorithm in generalized unit disk graphs with an improved running time of $O(\log \Delta \log^* n)$ was given in [229]. Also, a distributed implementation of the centralized PTAS for computing minimum dominating sets in unit disk graphs [303] was developed in [230]. Other distributed algorithms for clustering in unit disk graphs include [310, 388] and an empirical comparison of some of the above mentioned clustering algorithms has been conducted in [33].

4

MAC Layer and Coloring

Steffen Mecke

4.1 Introduction

The *Medium Access Layer* is part of the *Data Link Layer* (layer 2 of the OSI model) and sits directly on top of the Physical Layer (layer 1). Its purpose is to manage access to the shared wireless medium. If transmissions from two different nodes arrive simultaneously at a receiver, neither of them is received due to interference.

In this section we start by describing the problem in more detail and give a short and necessarily incomplete overview of some of the approaches that are widely used by the network community, such as CSMA, CSMA/CA, and TDMA (see [249] for a more detailed survey). Later on, we will introduce some other approaches for managing concurrent transmissions, namely scheduling, frequency assignment and coloring approaches. We will conclude the introductory section by introducing several variants of coloring problems and try to argument why coloring problems are an important and useful subject to study in order to understand and solve MAC layer issues. In Section 4.2 we present an overview of some distributed coloring algorithms. We start with simple heuristics and try to gradually develop more and more sophisticated methods. We then describe two completely different approaches to efficiently solve coloring problems in a nearly optimal fashion. While the second approach is based on a non-constructive combinatorial proof, the first one, called deterministic coin tossing, can lend itself to practical solutions to the coloring problem.

4.1.1 MAC Protocols

A *Medium Access Control (MAC)* protocol (sometimes also called a *Distributed Coordination Function* or *DCF*) decides when competing nodes may access the shared medium, i.e. the radio channel, and tries to ensure that no two nodes are interfering with each other's transmissions. In the event of a

D. Wagner and R. Wattenhofer (Eds.): Algorithms for Sensor and Ad Hoc Networks, LNCS 4621, pp. 63–80, 2007.

collision, a MAC protocol may deal with it through some contention resolution algorithm, for example by resending the message later at a randomly selected time. Alternatively, the MAC protocol may simply discard the message and leave the retransmission—if any—up to the higher layers in the protocol stack. MAC protocols for wireless networks have been studied since the 1970s, but the successful introduction of wireless LANs in the late 1990s has accelerated the pace of developments.

The assessment of the quality of a solution to this problem must take into account many aspects. Among the possible target functions are maximal channel utilization (collisions), latency, energy usage (idle network, per bit, per node), number of channels (frequencies), memory usage, simplicity, and others. These goals are linked with each other, some are conflicting, some are correlated. For example, a good channel utilization usually comes with low average latencies; however, sometimes at the price of fairness, if some nodes are allowed to send more often than others. The number of used channels is often assumed to be one, nowadays (but see also Section 4.1.3). Either there *is* only one channel (due to technical or legal restrictions) or certain tricks can be used to make multiple channels appear as one, like frequency division multiple access (*FDMA*). Wireless sensor nodes are often very restricted in their capabilities, that is, they have slow processors and little and slow memory. Therefore, simple algorithms are essential because sensor nodes cannot afford to do extensive computations nor store much information about the topology of the network. It is important to keep this in mind although algorithmic measures, like "number of messages sent", often do not take this into account. Energy usage, finally, is one of the most important and most complex requirements. MAC schemes which guarantee low latency or high bandwidth often do this at the cost of high energy usage. If energy requirements are tight, it is often advisable to sacrifice some of the other goals in order to permit nodes to take advantage of sleep modes. Recently, many attempts have been made to balance these conflicting goals in one or the other direction. How much energy can be saved often depends highly on the requirements of applications, such as predominant communication patterns and the structure of the network, as well as of the concrete hardware. It is therefore difficult, if not impossible, to speak of *the* optimal MAC scheme. Instead, users or system designers have to pick the right protocol for their specific application.

There are two main families of MAC protocols: contention based and scheduled protocols. In contention based protocols, nodes contend for the medium whenever they need to send a packet. In schedule based approaches, sending is done according to a predefined schedule. Most existing schedule based protocols also have a contention based element that allows newly arriving nodes to join the network.

One of the first protocols was *ALOHA* [10, 333]. It is extremely simple: If a node needs to send a packet, it simply sends it as soon as it is generated. If the transmission is received, the receiver sends an acknowledgment message (*ACK*). In the case of a collision, no acknowledgment is generated, and the

sender retries after a random period. This mechanism, however, results in a
very poor channel usage if there are many nodes, due to many collisions.

CSMA and CSMA/CA. The *Carrier Sense Multiple Access (CSMA)*
method ([365]) tries to avoid collisions. When a node wants to send a message,
it first listens if another node is transmitting. If it does not sense any traffic,
it assumes that the channel is clear and starts transmitting. If the medium is
busy, it waits until it becomes free, waits for a random back-off time (in order
to avoid starting its transmission at the same time as other waiting nodes)
and starts sending after this time has elapsed. For several reasons this method
still cannot avoid collisions entirely. One key problem is the so-called *hidden
terminal* problem (see Fig. 4.1). If node   and node C both want to send to
node B, but they cannot hear each other, they are both going to sense the
carrier to be idle and start sending, which leads to a collision at B. This problem is addressed by the Medium Access with Collision Avoidance (*MACA*,
[209]) protocol. It introduces a three-way handshake: The sender (   ) initiates
it by transmitting a (short) *RTS* (Request-To-Send or Ready-To-Send) message. The receiver (B) responds with a Clear-To-Send (*CTS*) packet, which
informs C of the upcoming transfer. Nodes overhearing an RTS or CTS message do not transmit. The data transfer from   to B is now "guaranteed" to
be collision free. There can still be collisions if two nodes are sending RTS
packages at the same time, for example, but these packages are assumed to be
so small that the probability of such a collision is assumed always to be negligible. After   's transmission is finished, B sends an *ACK* message, notifying
all listening nodes (C) that the medium is now idle again.

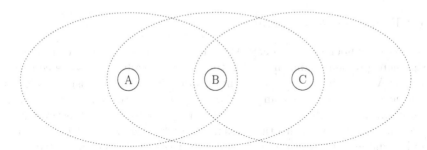

Fig. 4.1. The hidden terminal problem.

This *CSMA/CA* (Carrier Sense Multiple Access with Collision Avoidance)
mechanism is widely used today, for example by the IEEE 802.11 wireless
LAN protocol family ([191]). In practice and theory ([35]), a channel usage
(throughput) of up to 80 % has been found. Numerous similar protocols have
been proposed (for example, B-MAC, S-MAC, T-MAC, WiseMAC, ...) to
address the special needs of wireless sensor networks, especially regarding
energy efficiency.

TDMA. While CSMA-based protocols have been widely used in ad hoc, multi-hop networks, there is an entirely different approach. If there is a central authority (*access point*, such as in WLAN and mobile phone networks), traffic can be regulated by broadcasting a schedule that specifies when and for how long each node may transmit. The most important approach is Time-Division Multiple Access (*TDMA*). TDMA systems divide the channel into short slots which are grouped into larger frames. The access point decides which slot is to be used by which node. If nodes do not usually communicate with each other, but mainly with the access point (as in wireless LANs), scheduling is especially easy.

TDMA protocols can reduce protocol overhead and facilitate power saving mechanisms. The main drawback is the increased cost (in terms of communication, computational power, and other resources) at the access point. In many scenarios for ad hoc and sensor networks, there is no infrastructure like access points or base stations. In that case, the nodes either have to elect access points ($\leadsto$ clustering, cf. Chapter 3) or agree on a schedule. One such approach is proposed in [371].

4.1.2 Scheduling

There have been entirely different approaches to avoid collisions of transmissions on the shared medium. One of them is to optimally schedule all the traffic in a network at once, but this approach is hardly feasible for ad hoc networks because it uses global knowledge and centralized algorithms. See, e.g., [112], [18], or [176] for results on scheduling and also Section 6.4.

4.1.3 Frequency Assignment

There is a number of Frequency Assignment problems (also called channel assignment problems) known in literature. They usually share some common features: A set of direct wireless communication connections must be assigned frequencies such that the transmissions between any two endpoints of each connection are possible. The frequencies assigned to two connections may cause interference at each other, resulting in loss of transmissions. In the simplest model, two transmissions interfere if they have the same frequency and the senders are geographically close to the receiver. More complex models take into account that different pairs of frequencies can cause different levels of interference and/or consider more detailed models of radio propagation. The usual goal is to minimize the number of frequencies used and the amount of interference caused by the assignment. Techniques to solve these problems include mathematical optimization as well as graph-theoretic methods. There are also dynamic versions of frequency assignment problems.

Most frequency assignment approaches have been developed in the context of cellular networks. Usually, they either attempt to cover a certain area such that every point is served by a base station or they manage traffic between base stations. In both cases, the base stations are seen as fixed over a long period of time. Solutions can be searched for in advance, and they are valid over a long period of time. Also, many details of the scenario are known in advance, for example the exact profile of the terrain. All these points usually do not apply to sensor and ad hoc networks. In most scenarios neither the terrain nor traffic patterns are known in advance and the goal is not to manage traffic between base stations and clients. Therefore, many of the sophisticated tools developed for frequency assignment are not applicable for ad hoc networks.

4.1.4 Coloring Problems

Many frequency assignment and other problems can be formulated as coloring problems. They come in different flavors. All of the following problems are minimization problems. They differ in what objects are to color and subject to which constraints. Here we will state the complexity status of some of them. See the chapter notes (Section 4.4) for references.

The most classical coloring problem is the *vertex coloring* problem. Given a graph, a vertex coloring is an assignment $\chi\colon V \to \mathbb{N}$ of integer numbers (colors) to vertices such that no two adjacent vertices have the same color. The vertex coloring problem consists of finding the minimum number of colors needed to color a given graph. It is $\mathcal{NP}$-hard and, for general graphs, it is even hard to approximate an optimal solution. An *edge coloring* is an assignment of colors to edges such that edges sharing a common endpoint have the same color. The problem of finding a minimum edge coloring is also $\mathcal{NP}$-complete and can be reduced to vertex coloring—but edge coloring is easier to approximate.

A generalization of vertex coloring is *distance-d-coloring*. A coloring is a distance-d-coloring if for every vertex v all vertices at distance at most d to v have a color which is different from v's color. A standard vertex coloring is a distance-1-coloring. Finding a distance-d-coloring in a graph $G = (V, E)$ is equivalent to finding a distance-1-coloring in the graph $G' = (V, E')$, where two nodes are adjacent in G' if and only if there distance in G is at most d.

If we allow a vertex to have at most D neighbors of the same color, we have the problem of *coloring with defect D*. Another generalization of the coloring problem is *T-coloring*: for a set $T \subset \mathbb{N}$, a T-coloring is a coloring f that satisfies the additional constraint that $|f(x) - f(y)| \notin T$ for any edge (x, y). In this case, not only the number $\chi_T(G)$ of used colors is of interest, but also the range, from which the colors are chosen (denoted by $\mathrm{span}_T(G)$). An ordinary vertex coloring is a $\{0\}$-coloring. There is always a 0-coloring with $\chi_T(G) = \mathrm{span}_T(G)$.

A different problem, arising in cellular networks, is *CF-coloring* (conflict-free coloring). Given a fixed domain X (e.g., the plane, $\mathbb{R}^2$) and a collection $\mathcal{S}$ of subsets of X (e.g., disks whose centers correspond to base stations), a function $\chi\colon \mathcal{S} \to \mathbb{N}$ is a CF-coloring of $\mathcal{S}$ if for every $x \in \bigcup \mathcal{S}$, there exists a color $i \in \mathbb{N}$ such that $\{S \in \mathcal{S}\colon x \in S \text{ and } \chi(S) = i\}$ contains a single subset

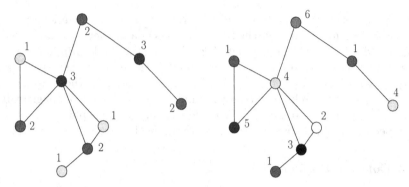

Fig. 4.2. A coloring with 3 colors and a distance-2-coloring with 6 colors of a graph

$S \in$. In other words, the goal is to find an assignment of frequencies to base stations such that for every point there is a frequency on which only a single base station can be received.

All these coloring problems provide different challenges and require different solutions. Compared to the previously described approaches, like MAC protocols or scheduling, coloring problems share the advantage that they provide solutions and insights to the MAC layer problems at a higher level of abstraction. They are less restricted to special hardware or special settings in their pure form but they can be adapted to fit many relevant scenarios. Coloring problems can be a common abstraction for many of the previously discussed problems. Distributed coloring algorithms could be used to replace or at least to improve the classical MAC protocols based on CSMA/CA. They can be used by a base stations or cluster heads to organize TDMA schedules. They can solve frequency assignment problems in static scenarios or they can even be used to provide an approximation to scheduling problems. Note that the interpretation of colors can also differ from problem to problem. Sometimes, different colors can represent time slots (TDMA), sometimes frequencies (FDMA) or sometimes a mixture of both. Coloring algorithms can provide fundamental insight and hard guarantees, although to be of practical use, they must be adapted to be useful for the solution of real-world problems.

In the rest of this chapter, we will concentrate on the standard vertex coloring problem, because it is the most natural problem and we hope that it can serve as a starting point for more sophisticated problem solutions for different scenarios. Note that a vertex coloring does not provide a MAC-layer at once. A distance-2-coloring is necessary for this purpose. We leave it as an open question how to modify algorithms for distance-1-colorings in order to compute distance-2-colorings.

4.2 Algorithms for Vertex Coloring

For successful coloring, we need some kind of symmetry breaking mechanism or even the simplest problems will become unsolvable. We could do this by randomization, but usually we will assume that every node has a unique ID from the range $1, \ldots, N$.

4.2.1 Lower Bounds

Vertex coloring is one of the classical graph problems. There are numerous results on the (in)-tractability of vertex coloring and in this subsection we will state a few of them. It is not only $\mathcal{NP}$-complete but also hard to approximate. If $\mathcal{P} \neq \mathcal{NP}$, it is not approximable within $O(|V|^{1/7-\varepsilon})$. That is, there is no polynomial algorithm which is guaranteed to find a coloring using $k \cdot$ OPT colors with $k \in O(|V|^{1/7-\varepsilon})$ for any $\varepsilon > 0$, where OPT is always the value of an optimal solution. Other coloring problems are, in general, not easier. Coloring with defect D, for example, is not approximable within $|V|^\varepsilon$ for some $\varepsilon > 0$.

For special graph classes, however, the problem becomes easier. For example, there is a polynomial algorithm for planar graphs with an *absolute* error guarantee of 1. For unit disk graphs the problem is approximable within 3 but not within 1.33.

In many ad hoc network scenarios, the maximum degree can be assumed to be bounded by a (not too large) constant; so it makes sense to concentrate on this case. In the centralized case, it is easy to find a coloring with $\Delta + 1$ colors for any graph (where Δ is the maximum degree of the graph). Algorithm 6 finds such a coloring in linear time. However, in a *local* computation model it is not possible to find an optimum coloring if we allow only $o(\mathrm{diam}(G))$ time. The following section, nevertheless, will introduce heuristics for greedy coloring in a distributed setting.

Algorithm 6: Sequential Greedy

---

for $i := 1 \ldots N$ **do**
   ⌊ $c(v_i) :=$ smallest color not used for neighbors of v_i

---

We will conclude with one last negative result:

Definition 4.2.1. *Let* $\log^{(1)}(n) := \log(n)$ *and* $\log^{(k)}(n) := \log \log^{(k-1)}(n)$ *for* $k > 1$. *Then* $\log^*$ *is the smallest* k *for which* $\log^{(k)}(n) < 1$.

There is a time lower bound of $\Omega(\log^* N)$ for finding a 3-coloring in a ring (see also Chapter 7). This bound, of course, also applies to general graphs. Section 4.2.3 describes algorithms with running times close to this lower bound.

4.2.2 Distributed Greedy Heuristics

The simplest algorithms for coloring are greedy algorithms. They have originally been developed for a centralized setting but some of them can easily be distributed. While having very bad worst case performance, their advantages are their simplicity and a good performance on "average" instances.

Algorithm 7: MIS color at vertex v

---

while v *is not colored* **do**
  calculate maximal independent set MIS of uncolored vertices
  if v *is in MIS* **then**
    choose color not used for neighbors of v
    remove v from the graph

Maximal Independent Sets. The Maximal Independent Set (MIS) coloring algorithm (Algorithm 7) colors the graph by repeatedly finding a largest possible independent set of uncolored vertices (i.e., vertices which are no neighbors of each other) in the graph. All vertices in such a set are given a color different from their neighbors' colors and are then removed from the graph. This process is iterated until all vertices have been colored.

Algorithm 7 leaves some degrees of freedom. Any algorithm for constructing maximal independent sets can be used. See Chapter 3 for a more exhaustive survey on distributed MIS-algorithms. It is not even necessary to compute a *maximal* independent set. Any independent set is sufficient. A further variation is either to give all vertices of the MIS the same color (difficult in a distributed setting) or just to choose, for every vertex in the MIS, a color which is different from its neighbors' colors.

The quality of the MIS coloring depends heavily on the choice of the MIS algorithm. One of the simplest distributed algorithms for finding an MIS is Algorithm 8. It employs a (local) leader election mechanism to determine which nodes are joining the independent set. But there are also (more) deterministic methods:

Largest Degree First. Another choice to find an independent set (not necessarily maximal) is Algorithm 9. It assigns the degree of a vertex as its weight.

Algorithm 8: Monte Carlo MIS at vertex v

---

while v *not in MIS and neither of v's neighbors in MIS* **do**
  send a random number ρ to all neighbors
  if ρ *is local maximum* **then**
    join MIS and broadcast decision

Algorithm 9: Largest Degree First MIS at vertex v

---

while v *not in MIS and neither of v's neighbors in MIS* **do**
  send $\rho := \deg(v)$ to all neighbors
  if ρ *is local maximum* **then**
    $\lfloor$ join MIS and broadcast decision

Vertices whose weights are local maxima among their neighbors are chosen and given the smallest available color among their neighbors in Algorithm 7. If two neighboring vertices have the same degree, randomization can be used to break ties. In this manner the vertices are colored in order of decreasing degree. The process is iterated until the graph is successfully colored.

Smallest Degree Last. This method tries to improve on the previous one by using a more sophisticated system of weights. It consists of two phases, a weighting phase and a coloring phase.

In the weighting phase all vertices with the smallest degree d are given a weight w. After a vertex has been given a weight, it is removed from the graph, thus changing the degree of its neighbors. The algorithm repeatedly removes vertices and increases w. When there are no more vertices of degree d, the algorithms looks for vertices of degree $d + 1$ and so on.

The second phase works like the Largest Degree First method but with the weights from the first phase instead of the vertex degrees. The starting value of w and how much it is increased is not important, but it is clear from the procedure that vertices in denser regions of the graph get higher weights and are therefore colored earlier. This approach tries to color dense portions of the network (also known as *kernels*) first and the less strongly connected components later. Note that this last algorithm has a parallel but not a straightforward, efficient, distributed implementation.

All of the above coloring algorithms can compute an $O(\Delta)$-coloring but there is little to say about their worst-case time complexity. It is usually "arbitrarily" bad: it can take $\Theta(N)$ rounds to compute a valid coloring. There are also very few positive results on the quality of the solutions (e.g., approximation guarantees) of greedy colorings. They have, however, been found to often work reasonably well in practice. In the following subsections, we give descriptions of distributed algorithms that color any graph with $\Delta + 1$ colors (which is at least an $O(\Delta)$-approximation). The fastest such algorithm finishes in $O(\Delta \log^*(N))$ time.

4.2.3 Coloring by Deterministic Coin Tossing

In this subsection we want to describe a more sophisticated algorithm which calculates an $O(\Delta)$-coloring of an arbitrary graph in a distributed fashion. It is divided into four parts. The first part is an algorithm for computing a 6-coloring of a rooted tree in only $\log^* N$ rounds. The second part is a reduction of this 6-coloring to a 3-coloring. The third part computes a decomposition of an arbitrary graph into Δ pseudo-forests. In the last part, 3-colorings for the Δ pseudo-forests are combined to a coloring of the original graph.

Algorithm 10: 6-coloring rooted trees at vertex v

---

1 $L := \lceil \log N \rceil$
2 $C_v := \mathsf{ID}$
3 **while** $L > 3$ *(and then one more time)* **do**
4 $\quad$ send (L, C_v) to all children; receive $(L', C_{\mathrm{parent}(v)})$
5 $\quad$ **if** $L = L'$ **then**
6 $\quad\quad$ **if** v *is root* **then**
7 $\quad\quad\quad$ $i_v := 0;\ b_v := \mathrm{bit}_0(C_v)$
8 $\quad\quad$ **else**
9 $\quad\quad\quad$ $i_v := \min\{i|\ \mathrm{bit}_i(C_v) \neq \mathrm{bit}_i(C_{\mathrm{parent}(v)})\}$
10 $\quad\quad\quad$ $b_v := \mathrm{bit}_{i_v}(C_v)$
11 $\quad$ $C_v := i_v \cdot b_v$
12 $\quad$ $L := \lceil \log L \rceil + 1$

6-Coloring of Trees. Algorithm 10 computes a 6-coloring for a rooted tree. We will assume in the following that all edges are directed from the root to the leaves. Initially, every node is assigned its ID as color. The algorithm works in rounds. Line 5 is used to synchronize the rounds. In every round, a vertex first compares its ID with the ID of its parent. It computes the position of the lowest bit in which they differ. The vertex's new color is this position concatenated with the bit itself (we denote this bitwise concatenation by $i_v \cdot b_v$). Note that this new color has a size of at most $O(\log N)$. This is iterated as long as possible. Fig. 4.3 shows an example.

Lemma 4.2.2. *Algorithm 10 computes a valid coloring.*

Proof. We prove this by induction. At the beginning, all colors are different, as the node IDs are different by assumption. Assume that the coloring C is valid at the beginning of an iteration. We will show that at the end of the iteration it is also valid. Let v and w be two adjacent vertices with v being the parent of w. By the algorithm, w chooses some index i such that the ith bit $\mathrm{bit}_i(C_w) \neq \mathrm{bit}_i(C_v)$ and v chooses some index j such that $\mathrm{bit}_j(C_v) \neq \mathrm{bit}_j(C_{\mathrm{parent}(v)})$. The new color of w is $i \cdot \mathrm{bit}_i(C_w)$ and the new color of v is $j \cdot \mathrm{bit}_j(C_v)$. Then, either $i \neq j$ (and we are done) or $i = j$, but then $\mathrm{bit}_i(C_w) \neq \mathrm{bit}_i(C_v) = \mathrm{bit}_j(C_v)$ and again the new colors are different. Note that this also holds if v is the root, because we take bit zero as a pivot. $\square$

Lemma 4.2.3. *Algorithm 10 terminates with a 6-coloring in $\log^* N$ rounds.*

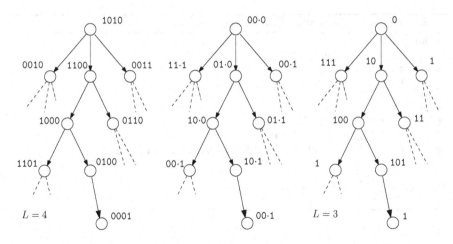

Fig. 4.3. Example of 6-coloring a tree. The node labels in the left figure are the (binary) node IDs in the beginning ($L = 4$), the labels after the first step are in the middle, the final state is to the right. An additional iteration would reduce the number of colors by one, because the color 111 could no longer occur.

Proof. The proof is again by induction. Let ℓ_k denote the number of bits in the representation of colors after k iterations. For $k = 1$ we have

$$\ell_1 = \lceil \log \ell_0 \rceil + 1 \le 2\lceil \log \ell_0 \rceil$$

if $\lceil \ell_0 \rceil \ge 1$. If for some $k > 1$ we have $\ell_{k-1} \le 2\lceil \log^{(k-1)} \ell_0 \rceil$ and $\lceil \log^{(k)} \ell_0 \rceil \ge 2$, then

$$\begin{aligned}
\ell_k &= \lceil \log \ell_{k-1} \rceil + 1 \\
&\le \lceil \log(2\log^{(k-1)} \ell_0) \rceil + 1 \\
&\le 2\lceil \log^{(k)} \ell_0 \rceil .
\end{aligned}$$

Therefore, as long as $\lceil \log^{(k)} \ell_0 \rceil \ge 2$,

$$\ell_k \le 2\lceil \log^{(k)} \ell_0 \rceil .$$

Hence, the number ℓ_k of bits in the representation of colors decreases until, after $O(\log^* N)$ iterations, $\lceil \log^{(k)} \ell_0 \rceil$ becomes 1 and ℓ_k reaches the value of 3. Another iteration produces a 6-coloring: 3 possible values of the index i and 2 possible values of the bit b. The algorithm terminates at this point.   □

Note that N (or a common upper bound to N) must be known by all vertices in advance in order for Algorithm 10 to be correct.

3-Coloring of Trees. Algorithm 11 converts the 6-coloring into a 3-coloring. It does this by successively "shifting down" the colors and recoloring the vertices

---

Algorithm 11: 3-coloring from 6-coloring

```
1  for c := 6, 5, 4 do
2  │  send C_v to children
3  │  if v is root then
4  │  │  └  C_v := smallest color different from old color
5  │  else
6  │  │  send C_parent(v) to children
7  │  │  if C_parent(v) = c then
8  │  │  │  └  C_v := smallest color different from C_v and C_grandparent(v)
9  │  │  else
10 │  │  │  └  C_v := C_parent(v)
```

---

with the highest color. See Fig. 4.4 for an example. Shifting down means that every vertex takes the color of its parent vertex and the root takes the lowest color different from its previous color. Afterwards, for every vertex all its children have the same color and its parent has a second, maybe different color. So every node has at most two different colors in its neighborhood. Now, vertices with the highest color choose the lowest possible color as their new, legal color. After the first shifting, every vertex with color 6 selects the smallest color (among $\{1, 2, 3\}$) that is not used by its parent and its children. After the second shifting, all vertices with color 5 are recolored and after the third shifting, all vertices with color 4 are recolored. In this way the 6-coloring is converted into a 3-coloring in a constant number of rounds.

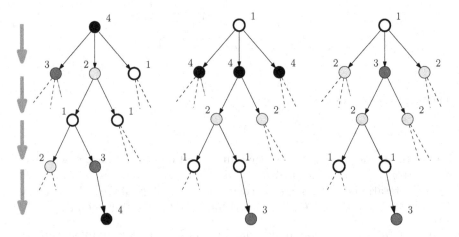

Fig. 4.4. Example of *shifting down* a coloring. To the left, a 4-coloring of the tree, in the middle the coloring after shifting down colors and to the right the final coloring after recoloring of nodes with color 4.

Coloring of Arbitrary Graphs. The technique of Algorithm 10 can be generalized to other classes of graphs, for example to rings, pseudo-forests, but also to arbitrary graphs. However, on general graphs this results in a coloring with an exponential (in Δ) number of colors.

We will describe a better approach in the following. The idea is to decompose the graph into a collection of pseudo-forests, compute a separate coloring for each pseudo-forest, and finally compose these colorings into a valid coloring for the original graph.

Forest Decomposition. A *pseudo-forest* is a directed graph where every vertex has in-degree one. It is quite straightforward to see that Algorithm 10 and 11 work for pseudo-forests, too, as they only use the fact that every vertex has a unique parent. Algorithm 12 computes a pseudo-forest decomposition of an arbitrary graph. Every node sorts its $O(\Delta)$ adjacent edges in an arbitrary fashion and then selects (at most) one edge for every one of the Δ pseudo-trees. Fig. 4.5 shows an example of a pseudo-forest decomposition.

Algorithm 12: Tree Decomposition at vertex v

---

1 sort adjacent edges arbitrarily: $e_1, \ldots, e_\Delta$
2 **for** $i := 1, \ldots, \Delta$ **do**
3   **if** $e_i = \{v, w\}$ *is unmarked* **then**
4   mark e_i
5   direct e_i from w to v and add edge (w, v) with number i to forest F_i,
6   **if** w *has chosen the same edge for i* **then**
7   resolve conflict arbitrarily

Merging Colorings. Algorithm 13 computes a $(\Delta + 1)$-coloring of an arbitrary graph. We start by computing a pseudo-forest decomposition, color every forest separately, and finally merge the Δ many 3-colorings into a single coloring. At the end, C^Δ is a legal $(\Delta + 1)$-coloring.

Before iteration i of the outer loop, C^{i-1} is a valid coloring in the graph induced by the edges in $F_1, \ldots, F_{i-1}$. Moreover, C^i is a coloring of the forest F_i, so the pair of colorings (C^{i-1}, C^i) is a valid coloring in the graph induced by $F_1, \ldots, F_i$. In the innermost loop, the nodes with color (j, c) (which form an independent set) are colored simultaneously with a color from $\{0, \ldots, \Delta\}$. This produces a coloring of $F_1, \ldots, F_i$ with one color per node.

Putting It All Together. Table 4.1 summarizes the running times and message complexities of the above algorithms for different models of computation. Under the synchronous model, we only need to count the number of communication rounds. Computing a forest decomposition obviously needs a constant number of rounds. Only a constant number of messages is sent over each edge.

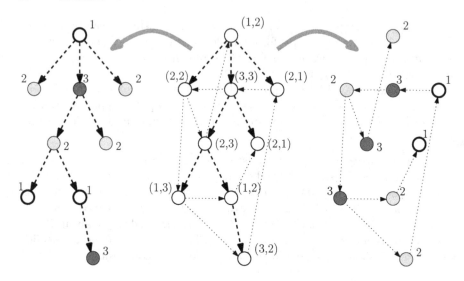

Fig. 4.5. A decomposition into two pseudo-forests (dotted and dashed edges, respectively) with two respective 3-colorings (numbers). Note that the pair of 3-colorings is a legal 9-coloring for the composed graph.

Algorithm 13: (+ 1)-color at vertex v

---

1 compute a (pseudo-)forest decomposition $F_1 \cup \cdots \cup F_\Delta = G$
2 **for** $i := 1, \ldots, \Delta$ **do**
3 $\quad\lfloor\;\; C_v^i := $ 3-coloring of F_i

4 send C_v^1 to all neighbors in G
5 **for** $i := 2, \ldots, \Delta$ **do**
6 $\quad$ **for** $j := 1, \ldots, \Delta + 1$ **do**
7 $\quad\quad$ **for** $c := 2, 3$ **do**
8 $\quad\quad\quad$ **if** $C_v^{i-1} = j$ and $C_v^i = c$ **then**
9 $\quad\quad\quad\quad$ $C_v^i := $ minimal legal color w. resp. to edges from $F_1 \cup \cdots \cup F_i$
10 $\quad\quad\quad\quad$ send (C_v^i, i, j, c) to neighbors

If a node can reach all its neighbors at the same time, via a broadcast message, only one message is required per node.

We have proven that the computation of a 6-coloring of a pseudo-tree needs $\log^* N$ rounds. In each round, every node sends $O(\ )$ messages of size $O(\log n)$. There is one message per round for every edge. This adds up to $|E| \log^* N$ messages per tree or $|E| \log^* N$ messages in total.

Merging the colors obviously works in ^{2} rounds. But a node only has to send messages once for every tree whose coloring is merged. That makes a total of $|E|$ messages in the unicast model or $|V|$ in the multicast model.

Step	Time complexity Synchronous	Message complexity	
		Unicast	Multicast
Forest decomp.	$O(1)$	$O(|E|)$	$O(|V|)$
6-coloring	$O(\log^* N)$	$O(\Delta|E|\log^* N)$	$O(\Delta|E|\log^* N)$
Merging step	$O(\Delta^2)$	$O(\Delta|E|)$	$O(\Delta|V|)$
Total	$O(\Delta^2 + \log^* N)$	$O(\Delta|E|\log^* N)$	$O(\Delta|E|\log^* N)$

Table 4.1. Running times of Algorithm 13

4.2.4 Combinatorial Coloring Algorithms

We will now briefly describe another approach to obtain a coloring with few colors which uses the following combinatorial tools. It is a beautiful result which would lead to a very efficient coloring algorithm if it didn't have the flaw that the proof of Lemma 4.2.4, on which it heavily depends, is non-constructive.

Lemma 4.2.4 ([114], Theorem 3.1). *For integers $n > \Delta$, there is a family J of n subsets $F_0, \ldots, F_n \subseteq \{1, \ldots, 5\lceil \Delta^2 \log n\rceil\}$ such that if $F_0, \ldots, F_\Delta \in J$, then*

$$F_0 \not\subseteq \bigcup_{i=1}^{\Delta} F_i \ .$$

Lemma 4.2.5 ([114], Example 3.2). *Let $q = p^k$ be a prime power. Then there is a collection J of q^3 subsets of $\{1, \ldots, q^2\}$ such that for any $(q-1)/2$ sets $F_0, \ldots, F_{(q-1)/2} \in J$ then*

$$F_0 \not\subseteq \bigcup_{i=1}^{(q-1)/2} F_i \ .$$

Using the first of these lemmas we can prove the following result:

Theorem 4.2.6. *Let G be a graph of order n and maximum degree Δ. It is possible to color G with $5\Delta^2 \log n$ colors in one unit of time.*

Proof. Fix a family $J = \{F_1, \ldots, F_n\}$ as in the Lemma 4.2.4. Consider a vertex with label j whose neighbors' labels are $j_1, \ldots, j_d$ with $d \leq \Delta$. Since

$$F_j \not\subseteq \bigcup_{i=1}^{d} F_{j_i} \ ,$$

there is a $1 \leq x \leq 5\lceil \Delta^2 \log n\rceil$ with $x \in F_j \backslash (\bigcup_{i=1}^{d} F_{j_i})$. The color of this vertex is x. It is obvious by Lemma 4.2.4 that this is a proper $O(\Delta^2 \log n)$-coloring of G. □

With the same proof, we can achieve the more general

Theorem 4.2.7. *Let G be a graph whose vertices are properly colored with k colors. It is possible to color G with $5\Delta^2 \log k$ colors in one unit of time.*

By iterating the procedure implied by Theorem 4.2.7 we can obtain a $10\Delta^2 \log k$-coloring in time $O(\log^* n)$. To eliminate the $\log \Delta$ term we use Lemma 4.2.5. Select q to be the smallest prime power with $q \geq 2\Delta + 1$. There is certainly one with $q \leq 4\Delta + 1$. In the same fashion as in the proof of Theorem 4.2.6 we can now transform an $O(\Delta^3)$-coloring into an $O(\Delta^2)$-coloring. We have now obtained

Theorem 4.2.8. *Let G be a labeled graph of order n with largest degree Δ. It is possible to color G with $O(\Delta^2)$ colors in time $O(\log^* n)$.*

As mentioned before, this result relies on the non-constructive proof of Lemma 4.2.4. The situation is, however, not hopeless. If Δ is known in advance the we could pre-compute the subsets F_i and hard-code one of them into every node. The sensor node would then be able to run the algorithm implied by Theorem 4.2.6. To implement the result of Theorem 4.2.7, however, every node would have to store at least $\Theta(5\Delta^2 \log n)$ subsets, one for every color it chooses in the first iteration. If this is practically feasible or not, remains to be seen.

It has been claimed that Linial's result could be improved to find a $O(\Delta)$-coloring in time $O(\log^*(n/\Delta))$. See the chapter notes, Section 4.4, for a reference. With this, there would be an approach which is competitive with the one from the previous section.

4.3 Conclusion

There is still a large gap between MAC protocols for wireless networks which are in use today on the one hand and the large array of existing algorithmic tools on the other hand. Most MAC protocols are quite "ad hoc" in nature (especially contention based protocols) and do not explicitly take into account the structure of the communication graph. Schedule based protocols, on the other hand, often waste bandwidth because they do not adapt to the current traffic pattern. There have been some attempts to overcome these problems recently, but usually in a heuristic fashion that does not give any guarantees on their performance, which is usually measured by simulation only.

Other approaches, like frequency assignment or scheduling, either need strong computational resources, that are usually not available in sensor networks, or too detailed information, for example on traffic or on the terrain. Coloring problems are able to model many of the aspects of the MAC layer and are yet simple enough to be handled by the sensor network in an on-line fashion. However, it is an open problem how to model the constraints of the communication graph into the right coloring problem and its solution back into a schedule for the network. A distance-2-coloring, for instance, theoretically solves the problem of the assignment of time slots to concurring nodes, but, if it is not adapted further, it wastes bandwidth because it does not reassign unused communication slots. Whether this solution is competitive to other MAC protocols, for example in terms of throughput, is uncertain.

MAC protocols and simple graph coloring are just the beginning of the solution of MAC layer issues. We haven't addressed several important issues of the MAC layer in wireless networks, for example energy management or dynamic versions of coloring. The issues in MAC layer design are also strongly connected to the topics of topology control (Chapter 5) and interference (Chapter 6). They can be solved independently by first establishing a topology with low interference and hoping that this will lead to less collision. It would be better, of course, to design a topology control algorithm that takes the MAC layer into account or, vice versa, to design a MAC protocol that works well for a certain type of topology. Other effects of cross-layer optimization (e. g. by designing the MAC layer together with a routing protocol) are also thinkable, but widely unexplored.

Finally there is the issue of the "chicken-and-egg" problem: A functioning MAC layer is a fundamental prerequisite for higher layers and, especially, for classic ditributed algorithms, which use some kind of message passing model. As the coloring algorithms use this model themselves, we run into a serious problem here: We need a MAC layer in order to set up a MAC layer. The pragmatic way to address this problem is to use a simpler MAC layer approach, like CSMA/CA first and run the algorithm using this MAC layer. Once the necessary computations are finished, we can switch to the new, hopefully more efficient, protocol. It would be desirable, however, to find a more elegant solution. One approach could be to use the so-called *unstructured radio network* model. This model has already been discussed in Chapter 3, sec. 3.3.5. There have been successful attempts to compute good colorings (with $O(\Delta)$ colors) very efficiently (every node chooses a color in time $O(\Delta \log n)$) even for this harsh computational model.

4.4 Chapter Notes

Works on MAC layer issues, among many others mentioned in the introduction, are [249] and [325]. For a survey on frequency assignment, see also [300].

Vertex coloring is one of the classical graph theoretic problems of computer science. Results on complexity and approximation have been around for a long time, see the books by Garey and Johnson ([150]) and by Ausiello et al. ([23]) and also [377]. for references. For further results on non-approximabilty of coloring is see [34]. Complexity results on unit disk graphs can be found in ([276]). Among the works on defective coloring are [87], and the surveys [138] and [396]. For T-coloring consider [177, 332].

Heuristics for graph coloring can be found in [11]. For results on independent sets and independent set coloring see [266] and also Chapter 3. Another work on greedy coloring heuristics is [180].

Deterministic Coin Tossing was first proposed by Cole and Vishkin ([81]) for 3-coloring of a ring. Their idea has been adopted by Goldberg, Plotkin and Shannon ([163]) to color arbitrary graphs. Their results have been developed in the context of parallel computing and we have adapted them to the distributed scenario. Coloring by families of finite sets has been proposed by Linial ([260]). The necessary combinatorial tools are described in [114]. De Marco and Pelc ([98]) claim to have improved on Linial's result and to be able to find a $O(\Delta)$-coloring in time $O(\log^*(n/\Delta))$. Recent results on lower and upper bounds for graph coloring in distributed settings have been presented in [239].

For results on coloring of unit disk graphs, but in the stronger *unstructured radio network* model, see [293] and Chapter 3.

5

Topology Control

Kevin Buchin and Maike Buchin

5.1 Introduction

Information between two nodes in a network is sent based on the network topology, the structure of links connecting pairs of nodes of a network. The task of topology control is to choose a connecting subset from all possible links such that the overall network performance is good. For instance, a goal of topology control is to reduce the number of links to make routing on the topology faster and easier.

In this chapter we assume that the network topology can be controlled by varying the transmission radii of the nodes and selecting and discarding possible links. Topology control was first considered by Tagaki and Kleinrock [355]. The effect of topology control on the network is demonstrated in Figure 5.1 (taken from [392]).

In the following we consider localized algorithms, i.e., algorithms where decisions for establishing connections are done locally by the nodes. Each

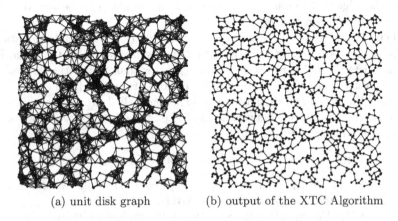

(a) unit disk graph   (b) output of the XTC Algorithm

Fig. 5.1. A network without and with topology control applied.

D. Wagner and R. Wattenhofer (Eds.): Algorithms for Sensor and Ad Hoc Networks, LNCS 4621, pp. 81–98, 2007.

node independently explores its surrounding region and establishes connections with neighboring nodes in its transmission range. A localized algorithm should be sufficiently simple and efficient such that it can be carried out by the mobile nodes.

Alternatively one can consider centralized algorithms for establishing the network topology. These gather all connectivity and location information in one node, construct the topology of the whole network, and propagate it to all other nodes. We do not consider centralized algorithms since a localized construction and maintenance is preferred because of the limited resources and the mobility of the nodes.

Topology control aims to conserve energy as the power in an ad hoc network is limited. The topology should allow for efficient routing and improve the overall network performance. These goals can be formalized in different ways. In Section 5.2 we give an overview of quality criteria that influence the network performance based on the topology.

The topology control problem is well described in a graph-theoretic setting. Consider the bidirected graph $G = (V, E)$, where the node set V consists of the nodes of the network, and an edge (u, v) between the nodes u and v exists if the nodes u and v lie in each other's transmission range. If the nodes have differing transmission radii, this yields a directed graph where edges exist if one node lies in the transmission range of the other (and not necessarily vice versa). We will assume uniform maximal transmission ranges.

The task of topology control can now be formulated as follows: Given the graph $G = (V, E)$ compute a subgraph $G' = (V, E'), E' \subseteq E$ which allows for energy efficient routing and increases the network performance and lifetime. These goals can be measured by several criteria, such as connectivity and sparseness of G', which are described in the next section.

Based on the energy model used, one can also consider *edge weights* for the graph G. I.e., consider the weight function w which assigns to each edge e in E a weight $w(e)$, which is the energy needed to transport information along the edge e. The energy cost of a *path* in the graph is the sum over the costs of the edges in the path.

Typical edge weights are given by the link metric, the Euclidean distance between the endpoints of the edge or an energy metric which takes the Euclidean distance and raises it to a power greater than 1. For the link metric all edge weights are one, i.e., the number of hops are counted. An energy metric raises the Euclidean distance to an exponent called *attenuation exponent* which is typically a value between 2 and 5. I.e., the edge e between two nodes u and v obtains one of the weights

$$\mathrm{w_{link}}(e) := 1, \quad \mathrm{w_{dist}}(e) := d_2(u, v), \quad \text{or} \quad \mathrm{w_{energy}(a)}(e) := d_2(u, v)^a$$

where $a > 1$ and $\mathrm{w_{link}}, \mathrm{w_{dist}}$, and $\mathrm{w_{energy}(a)}$ denote the edge weights for the link, distance, and energy metric, respectively.

Taking the Euclidean distance between the endpoints requires that the graph is embedded in the Euclidean plane or space, i.e., each node has coordinates. In general, nodes need not know their coordinates. Some algorithms require that they know their coordinates (e.g., assume that all nodes have a GPS), other algorithms require that the nodes can determine the distances or directions to nodes in their neighborhood and some algorithms do not use any distance or direction information at all. Often algorithms assume nodes to have distinctive IDs.

Many localized algorithms for topology control compute a locally defined geometric graph. Therefore we give an overview of locally defined geometric graphs in Section 5.3 and in Section 5.4 describe some algorithms computing different locally defined graphs.

5.2 Quality Criteria

In this section we consider different criteria that may be used to measure the quality of a network topology.

Connectivity. Perhaps the most basic requirement for a topology is *connectivity*, i.e., that there is a path between any pair of vertices in V. More specifically, one requires that any two nodes that are connected by a path in G should also be connected by a path in G'. Sometimes the stronger notion of *k-vertex-connectivity* or *k-edge-connectivity* is required. A graph is k-vertex-connected (k-edge-connected) if at least k vertices (edges) need to be removed for the graph to become disconnected. The reason to ask for connectivity is simple: A topology G' that is less connected than G does not have the same capabilities to transfer information. Asking for k-connectivity might lower the risk of path loss due to interference.

Sometimes instead of total connectivity a *giant connected component* suffices. A connected component is a maximal connected subgraph. A giant connected component is a connected component of linear size, i.e., containing cn vertices, where n is the number of vertices in V and $0 < c \leq 1$ a constant. For instance if the task of an ad hoc network is to compute an average value over the network, a giant connected component containing all but few of the vertices will yield a value very close to that of the total network. For obtaining a giant connected component instead of total connectivity, often a much smaller transmission radius is sufficient [323], and therefore much less energy.

Symmetry. Another basic requirement for a topology is *symmetry*, which means that if (u, v) is an edge in G', so is (v, u). In a graph-theoretic sense this means that the computed graph G' should be bidirectional. Many routing algorithms require symmetry. Without symmetry, communication in the network becomes more complicated, e.g., it may not be possible to send back an acknowledgement (ACK) for received data.

Stretch Factors. A main goal of topology control is high *energy-efficiency* to increase the network lifetime. Depending on the energy model used, energy-efficiency can be measured in different ways, typically using one of the following *stretch factors*: *energy stretch factor*, *hop stretch factor*, or *distance*

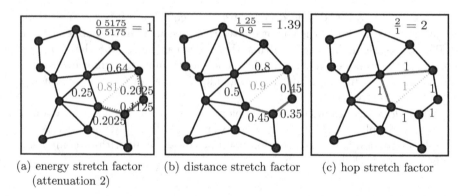

(a) energy stretch factor   (b) distance stretch factor   (c) hop stretch factor
(attenuation 2)

Fig. 5.2. Example for stretch factors for a pair of nodes.

stretch factor, see Figure 5.2. A stretch factor is the largest ratio of a quantity measured in a subgraph and in the full communication graph.

The energy stretch factor of a subgraph G' in G is the largest ratio for any pair of vertices in V between the energy of the minimum energy path in G' and in G. Formally, the energy stretch factor is $\max_{E} \frac{G'(\ )}{G(\ )}$ where $E\ (u,v)$ denotes the energy of the minimum energy path between u and v in G. The energy of a path is the sum of the energies of the edges, where the energy of an edge is measured according to the chosen energy model.

The hop stretch factor is the largest ratio of number of hops, i.e., number of edges, in G' and G, respectively, between any two vertices in V.

The distance stretch factor is the largest ratio of the Euclidean lengths of shortest paths in G' and G, respectively, for any pair of vertices in V. The problem of designing graphs with low distance stretch factor has been extensively studied by network designers. If an attenuation exponent larger than 1 is used in the energy model, a constant distance stretch factor implies a constant energy stretch factor.

A subgraph G' of G that has a constant distance stretch factor is a *spanner* of G. A spanner is a subgraph G' of a graph G in which the distance between any two nodes in G' is within a constant factor of the distance in G.

Sparseness. The topology determined by a topology control algorithm should be simple and easy to maintain. These subjective goals are influenced by the *sparseness* of the graph. A graph is called sparse if it has few edges: The number of edges is linear in the number of nodes, i.e., the average vertex degree is constant. The *degree* of a vertex is the number of of edges adjacent to the vertex. The maximal vertex degree may still be high in a sparse graph. A stronger requirement is therefore that all vertex degrees are bounded by a constant.

Due to the low complexity of sparse graphs, routing algorithms are more time and power efficient in sparse graphs. There is a tradeoff between connectivity and sparseness: A sparse graph is likely to have low connectivity.

Throughput. Topology control aims at a topology with large *throughput*, i.e., where as much data as possible can be sent over the network. A measure for the throughput of a network is its *bit-meter* capacity. We say that the network transports one bit-meter if one bit has been transported over a distance of one meter. The network's bit-meter capacity is the amount of bit-meters that can be transported in one second.

The bit-meter capacity measures the optimal throughput of a network. For measuring the worst case throughput, *throughput competitiveness* can be used. This is the largest number ϕ, $0 \leq \phi \leq 1$, such that for any pair of vertices (v, w), if a flow of r can be routed in G from v to w then a flow of ϕr can be routed in G' from v to w.

Interference. Interference occurs when a node cannot receive messages because several nodes in its transmission range send at the same time. Topology control handling interference is discussed in Chapter 6.

Adaptability. A further goal of topology control is to determine a topology that is robust to mobility, i.e., that can react quickly to changing network conditions. This is measured by the adaptability of a network, which is the maximum number of nodes that need to change their topology information as a result of the movement of a node. All locally defined graphs have reasonable adaptability because in case a node moves the nodes in its neighborhood can locally recompute their edge sets.

Planarity. Some geographic routing algorithms require that the embedding of the topology in the plane does not contain crossing edges, i.e., it is a planar straight-line graph.

5.3 Locally Defined Geometric Graphs and Further Proximity Graphs

A localized algorithm for topology control computes a subgraph $G' = (V, E')$ of a given graph $G = (V, E)$ based on local decisions. This means that the decision whether an edge at a node $v \in V$ is selected to be in E' is made by considering the information of all nodes in a k-hop neighborhood of v, i.e., all nodes $w \in V$ for which there is a path starting at v and ending at w containing at most k edges for a constant k. For many topology control algorithms k equals one or two, i.e., only the information of the neighbors or of the neighbors' one-hop neighborhood is used.

If the nodes are assumed to lie in some geometric space, typically the Euclidean plane $\mathbb{R}^2$, this problem is closely related to geometric graphs which are defined by local properties. Many distributed topology control algorithms compute such a locally definable graph, or a variant of such a graph or combine two locally definable graphs. In this section we give an overview of locally definable graphs, many of which have been used for topology control, explicitly or implicitly. We consider here only geometric graphs in the two-dimensional Euclidean plane $\mathbb{R}^2$.

An example of a locally definable graph is the *Gabriel graph (GG)*: Given a set of vertices V, two points from V are connected by an edge if the disk with these points as a diameter contains no further points from V, see Figure 5.3. If the set of possible edges is restricted to the edges of a graph G, we call this the restricted Gabriel graph denoted by $GG(G)$ (and analogously for other graphs).

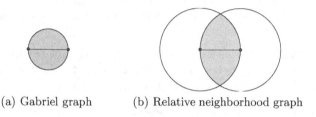

(a) Gabriel graph   (b) Relative neighborhood graph

Fig. 5.3. Empty neighborhoods of an edge.

In the Gabriel graph edges are defined by an *empty neighborhood*, i.e., an edge is in the graph if a certain neighborhood of the edge does not contain any vertices. For the Gabriel graph the empty neighborhood of an edge is the disk which has the edge as a diameter. Such graphs are considered in detail in Section 5.3.1. These graphs are typically sparse, i.e., have a low total number of edges though the maximum vertex degree might be high. A shortcoming of these graphs is that they might have a high distance stretch factor.

A better stretch factor can be achieved by connecting a point to a nearby point in all directions. An example is the *Yao graph*: For every point the plane is divided into a set of cones with a fixed angle with the point as base and the point is connected with the nearest neighbor in every cone. These kinds of graphs are described in Section 5.3.2.

Closely related to locally definable graphs are *proximity graphs*, i.e., geometric graphs with edges between vertices that are by some means close to each other. These graphs are discussed in Section 5.3.3. The Gabriel graph is such a graph. But not all proximity graphs are locally definable. For instance a *minimal spanning tree* is a minimal (with respect to edge weight) connected graph on a set of points. Another important proximity graph is the *Delaunay triangulation*. It can be defined by a local property of the triangles. This property does not yield a local property for the edges. Therefore, localized variants of the Delaunay triangulation are considered.

In Section 5.3.4 we consider graphs which have an *exclusion region*. An exclusion region is similar to an empty neighborhood: a region close to an edge that needs to be empty for the edge to be in the graph. For an exclusion region, in contrast to an empty neighborhood, not the whole region needs to be empty, but the region is divided into two halves by the edge and one of the halves needs to be empty. E.g., a possible exclusion region is the empty neighborhood of the Gabriel graph, i.e., the disk which has the edge as diameter, and an edge is then in the graph if one of the half-disks is empty.

Figure 5.4 shows examples for most of the graphs that are described in the rest of this section. The graphs are shown in order of inclusion. Note that the graph inclusion $H \subset G$ implies that the graph H has smaller degree and larger stretch factors than the graph G. Many of the properties of the graphs in this section are also described in the references given in the chapter notes [199][255]. At the end of this section the most important graphs and their properties are summarized in Table 5.3.4.

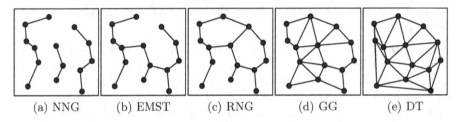

| (a) NNG | (b) EMST | (c) RNG | (d) GG | (e) DT |

Fig. 5.4. Examples of locally defined geometric graphs in order of inclusion: nearest neighbor graph (NNG), Euclidean minimum spanning tree (EMST), relative neighborhood graph (RNG), Gabriel graph (GG), and Delaunay triangulation (DT).

In the following n always denotes the number of vertices, i.e., $n = |V|$. Note that a connected and planar graph has at least $n - 1$ edges (because of connectivity) and at most $3n - 6$ (using Euler's formula). In particular, a connected and planar graph is sparse.

5.3.1 Graphs with Empty Neighborhoods

Gabriel Graph (GG). The empty neighborhood of an edge in the Gabriel graph is the disk with the edge as a diameter (see Figure 5.3). Formally, an edge (v, w) exists if there is no node u such that $d(v, u)^2 + d(w, u)^2 \leq d(v, w)^2$. I.e., there is no vertex u s.t. (u, v, w) form a triangle with an angle larger than $\pi/2$ at u.

The Gabriel graph is a bidirectional and connected graph. It is planar and also sparse. The distance stretch factor of the Gabriel graph is $\sqrt{n-1}$ and it is not a spanner. The Gabriel graph has optimal energy stretch factor 1; the maximum vertex degree is $n - 1$.

Relative Neighborhood Graph (RNG). The empty neighborhood of an edge in the relative neighborhood graph is the intersection of the two disks centered at the vertices of the edge with their radii equal to the distance between the two vertices (see Figure 5.3). This intersection is called the *relative neighborhood* of the edge. One can also say that the two connected vertices are relatively close, i.e., there is no node closer to both vertices than they are to each other. Formally, an edge (v, w) exists if

$$d(v, w) \leq \max_{u \in V \setminus \{v, w\}} (d(v, u), d(w, u)).$$

The relative neighborhood graph is a bidirectional and connected graph. It is planar and also sparse. Its distance stretch factor is $n - 1$, i.e., it is not a spanner, and its energy stretch factor is also $n - 1$. The maximum vertex degree is $n - 1$ but can be bounded by 5 if no two points have the same distance to a third point. If two points do have the same distance to a third point, we will call this a *tie* and discuss it further in Section 5.4.3, where an algorithm using the relative neighborhood graph is described. If one uses *tie breaking*, i.e., choosing only one edge in the case of a tie, the maximum vertex degree can be bounded by 6.

β-Skeletons. β-skeletons are a parameterized family of graphs, which are parameterized by a positive, real number β. There are two versions of β-skeletons: lune-based and circle-based. In both versions an edge is established if its β-neighborhood is empty. For $0 < \beta \leq 1$ the β-neighborhoods of lune- and circle-based β-neighborhoods are the same. For larger β, the β-neighborhoods of lune-based β-skeletons are the intersection of two circles whereas the β-neighborhoods of circle-based β-skeletons are the union of two (different) circles. As an example we now give the definition of the β-neighborhoods for lune-based β-skeletons for $\beta \geq 1$ (For the other cases see [217]).

The β-neighborhood of an edge in a lune-based β-skeleton is the intersection of two circles. For $\beta \geq 1$ the circles have centers lying on the line defined by the edge and moving away from the edge with growing β. More specifically (see Figure 5.5),

-    for $\beta = 1$ both circle centers are the midpoint of the edge,
-    for $1 < \beta < 2$ the centers move from the mid- to the endpoints of the edge,
-    for $\beta = 2$ the centers are the endpoints, and
-    for $\beta > 2$ the centers move further away from the edge.

The radii of the circles grow with β and they are chosen such that always the opposite endpoint of the edge lies on the boundary of the circle. Formally, the lune-based β-neighborhood of an edge (v, w) and $\beta \geq 1$ is

$$B\left((1 - \frac{\beta}{2})v + \frac{\beta}{2}w, \frac{\beta}{2}d(v, w)\right) \cap B\left((1 - \frac{\beta}{2})w + \frac{\beta}{2}v, \frac{\beta}{2}d(v, w)\right)$$

where $B(c, r)$ denotes the disk with center c and radius r and $d(v, w)$ is the Euclidean distance between the vertices v and w.

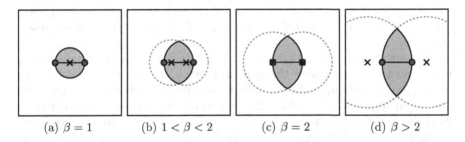

(a) $\beta = 1$ (b) $1 < \beta < 2$ (c) $\beta = 2$ (d) $\beta > 2$

Fig. 5.5. β-Skeletons.

Both lune- and circle-based β-skeletons are a generalization of the Gabriel graph. Lune-based β-skeletons are also a generalization of the relative neighborhood graph. I.e., for $\beta = 1$ both β-skeletons equal the Gabriel graph and for $\beta = 2$ the lune-based β-skeleton is the relative neighborhood graph (see Figure 5.5).

Another parameterized family of graphs are γ-neighborhood graphs [376], which generalize circle-based β-skeletons and the Delaunay triangulation.

5.3.2 Cone-Based Graphs

Yao Graph (YG). The Yao graph is a graph parameterized by a constant $k \geq 6$. For each vertex, the plane is divided into k cones of equal angle and the vertex is connected to the nearest vertex in each cone (see Figure 5.6(a) left). There may be several closest vertices in one cone, which we again call a *tie*. In this case *tie breaking* can be used, i.e., choosing one or more of the closest vertices to connect to. In the original definition [400], an arbitrary closest vertex was selected. An alternative is to connect to all closest vertices.

This yields a weakly connected directed graph, "weakly" meaning that — disregarding edge directionality — all nodes are connected. The bidirectional graph, in which edge directions are ignored, is called the *undirected Yao graph* (see Figure 5.6(a) right). The Yao graph is typically not planar, but for fixed

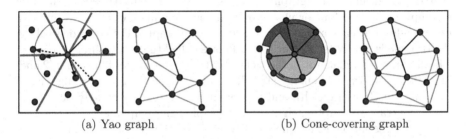

(a) Yao graph   (b) Cone-covering graph

Fig. 5.6. Cone-based graphs.

k it is sparse. It is a spanner with distance stretch factor $1/(1 - 2\sin(\pi/k))$ and energy stretch factor $1/(1 - (2\sin(\pi/k))^\alpha)$ where α is the attenuation exponent.

The maximum vertex degree is $n - 1$, but the outdegree can be bounded by k if no two vertices have the same distance to a third vertex, or if, as in the original definition, in the case of a tie one vertex is chosen. A graph with constant distance stretch factor and bounded vertex degree can be constructed from the Yao graph [19]. The Yao graph contains the relative neighborhood graph if in case of a tie one connects to all closest vertices or if it is assumed that no two points have the same distance to a third vertex.

Θ-Graph. The Θ-graph is often used synonymously to the Yao graph. It is originally defined sightly different [212]: Instead of connecting to the nearest vertex in each cone, one connects to the vertex for which the projection on the axis of the cone is the shortest. The Θ-graph has similar properties as the Yao graph.

Cone-Covering Graph. A graph related to the Yao graph that is the basis of an algorithm described in Section 5.4.1 is the following: Each vertex processes its neighbors by their distance, starting with its nearest neighbor. Edges to neighbors are established until the α-cones of the processed neighbors cover the complete angular range around the vertex (see Figure 5.6(b) left). This graph has similar properties as the Yao graph but typically the longest edge is shorter (and therefore the necessary transmission range smaller). To guarantee no empty cone of angle greater than α, the angles for the Yao graph need to be at most $\alpha/2$, i.e., it connects to $k \geq 4\pi/\alpha$ nodes. The degree of the cone-covering graph can be arbitrarily high, e.g., for points lying on a spiral. It can be efficiently constructed by incrementally increasing the transmission radius.

5.3.3 Further Proximity Graphs

Delaunay Triangulation (DT) and Variants. Two vertices are connected by an edge in the Delaunay triangulation if there exists an empty circle through the two vertices. The Delaunay triangulation is a bidirectional, connected, planar and sparse graph. All faces of the graph are triangles (with the assumption that no four or more points lie on a common circle). Formulated for triangles, one can say that a triangle is in the graph if the circle defined by the triangle is empty (see Figure 5.7). This is called the *empty circle property*. The distance stretch factor of the Delaunay triangulation lies between $\pi/2$ and $(4\sqrt{3}/9)\pi \approx 0.77\pi$. The energy stretch factor is 1 (the Delaunay triangulation contains the Gabriel graph), and the maximum vertex degree may be $n - 1$.

In contrast to empty neighborhood graphs, the empty area in the Delaunay triangulation is not necessarily close to the points, but may reach arbitrarily far away. Therefore the Delaunay triangulation is not locally constructible in the sense that one may need to check vertices arbitrarily far away

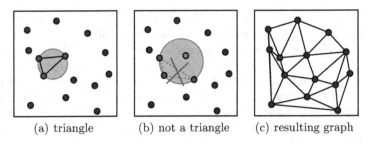

(a) triangle   (b) not a triangle   (c) resulting graph

Fig. 5.7. Delaunay triangulation. A triangle is in the Delaunay triangulation if its circumcircle is empty.

from an edge. Therefore localized versions of the Delaunay triangulation have been defined: the k-localized Delaunay triangulation, the restricted Delaunay triangulation, and the partial Delaunay triangulation. We describe here the k-localized Delaunay triangulation for which an algorithm will be described in Section 5.4.2.

For the k-localized Delaunay triangulation one considers the *k-localized empty circle property*, which says that no point of the *k-neighborhood* lies in the circumcircle of the triangle. A k-neighbor is a vertex to which a path of length k exists in the original graph. The k-localized Delaunay triangulation is the Gabriel graph plus all triangles fulfilling the k-localized empty circle property. For $k = 1$ the k-localized Delaunay triangulation is not necessarily planar but it can be converted into a planar spanner by further processing. For $k \geq 2$ it is a planar spanner. The k-localized Delaunay triangulation contains the Delaunay triangulation, therefore the distance stretch factor is at most that of the Delaunay triangulation and the energy stretch factor is again 1. The maximum vertex degree is also $n - 1$.

Euclidean Minimum Spanning Tree (EMST). A Euclidean minimum spanning tree is a graph that connects all vertices and whose total Euclidean edge length is minimized. It is a *tree*, i.e., a graph without cycles. It is connected and sparse but its energy stretch factor and distance stretch factor are $n - 1$, i.e., it is not a spanner. Note the difference between spanning and spanner: A spanning tree connects all vertices, a spanner is a graph with constant distance stretch factor.

The maximum vertex degree in a Euclidean minimum spanning tree is 6. For a given set of vertices, the Euclidean minimum spanning tree need not be unique. Many useful properties of the Euclidean minimum spanning tree have been proven by Gilbert and Pollack [160].

Nearest Neighbor Graph (NNG). The nearest neighbor graph is perhaps the most basic proximity graph. Each vertex is connected to its nearest neighbor. The nearest neighbor graph is a directed graph. It is in general not connected, and therefore has not been used for topology control. It is planar and sparse but not a spanner. It is contained in a Euclidean minimum spanning tree and thus has maximum vertex degree 6.

In the k–Nearest Neighbor graph each node is connected to its k nearest neighbors. Blough et al. [39] use the graph containing only the bidirectional edges, i.e., both endpoints are in the k–neighborhood of the other end point, to achieve a topology where each node has a bounded (by k) number of nodes in its transmission range. For nodes distributed uniformly at random in a square and a suitable constant λ the resulting graph is connected with high probability if $k \leq \lambda \log n$.

5.3.4 Graphs with Exclusion Regions

Even if a graph is not locally definable there might be a non-trivial larger graph containing it which is locally definable. In particular it might have an exclusion region: a region around an edge such that a necessary condition for the edge to be in the graph is that the region is empty on one side of the edge. In general the number of edges obtained in such a way might be quadratic, i.e., the resulting graph is not sparse. But the edges might be of interest as a candidate set for further processing.

An example of a graph with an exclusion region is the Delaunay triangulation. Its exclusion region is the disk which has the edge as a diameter (the same area defining the empty neighborhood for the Gabriel graph, but now we allow vertices to lie in one half of it). Again, the exclusion region only gives a necessary – not a sufficient – condition for an edge to be in the Delaunay triangulation.

Also the minimum weight triangulation (triangulation with minimum total edge length) allows for an exclusion region, as well as the minimum dilation triangulation (triangulation with minimum maximal stretch factor between any two nodes) and the greedy triangulation (incrementally constructed triangulation adding shortest edges first if possible). An overview over graphs with exclusion regions is given by Drysdale et al. [193].

The following table summarizes the most important graphs and their properties. The bounds in the table are worst case bounds.

5.4 Localized Algorithms

We describe here three localized topology control algorithms. Each algorithm makes use of one of the graphs described in the previous section: the cone-covering graph, a localized version of the Delaunay triangulation and the relative neighborhood graph.

5.4.1 Cone Based Topology Control (CBTC)

We describe a cone based algorithm proposed by Wattenhofer et al. in [391] and revised and extended by Li et al. in [254]. The algorithm consists of two

graph property	GG	RNG	YG	NNG	EMST	DT
connected	yes	yes	weakly	no	yes	yes
planar	yes	yes	no	yes	yes	yes
sparse	yes	yes	yes	yes	yes	yes
maximum vertex degree	$n-1$	$n-1$ $(5,6)$	$n-1$	6	6	$n-1$
distance stretch factor	$\sqrt{n-1}$	$n-1$	$\frac{1}{1-2\sin\frac{\pi}{k}}$	–	$n-1$	$\leq \frac{4\sqrt{3}}{9}\pi$
energy stretch factor	1	$n-1$	$\left(\frac{1}{1-2\sin\frac{\pi}{k}}\right)^a$	–	$n-1$	1
spanner	no	no	yes	no	no	yes

Table 5.2. Graph properties of the Gabriel graph (GG), the relative neighborhood graph (RNG), the Yao graph (YG), the nearest neighbor graph (NNG), the Euclidean minimum spanning tree (EMST) and the Delaunay triangulation (DT). The a in the energy stretch factor of the Yao graph denotes the attenuation exponent.

phases: a basic step (establishing a candidate edge set) and several possible improvements (removing unnecessary edges) as a second step. An example for the resulting topology is shown in Figure 5.8 (taken from [254]).

Algorithm 14: Cone Based Topology Control

---

1   Neighbors $N_u \leftarrow \emptyset$
2   Directions $D_u \leftarrow \emptyset$
3   Transmission power $p_u \leftarrow p_{min}$

4   **while** $p_u < p_{max}$ & $\exists$ *empty α-cone in* D_u **do**
5   Increase p_u
6   Broadcast and gather Acks
7      $N_u \leftarrow N_u \cup \{v : \text{discovered } v\}$
8      $D_u \leftarrow D_u \cup \{dir(v) : \text{discovered } v\}$

Basic Step. The basic algorithm 14 is easy to state: For a predefined angle α the algorithm finds for every node the smallest circular neighborhood such that there is a point in any cone of angle α starting at the node. I.e., in the basic step the cone-covering graph described in Section 5.3.2 is computed.

For this, each node incrementally increases its transmission radius and records the nodes it reaches. Whenever it reaches a new node it checks whether the cone defined by this node is already covered by previously established edges. If not, a new edge to this node is established. Note that nodes only need to know the distances and directions of their neighbors, and not their exact positions.

Then the symmetric closure of the up to now directed graph is computed. I.e., when each node has covered its angular range with cones or reached its maximal transmission range, it contacts all its neighbors that it has established edges to, and if necessary the neighbors add those edges. This symmetric closure is connected for any $\alpha \leq 5\pi/6$. For larger α it is not necessarily connected.

Remarks on the Connectivity:. In [254] it is proven, that the symmetric closure is connected. But actually a stronger result holds: It contains a Euclidean minimum spanning tree. This follows from Lemma II.2 in [254]:

Let R be the maximum transmission range and G_R the graph with edge set E_R containing all edges (u,v) lying in each others transmission range. Let E_α be the set of edges chosen by the algorithm. If $\alpha \leq 5\pi/6$, and u and v are nodes in V such that $(u,v) \in E_R$ (that is, (u,v) is an edge in the graph G_R, so that $d(u,v) \leq R$), then either $(u,v) \in E_\alpha$ or there exist $u',v' \in V$ such that (a) $d(u',v') < d(u,v)$, (b) either $u' = u$ or $(u,u') \in E_\alpha$, and (c) either $v' = v$ or $(v,v') \in E_\alpha$.

First observe that, using the unit disk graph model, if the maximal reachable graph (all possible edges established) is connected, it contains a Euclidean minimum spanning tree. Using Lemma II.2 one can show that in the basic step no Euclidean minimum spanning tree edges are removed: Assume a Euclidean minimum spanning tree edge is removed, then by the lemma it can be replaced by a shorter edge which is a contradiction to it being a Euclidean minimum spanning tree edge.

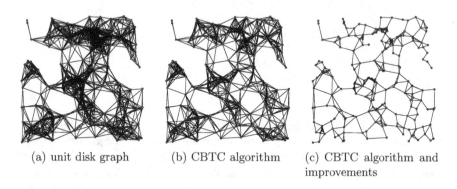

(a) unit disk graph   (b) CBTC algorithm   (c) CBTC algorithm and improvements

Fig. 5.8. Topologies computed by the CBTC algorithm.

Improvements. Several improvements which remove energy inefficient edges that have been established by the basic step are possible as a second step to the basic algorithm. All improvements preserve connectivity. In the original work [391][254] this is proved directly for each improvement but it can also be seen by the following argument:

1. The original topology computed by CBTC contains a minimum spanning tree (if the unit disk graph is connected).
2. Any minimum spanning tree is contained in the relative neighborhood graph.
3. The improvements delete only edges which are not in the relative neighborhood graph and therefore no edges of the minimum spanning tree.
4. After applying the improvements the graph still contains a minimum spanning tree and is thus connected.

Shrink Back. In the basic algorithm, nodes might not find a neighbor in every direction. I.e., there may be directions which are not covered by cones, simply because there are no neighbors in that direction. In this case the transmission radius will have been set to the maximum. This improvement reduces these maximum transmission radii to the smallest transmission radii that reach all chosen neighbors. The resulting graph will still contain a Euclidean minimum spanning tree and is therefore still connected.

Asymmetric Edge Removal. The key of this improvement is that for $\alpha \leq 2\pi/3$ a Euclidean minimum spanning tree is a subset of the directional edges, analogously to the case of the Yao graph with $k \geq 6$. This follows from the fact that the Euclidean minimum spanning tree is contained in the relative neighborhood graph. In other words: For $\alpha \leq 2\pi/3$ any outgoing edge of the cone-covering graph without corresponding ingoing edge is not necessary to maintain a Euclidean minimum spanning tree as subgraph. Thus removing asymmetric, i.e., directed, edges, does not remove any Euclidean minimum spanning tree edges and the resulting graph is still connected.

Pairwise Edge Removal. In this improvement edges are removed that are not contained in the relative neighborhood graph. This is done using the following observation: If a node u is connected to two vertices v and w, and if either (u, v) or (u, w) are the longest edge in the triangle (u, v, w), it can be removed. This is in fact a reformulation of the definition of the relative neighborhood graph. Again a Euclidean minimum spanning tree, which is contained in the relative neighborhood graph, will still be contained in the resulting graph, which therefore is connected.

Triangle Inequality Edge Removal. This is the original improvement given in [391], which uses the triangle inequality for edge removal. An edge (v, w) is removed if for some vertex u the edges (u, v) and (u, w) are in the graph, and $p(u, v) + p(u, w) \leq p(v, w)$, where $p(e)$ denotes the energy of the edge e. If p is the quadratic distance function, then all removed edges are not edges of the Gabriel graph.

5.4.2 Delaunay Based Topology Control

Several topology control algorithms compute a graph containing the Delaunay triangulation restricted to the full communication graph, i.e., a graph containing all edges that are in the Delaunay triangulation as well as in the communication graph. The resulting topologies have constant distance stretch factor in the case of the unit disk graph model. The algorithms discussed assume that location information is given at each node.

Gao et al. [148] define a *restricted Delaunay triangulation* as any planar graph containing the Delaunay triangulation restricted to the unit disk graph. They combine such a graph with node clustering (i.e., the restricted Delaunay triangulation is used to connect clusters) and show how it can be maintained when nodes move.

Li and his co-authors proposed several topology control algorithms based on the k–localized Delaunay triangulation [256] described in Section 5.3. For achieving a reasonable run-time, k is typically chosen as 1 or 2. The 1–localized Delaunay triangulation is not necessarily planar but a planar subgraph can be extracted. For $k \geq 2$ the k–localized Delaunay triangulation is always planar. In the following we describe algorithm 15 for computing the k–localized Delaunay triangulation.

Algorithm 15: k-Localized Delaunay Topology Control

---

1 Collect location information of k–hop neighbors
2 Find triangles incident to node in local Delaunay triangulation
3 Broadcast and receive triangle information
4 Check if incident triangles are confirmed by all three vertices
5 Add Gabriel graph edges
6 Add edges of verified triangles

Each node collects the location of its k–hop neighbors and computes the Delaunay triangulation of this set of points. Of the computed triangulation it discards all triangles not incident to itself. It then sends a list of its incident triangles to its neighbors and receives their list of incident triangles. Each node checks if its incident triangles are verified by its neighbors. I.e., each node u checks for a triangle uvw if the same triangle is also in the list of triangles of v and w. If not, the triangle is discarded. The communication cost of the verification step can be reduced to $O(n)$ by letting each triangle be checked only by the vertices at which it has an angle of at least $\pi/3$.

5.4.3 Relative Neighbor Topology Control (XTC)

This algorithm was proposed by Wattenhofer and Zollinger in [392]. If based on Euclidean distances it computes a subgraph of the relative neighborhood graph, and uses only relative positional information for this.

Algorithm. The XTC algorithm computes a subgraph of the relative neighborhood graph using only relative positional information, i.e., each node is required to sort its neighbors by their distance. For this XTC does not use the description of the relative neighborhood graph by empty neighborhoods, which would require full positional information. Instead XTC uses the description of the relative neighborhood graph that an edge is established between two vertices u and v if there is no vertex w that is closer to both u and v than u and v are to each other.

Algorithm 16: Relative Neighbor Topology Control (XTC)

---

1   Establish order $\prec_u$ over u's neighbors
2   Broadcast order
3   Receive orders
4   **forall** *neighbors v in increasing order according to $\prec_u$* **do**
5   **if** $\nexists w : w \prec_u v$ & $w \prec_v u$ **then**
6   Add v to final neighbor set

More explicitly, XTC (algorithm 16, from node u's perspective) does the following: First each node determines its one-hop neighbors, i.e., all other nodes in its transmission range. Then it sorts its neighbors by any reasonable order, for instance the link qualities to its neighbors. This order may correspond to the order on the Euclidean distances between the nodes, but need not. Instead they may also depend on obstacles between the nodes. Symmetric edge weights are assumed. In case of a tie, i.e., when two neighbors have the same "distance" to it, tie breaking is used for achieving a strict order. This will be explained in the next paragraph.

All nodes exchange their ordered neighbor lists with their neighbors. Then each node u processes its neighbor list in order to select those neighboring nodes that define the neighborhood in the topology control graph: For each node v on the neighbor list it checks whether there is a node w that comes before v on its own neighbor list and before itself on the neighbor list of v. I.e., it checks for a node w s.t. $w \prec_u v$ and $w \prec_v u$, where $\prec_v$ and $\prec_w$ denote the neighbor orders of v and w, respectively. If this is the case, v is not added to the topology defining neighborhood, otherwise it is.

Tie Breaking. A tie occurs if two nodes have the same distance to a third node. To break the tie means to assign distinct and consistent distances to all pairs of nodes. This situation can be described more generally for a weighted graph where the goal is to have distinct and consistent edge weights. Assuming that all nodes have distinctive IDs, this can be done as follows: Assign to each edge an extended, more general weight, which is the triple of the former edge weight, the smaller endpoint id and the larger endpoint id. A strict order is then given by the lexicographic order on these weight triples. For the XTC algorithm this means simply that if a node has two neighbors with equal distance it sorts them according to their IDs.

Properties of G_{XTC}. Let G_{XTC} denote the graph computed by XTC. G_{XTC} is symmetric by definition (the condition $u \prec_v w$ and $u \prec_w v$ is symmetric). G_{XTC} is connected in the general weighted graph model if the original graph was connected.

The following properties hold if Euclidean distances are used: The maximum vertex degree is 6, G_{XTC} is planar and a subgraph of the relative neighborhood graph. If also there are no ties, i.e., no two nodes that have the same distance to a third node, XTC exactly computes the relative neighborhood graph.

5.5 Chapter Notes

In this survey we discussed geometric graphs suitable for topology control and localized algorithms computing such graphs.

A survey on relative neighborhood graphs and its relatives is given by Jaromczyk and Toussaint [199]. The β-skeletons are proposed by Kirkpatrick and Radke [217]. Spanners, i.e., graphs with constant distance stretch factor, are discussed by Arya et al. [19] and Eppstein [113]. Drysdale et al. [193] give an overview of exclusion regions. Santi [338], [339] and Li [255] survey topology control for wireless ad hoc networks.

In Section 5.2 we gave criteria for topology control and in Section 5.3 described graphs which fulfill these criteria. In Section 5.4 we described algorithms computing some of these graphs. The criteria were described in a graph-theoretic sense and the reasoning why they are good for topology control is more or less heuristic. The criteria are also closely bound to the unit disk graph model and the assumption that the nodes lie in the Euclidean plane.

One of the main open problems in topology control is how realistic these assumptions are. In other words, what are realistic energy and interference models, and efficient topology control algorithms based on them? In particular, the impact of the mobility of nodes needs to be addressed. Several authors [39], [55] show that optimizing the given graph properties not necessarily implicates an improvement of the network performance. Approaches explicitly handling interference are further discussed in Chapter 6.

6

Interference and Signal-to-Noise-Ratio

Alexander Kröller

6.1 Introduction

In a wireless network, a signal sent from one node to another suffers from physical effects. It will be attenuated, where the amount of loss depends on the physical medium it passes through, the distance it travels, and many other influences. The signal cannot be decoded the signal if it is too weak at the receiver. So in order to reach a certain receiver, the sender has to make sure it sends the signal with sufficient power, such that it is still *strong enough* after the attenuation on its path to the receiver. More details on these concepts are given in Section 2.3 or any decent book on communications engineering, e.g., [336].

But what exactly means *strong enough?* Unfortunately, this does not just depend on the two nodes that wish to communicate. The receiver can decode the signal only if it is significantly stronger than all other signals the antenna picks up, for example noise produced by the receiver's circuitry. Because a signal sent by some node always travels over many paths in multiple directions, it may reach nodes other than the designated receiver, and counts as noise for them. Hence a transmission may fail just because some other nodes in the environment transmit signals at the same time – their signals *interfere* with the transmission.

Hence, it is an intrinsic property of wireless communications that the success of a transmission depends on the existence of simultaneous transmissions between other nodes. There is no simple solution to this situation, mainly because of the *hidden terminal problem:* There may be two senders that cannot talk to each other directly, but whose messages interfere, because the receivers are within communication range of both senders.

This chapter is about dealing with the effects of interference. Generally, the solution is twofold: First, messages should be sent with carefully chosen power levels to keep the interference at other nodes as little as possible. Second, messages that would interfere with each other should be sent at different times.

D. Wagner and R. Wattenhofer (Eds.): Algorithms for Sensor and Ad Hoc Networks, LNCS 4621, pp. 99–116, 2007.

Algorithmically, there are two possibilities. The first is simply to ignore the existence of interference. This is justified because the operating system of the nodes runs designated protocols that try to minimize interference effects and automatically retransmit messages that failed for whatever reasons. Section 4.1.1 reviews some of these protocols. Higher-level algorithms are then built on the assumption that message transmissions will always succeed, it just might take some time until the MAC layer protocol sorts everything out.

The other possibility is to consider interference as an additional constraint in the design of algorithms. The benefit is that such algorithms can use communication patterns that minimize the occurrence of interference. This chapter shows how this can be done. First, we introduce some basic models for interference. Section 6.3 shows how to restrict communication to a subset of the possible links that are less likely to interfere with each other. In Section 6.4, interference-free schedules for the links are discussed. Then, in Section 6.5, we analyze interference-aware routing.

6.2 Interference Models

When discussing failing radio transmission, one has to distinguish different types of failures: A *primary* conflict occurs when a node attempts to send multiple messages at once, or it tries to decode multiple messages. The limitation in sending is dealt with by the operating system software and usually does not need to be addressed explicitly in algorithms. So just the inability to receive has to be considered. A *secondary* conflict happens when a message is not sufficiently decoded by the intended receiver(s) due to interference.

We assume that a network (V, E) is given, where E consists of all potential communication links.

6.2.1 Physical Model

We first introduce the *physical* model, which is widely believed to resemble reality closely. Its introduction into algorithms for wireless networks is frequently addressed to [174].

Assume a node v transmits a message with power $P_v > 0$. A node w at distance $d(v, w)$ will receive the signal if the *signal strength* $SS(v, w)$ is sufficiently high:

$$SS(v, w) := \frac{P_v}{d(v, w)^\alpha} \geq \beta . \tag{6.1}$$

Here, α is a model parameter that reflects the environment. It is usually assumed that α is a position-independent constant with a known value $2 \leq \alpha \leq 5$. β is a constant that models characteristics of the receiver. Two direct observations can be made:

1. The necessary sending power to send a signal over distance d is $\Omega(d^\alpha)$.
2. The area in which a message can be received is the ball $B(v, (P_v/\beta)^{1/\alpha})$.

Note that (1.) is also observed in reality, while (2.) is not. This can be overcome by using direction-dependent signal strength functions instead. See also Section 2.3 for more details.

If a transmission is successful according to (6.1), it may still get lost due to interference. The receiver w can decode the message successfully only if the signal is significantly stronger than ambient noise and other interfering signals. This is modeled by the *Signal to Interference and Noise Ratio* (SINR) inequality

$$\frac{SS(v, w)}{N + \sum_{u \in U} SS(u, w)} \geq \gamma , \tag{6.2}$$

where $N \geq 0$ is the noise induced by the environment and the electrical circuit of w, and U is the set of nodes that are transmitting in parallel to v, each with its own sending power P_u. Here, $\gamma \geq 1$ is a constant depending on hardware and software characteristics of the receiving node.

The SINR model can also be expressed as a *weighted conflict graph*: Consider two communication links $e_1 = v_1 w_1$ and $e_2 = v_2 w_2$. We define a weight $s(e_1, e_2)$ that describes how much interference e_1 causes on e_2 by

$$s(e_1, e_2) = \begin{cases} 0 & \text{if } e_1 = e_2 \\ \infty & \text{if } e_1 \neq e_2 \text{ with primary conflict} \\ \frac{SS(v_1, w_2)}{\frac{1}{\gamma} SS(v_2, w_2) - N} & \text{otherwise} \end{cases} . \tag{6.3}$$

This defines a graph whose nodes are the edges of the original graphs. The weighted edges of the conflict graph define where interference occurs. Now consider a set $F \subseteq E$ of simultaneous transmissions. Each transmission $f \in F$ will succeed according to the SINR inequality (6.2) if, and only if,

$$\sum_{f' \in F} s(f', f) \leq 1 . \tag{6.4}$$

This allows to formulate interference by linear constraints, assuming the sending powers P_v are fixed and known. We will exploit this to obtain linear programs for some problems in Sections 6.4 and 6.5.

6.2.2 Conflict Graph Models

There exists a whole class of interference models that construct a *conflict graph* (E, C). In such a graph, there is a node $e \in E$ for every communication edge in the network, and an edge $ef \in C$ whenever simultaneous transmission along e and f fails due to a primary conflict or interference. This is essentially a simplification of the previous model: It is just modeled whether *two* simultanous transmissions interfere with each other. The effect of accumulated noise when many nodes send simultaneously is ignored.

The benefit of using conflict graph models is that feasibility of parallel transmissions reduces to well-understood graph concepts: A set of transmissions can be done simultaneously if it is an independent (stable) set in (E, C). A coloring of $E' \subseteq E$ defines a scheduling of transmissions.

While E is usually assumed to consist of directed edges, C is not: Feasibility of a set of parallel transmissions does not depend on the orientation of the conflict edges in C.

We now introduce three common conflict graph models. Some of them assume that the nodes can not adjust their transmission power on a per-message basis, for them $\mathrm{rg}(v)$ denotes the transmission range of a node v. Note that $\mathrm{rg}(v) \geq d(v, w) \; \forall w : vw \in E$.

Assume node u sends a message to v, while another message is sent from u' to v' at the same time; without primary conflicts. Figure 6.1 shows for each model a situation where v cannot receive the message from u due to interference.

1. In the *directed distance-2* model, v successfully receives the message if it is no neighbor of u'. This depends on the network graph at the time of sending. There is a great amount of possible graph definitions (see Chapter 2); here, we concentrate on three variants.
   - The *general* variant assumes the network (V, E) is fixed, without any assumptions on geometry. The transmission fails if $u'v \in E$ (Figure 1(a)).
   - The variant with *fixed ranges* uses a static graph (V, E). Message failure occurs if (see Figure 1(b))

   $$d(u', v) \leq \mathrm{rg}(u') . \tag{6.5}$$

   - The variant with *adjustable sending powers* uses a disk graph, where each node may adjust its transmission range on a per-message basis. Hence, the message is lost if

   $$d(u', v') \geq d(u', v) . \tag{6.6}$$

   See Figure 1(c).
2. In the *directed σ-Tx* model for parameter $\sigma \geq 0$,

   $$(1 + \sigma)(\mathrm{rg}(u) + \mathrm{rg}(u')) \leq d(u, u') \tag{6.7}$$

   must hold for successful transmission. See Figure 1(d).
3. The *directed σ-Protocol* model with parameter $\sigma \geq 0$ requires

   $$d(u', v) \geq (1 + \sigma)d(u, v) \tag{6.8}$$

   for v to receive the message (Figure 1(e)).

Note that all of these models are simplifications of the SINR inequality (6.2) with different assumptions on the parameters and sending power schemes.

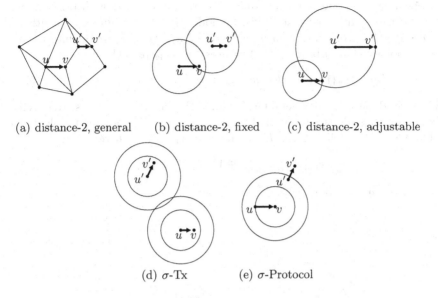

(a) distance-2, general   (b) distance-2, fixed   (c) distance-2, adjustable

(d) σ-Tx   (e) σ-Protocol

Fig. 6.1. Conflict graph models

Each of these models, including the SINR model, also exists in a *bidirected* variant: Here, it is assumed that a transmission requires some handshaking with the receiver, so both partners of the transmission are actually senders. In the bidirected variant, a message transmission is successful if it is successful according to the directed model even if the direction of any number of transmissions are reversed.

6.3 Low-Interference Topologies

This section deals with the problem of finding low-interference network topologies. We are looking for a topology that satisfies certain conditions (say, has property Ψ) while having low interference according to some interference measure χ. This is the MINIMUM INTERFERENCE TOPOLOGY Problem

$$(\text{MIT}_\Psi) \quad \min\{\chi(E') : E' \subseteq E, (V, E') \text{ has property } \Psi\} . \tag{6.9}$$

Note that this problem is closely related to Topology Control problems, see Chapter 5.

6.3.1 Objectives

We define and analyze three different interference measures, an edge-, a node-oriented and general one.

Assume two nodes are communicating over an edge uv with transmission radii adjusted on a per-packet basis. All nodes that are within range of either u or v may suffer from interference, see Figure 2(a). We measure the interference of a topology (V, E') by $\chi_{\text{edge}}(E') := \max_{e \in E'} \chi_{\text{edge}}(e)$, where

$$\chi_{\text{edge}}(uv) := |\{w \in V : d(u,w) \leq d(u,v) \vee d(v,w) \leq d(u,v)\}| . \qquad (6.10)$$

The second measure considers the likelihood that a node v receives interference from another node, assuming that sending power is fixed for the topology, see Figure 2(b). We define $\chi_{\text{node}}(E') := \max_{v \in V} \chi_{\text{node}}(v)$, where

$$\chi_{\text{node}}(v) := |\{w \in V : d(v,w) \leq r_w\}| \qquad (6.11)$$

and $r_w = \max\{d(w,x) : wx \in E'\}$.

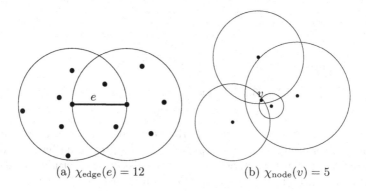

(a) $\chi_{\text{edge}}(e) = 12$ (b) $\chi_{\text{node}}(v) = 5$

Fig. 6.2. Interference measures χ_{edge} and χ_{node}

The third model is very general. We assume that each node v has a set of $k(v)$ possible transmission power levels. These correspond to a sequence $\varnothing = R_0^v \subsetneq R_1^v \subsetneq \ldots \subsetneq R_{k(v)}^v \subset V$ of receiver sets. Let $r : V \to \mathbb{N}$ with $0 \leq r(v) \leq k(v)\ \forall v \in V$ be a power assignment. Then r defines a topology

$$E_{\text{gen}}(r) := \{uv : u \in R_{r(v)}^v \wedge v \in R_{r(u)}^u\} . \qquad (6.12)$$

Note that all such topologies are subsets of $E := E_{\text{gen}}(k)$. For a topology $E' \subseteq E_{\text{gen}}(k)$, we define $r_{E'}$ to be the power assignment that chooses at every node the smallest possible power such that $E' \subseteq E_{\text{gen}}(r_{E'})$.

Let r be a given power assignment. Similar to (6.11), we define the interference at a node v by $\chi_{\text{gen}}(v, r) := |\{w : v \in R_{r(v)}^v\}|$. For this measure, we look at the *average* interference

$$\chi_{\text{gen}}(E') := \frac{1}{n} \sum_{v \in V} \chi_{\text{gen}}(v, r_{E'}) , \qquad (6.13)$$

as opposed to the worst-case considerations in the previous two measures.

6.3.2 Connected Topologies

We now analyze the Minimum Interference Connected Topology Problem (MIT$_{conn}$), where τ is the property of being connected. Often, topology control algorithms always connect each node to its nearest neighbor. Unfortunately, this leads to worst possible results:

Lemma 6.3.1. *[55, 378] Let* be edge *or* node. *Every algorithm for MIT$_{conn}$ that always includes the Nearest Neighbor Forest has an approximation factor in* (n).

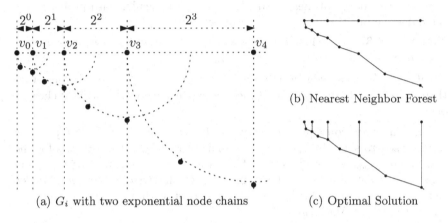

(a) G_i with two exponential node chains   (c) Optimal Solution

(b) Nearest Neighbor Forest

Fig. 6.3. Exponential Node Chains

Proof. Consider the graph G_i with $n = 3i + 1$ nodes and two exponential node chains (Figure 3(a)). It has a horizontal chain $v_0, \ldots, v_i$ where any two subsequent nodes v_j and v_{j+1} have a distance of 2^j. Every v_j, $= 1, \ldots, i$, has a partner node that is placed $2^{j-1} + \varepsilon$ below it, with a small $\varepsilon > 0$, and between two such partners there is another node at a carefully chosen position (see Figure 3(a)). The Nearest Neighbor Forest NNF(G_i) of G_i is shown in Figure 3(b). Its interference is

$$\text{edge}(\text{NNF}(G_i)) = \text{edge}(v_{i-1}v_i) = n - 1 = \ (n) \text{ resp.}$$
$$\text{node}(\text{NNF}(G_i)) = \text{node}(v_0) \geq i = \ (n) \ .$$

The optimal solution (Figure 3(c)) does not use the horizontal edges, resulting in a topology where both interference measures are $O(1)$.   □

Note that not doing topology control at all, i.e., letting $E' = E$, is also an $O(n)$-approximation. Observe that most structures that were presented in Chapter 5 include the NNF. Even worse, it turns out that no better structures can be constructed if we seek local algorithms:

Lemma 6.3.2. *[55] No local algorithm can have a better approximation factor than $\Omega(n)$ w.r.t. χ_{edge}.*

Here, an algorithm is called *local*, if the decision on whether to include an edge e depends only on a $O(1)$-hop neighborhood of e w.r.t. the graph (V, E).

Proof. Construct a graph with a long edge uv and $\Omega(n)$ nodes close to v, such that inclusion of uv in the topology covers all these nodes. Add a path of length $\Omega(n)$ that connects u and v. Select the node in the middle of the path, and decide randomly whether to break the path by removing this node or not. Deciding whether uv is in an optimal solution now requires to know whether the middle node at distance $\Omega(n)$ still exists.   □

However, this does not mean that MIT_{conn} is a particular hard problem. For the objective function χ_{edge} it can even be shown that

Theorem 6.3.3. *[55] A minimum spanning tree on (V, E) using edge weights $\chi_{\text{edge}}(e)$ is an optimal solution for MIT_{conn}.*

Proof. $\chi_{\text{edge}}(e)$ as an edge weight does not depend on the state of edges other than e. In a MST, one cannot replace an expensive edge by strictly cheaper ones.   □

As an immediate consequence, every greedy algorithm for the MST problem can be used to solve MIT_{conn}. So it remains to find good algorithms for the case when χ_{node} is the objective. This is an open problem, however, good algorithms exist if we restrict the problem to the *General Highway Model*: Here, we assume that (V, E) is a unit disk graph with all nodes located on a straight line. See Algorithm 17.

Algorithm 17: Algorithm for χ_{node} in one-dimensional networks

---

1 Sort nodes by position: Order $v_1, v_2, \ldots, v_n$ such that $x_1 < x_2 < \ldots x_n$
2 Partition nodes into disjoint segments $S_1, \ldots, S_{\lceil x_n - x_1 \rceil}$ of length 1:
  $v_i \in S_{\lfloor x_i - x_1 \rfloor + 1}$
3 From every segment $\{v_i, \ldots, v_j\}$, choose nodes $v_{i+k\lceil\sqrt{\Delta}\rceil}$, $k = 0, 1, 2, \ldots$ and
  v_j, and add them to H
4 Connect the nodes in H linearly according to their position, resulting in edge set E'
5 For each node $v \in V \setminus H$, choose a closest node $h \in H$, and add $\{v, h\}$ to E'

Theorem 6.3.4. *[378] Algorithm 17 computes a connected subgraph (V, E') with $\chi_{\text{node}}(E') = O(\sqrt{\Delta})$, where Δ is the maximum nodal degree in the network. There exist problem instances for which $\Omega(\sqrt{\Delta})$ is a lower bound on the objective.*

Although Algorithm 17 is optimal in some sense, there may be still problem instances where it performs bad. Fortunately there is a simple criterion to identify these instances and provide direct solutions for them:

Theorem 6.3.5. *[378] Let (V, E) be a network in the General Highway Model. Assume $V = \{v_1, \ldots, v_n\}$ with positions $x_1 < \ldots < x_n$. Let $E'' = \{\{v_i, v_{i+1}\} : i = 1, \ldots, n-1\}$.*

If $\chi_{\text{node}}(E'') \leq \sqrt{\Delta}$, then E'' is a $O(\sqrt[4]{\Delta})$-approximation for MIT_{conn}. Otherwise, Algorithm 17 outputs a solution with an approximation factor of $O(\sqrt[4]{\Delta})$.

Now we turn to the third interference measure, χ_{gen}. A set $S \subseteq E$ is a *star*, if the corresponding subgraph $(V(S), S)$ is a tree of depth 1. A star is called *active* w.r.t. a power assignment r, if all nodes in $V(S)$ are mutually disconnected in $(V, E_{\text{gen}}(r))$. Given r, we define the cost $c(S, r)$ of an active star by

$$c(S, r) = \frac{1}{|V(S)|} \sum_{v \in V(S): r'(v) \neq r(v)} |R^v_{r'(v)}|, \tag{6.14}$$

where $r' := r_{E_{\text{gen}}(r) \cup S}$ is the minimal power assignment when adding S to the current topology. MIT_{conn} can be approximated by Algorithm 18, which adds active stars in a greedy manner.

Algorithm 18: Greedy algorithm for MIT_{conn} with χ_{gen}

---

1 Let r be the trivial power assignment 0
2 **while** $(V, E_{\text{gen}}(r))$ *is not connected* **do**
3 $\quad$ Let $S := \arg\min\{c(S, r) : S$ is an active star$\}$
4 $\quad r \leftarrow r_{E_{\text{gen}}(r) \cup S}$

Proving correctness for this algorithm is trivial (a single edge connecting two network components defines a star). It can be shown that this algorithm performs very well even on this very general interference model:

Theorem 6.3.6. *[297]*

Algorithm 18 is an $O(\log n)$-approximation for MIT_{conn} with objective χ_{gen}.

Sketch of Proof. Using the dual of a linear covering formulation of MIT_{conn} with model χ_{gen}, it can be shown that the algorithm picks stars for which $\lambda c(S, r)/|S| \leq OPT$ holds, where λ is the number of connected components in $(V, E_{\text{gen}}(r))$ before S is added; afterwards it is $\lambda - |S|$. This leads to a recursion that proves the claim. $\qquad\qquad\qquad\qquad\qquad\qquad\qquad\qquad\square$

Because the model is generic, there might be room for improvement if we consider more realistic, e.g., geometric, networks. However, it turns out that the algorithm is most likely optimal, even in the restricted setting:

Lemma 6.3.7. *[297] Assume the receiver set sequences for χ_{gen} are induced by a discrete metric on the nodes. Then there exists no algorithm with an approximation factor better than $\Omega(\log n)$, unless $\mathcal{NP} \subseteq \text{DTIME}(n^{\log \log n})$.*

Sketch of Proof. There is an approximation-preserving transformation from the general Set Cover problem, for which the claimed lower bound is already known. See [297].

6.3.3 Spanners

This section deals with a stronger property Ψ than mere connectivity: We require the topology to be an Euclidean t-spanner. For that, we define the length of a path $P = (v_1, v_2, \ldots, v_k)$ by $\ell(P) := \sum_{i=1}^{k-1} d(v_i, v_{i+1})$. A topology (V, E') is a t-spanner w.r.t. (V, E) if for every u-v-path P in (V, E) there is a u-v-path P' in (V, E') with $\ell(P') \leq t\ell(P)$.

For the interference measure χ_{edge}, such topologies can be determined by Algorithm 19.

Algorithm 19: Algorithm for MIT_{span} with χ_{edge}

---

1 Set $E_{\text{todo}} \leftarrow E$, $E' \leftarrow \varnothing$
2 Let χ_{edge} be defined w.r.t. the changing (V, E_{todo})
3 **while** $E_{\text{todo}} \neq \varnothing$ **do**
4 $\quad$ $uv \leftarrow \arg\max\{\chi_{\text{edge}}(e) : e \in E_{\text{todo}}\}$
5 $\quad$ **while** $d^*(u, v)$ *w.r.t.* (V, E') *exceeds* $td(u, v)$ **do**
6 $\quad\quad$ $\chi' \leftarrow \min\{\chi_{\text{edge}}(e) : e \in E_{\text{todo}}\}$
7 $\quad\quad$ $E' \leftarrow \{e \in E : \chi_{\text{edge}}(e) \leq \chi'\}$
8 $\quad$ Remove uv and E' from E_{todo}
9 **return** E'

Theorem 6.3.8. *[55] Algorithm 19 computes an χ_{edge}-optimal Euclidean t-spanner.*

Proof. Note that E' is a t-spanner if, and only if, it contains a path P for every edge uv that has length $\ell(P) \leq td(u, v)$. The algorithm simply increases χ' until $\{e \in E : \chi_{\text{edge}}(e) \leq \chi'\}$ is a t-spanner. Because supersets of spanners are also spanners, every optimal solution must contain an edge e with $\chi_{\text{edge}}(e) \geq \chi'$. By definition of χ_{edge}, the final E' is optimal with objective value χ'. $\square$

The algorithm can be distributed easily: For each edge uv, one end-node collects its $t/2$-neighborhood, that is, all nodes that can be reached over a path of length $(t/2)d(u, v)$ from u or v. This can be done by local flooding. It then runs Algorithm 19 on this subgraph, and informs all edges it chose to become part of the spanner.

6.4 Topology Scheduling

A transmission schedule $S = \{S_1, \ldots, S_{T(S)}\}$ ($S_i \subseteq E \ \forall i$) describes a sequence of simultaneous transmission sets where each S_i is possible according to a chosen interference and transmission model. The indices $1, \ldots, T(S)$ are called *time slots*, and $T(S)$ is the *length* of the schedule. If a schedule is intended to be repeated, it is called *periodic*. A periodic schedule can also be described as an infinite sequence of sets, where $S_i = S_{i+T(S)} \ \forall i$.

6.4.1 Link and Broadcast Scheduling

Next we consider the LINK SCHEDULING (LS) and the BROADCAST SCHEDUL-ING Problem (BS): Given a topology $E' \subseteq E$, find a schedule S of minimum length $T(S)$ that contains all transmissions, i.e., $S_1 \cup \ldots \cup S_{T(S)} = E'$. Here, we assume the directed distance-2 interference model on general graphs, as defined in Section 6.2.2.

In link scheduling, each link is considered for itself. Broadcast scheduling assumes that all transmissions are broadcasts: A transmission must be scheduled in one slot with all other transmissions with the same sender. This problem is sometimes also called NODE SCHEDULING, because it is equivalent to assigning the slots to the nodes, where a node may send a message to any neighbor in its slots.

It is known that these kinds of problems are generally hard:

Theorem 6.4.1. *[326, 118, 112] Link and broadcast scheduling of general directed and undirected graphs with distance-2 interference is $\mathcal{NP}$-hard.*

However, there are algorithms to solve the link scheduling problem in trees optimally, there is a constant-factor approximation for planar graphs and an $O(\theta)$-approximation for general graphs, where θ is the thickness, i.e., the minimum size of a partitioning into planar subgraphs [326].

Now, we show how the Link Scheduling Problem with the SINR model can be solved using integer programming. We seek a short schedule for a given topology $E' \subseteq E$. Let $\hat{T}$ be an upper bound on the optimal schedule length, e.g., $\hat{T} = |E'|$. Let $M > |E'|$ be a large constant. Consider the following integer program [36]:

$$\min \sum_{i=1}^{\hat{T}} y_i \tag{6.15}$$

$$\text{s.t.} \sum_{i=1}^{\hat{T}} x_{e,i} \geq 1 \quad \forall e \in E' \tag{6.16}$$

$$x_{e,i} \leq y_i \qquad\qquad \forall i = 1, \ldots, \hat{T}, \forall e \in E' \tag{6.17}$$

$$\sum_{e \in E'} s(e, e') x_{e,i} \leq 1 + (1 - x_{e',i})M \qquad \forall i = 1, \ldots, \hat{T}, \forall e' \in E' \tag{6.18}$$

$$y \in \{0, 1\}^{\hat{T}} \tag{6.19}$$

$$x \in \{0, 1\}^{E \times \hat{T}} \tag{6.20}$$

Here, $y_i = 1$ for every slot that is used and $x_{e,i} = 1$ if edge e is scheduled in slot i. The objective (6.15) minimizes the schedule length. (6.16) requires that each edge is actually scheduled, and (6.17) ensures that if there is any edge scheduled in time slot i, it is counted in the objective function. Finally, (6.18) models the SINR constraint, as well as primary conflicts, using the weighted conflict graph (6.3) (this requires to use a constant $> M + 1$ where $s(e, e') = \infty$).

This is a direct formulation for (LS), but it is very large, and hence hard to solve in practice. A better formulation reduces (LS) to a Set Cover instance. Let $\{E_1, E_2, \ldots, E_k\}$ be the set of *all* transmission sets that are feasible w.r.t. the SINR constraint and primary conflicts. Then (LS) can be formulated as:

$$\min \sum_{j=1}^{k} z_j \tag{6.21}$$

$$\text{s.t.} \sum_{j:e \in E_j} z_j \geq 1 \quad \forall e \in E' \tag{6.22}$$

$$z \in \{0, 1\}^k \tag{6.23}$$

This is still too large to solve directly. By relaxing the integrality condition in (6.23), we obtain a linear program for fractional schedules; still with a possibly exponential number of variables. But this LP can be efficiently solved using column generation [36]: The dual separation problem can be formulated as a LP that produces feasible transmission sets.

The Broadcast Scheduling Problem can be solved the same way—both integer programs and the separation problem can be easily adopted for (BS).

6.4.2 Scheduling Complexity

So far, we considered scheduling a given topology. This leads to an interference measure χ that defines the interference of a topology as the minimum length of a transmission schedule. As in Section 6.3, we consider the problem

$$(\text{SC}_\Psi) \quad \begin{array}{l} \min T(S) \\ \text{s.t.} \quad S \text{ is a schedule for } E' \subseteq E, \\ (V, E') \text{ has property } \Psi \, . \end{array} \tag{6.24}$$

The minimum schedule length for a given property Ψ is called *scheduling complexity* of Ψ. [292] analyzes this problem where Ψ is connectivity. It uses the full SINR interference model, so there is the additional challenge to assign power levels.

In practice, power levels cannot be adjusted freely. Often, there is a predefined set of power schemes from which to choose, or even just a single, fixed one. So it makes sense to study the impact of power schemes on the scheduling complexity. We consider the scheduling complexity of property Ψ_a, which is the following relaxation of connectivity: A transmission set $E' \subseteq E$ has property Ψ_a, if every node has a nonzero out-degree. I.e., we consider schedules where every node is allowed to transmit at least one message.

There are two power schemes that are frequently used in MAC layer implementations: First the *uniform* scheme, where P_v is constant, independently of distance and presence of other transmissions. The second is the *linear* scheme, in which a node v sends messages to w with power $P_v = \varrho d(v, w)^\alpha$, where ϱ is constant. This model is based on solving the signal strength equation (6.1) while ignoring parallel transmissions.

Both power models turn out to be equally bad in the worst case:

Theorem 6.4.2. *[292] When the power scheme is uniform or linear, the scheduling complexity of property Ψ_a is worst possible, i.e., $\Omega(n)$.*

Sketch of Proof. The proof uses an exponential node chain, that is a sequence $v_0, \ldots, v_{n-1}$ of nodes on a line where $d(v_i, v_{i+1}) = 2^i$. It can be shown that at most $2^\alpha / \beta = O(1)$ nodes can be scheduled together. For the uniform power scheme, the interference at the receiver of the longest link is high because all parallel senders are at most two times farther than the sender for that link. With a linear scheme, the receiver over the shortest link receives intereference from all other senders.

In the network used in the proof, the link lengths are extremely varying. This turns out to be the reason for the bad results: We partition the links in the network into length classes

$$L_m := \{uv \in E : \lfloor \log d(u, v) \rfloor = m\} \tag{6.25}$$

and define the *diversity* of a network by

$$g(V) := |\{m : L_m \neq \varnothing\}| . \tag{6.26}$$

Theorem 6.4.3. *[292] The scheduling complexity of property Ψ_a with a linear power scheme is at most $O(g(V))$.*

Sketch of Proof. There is even a constructive algorithm to prove this theorem, which constructs $O(g(V))$ grids and schedules the links of each length class together [292].

So far we considered the property Ψ_a, which is just a relaxation of what we are really interested in: The scheduling complexity of connectivity. There is an algorithm that produces very short schedules for weak connectivity:

Theorem 6.4.4. *[292] The scheduling complexity of weak connectivity with free power assignments is at most $O(\log^4 n)$.*

Sketch of Proof. The algorithm first constructs a spanning tree that is directed towards a single root by iteratively adding nearest neighbor forests to the topology. Then, a scheduling and power scheme for the tree is constructed. There are two main ideas to make the schedule short:

- Links that are scheduled together are either of similar length or of very different length. This is done by grouping different length classes together for scheduling.
- The power scheme is non-linear, giving short links a slightly higher power than what a linear scheme suggests.

However, this is not applicable if a linear power scheme has to be used due to technical constraints. In this case, there is another algorithm that schedules links by their length class, leading to the following scheduling complexity:

Theorem 6.4.5. *[292] The scheduling complexity of weak connectivity under a linear power scheme is at most $O(g(V) \log n)$.*

Sketch of Proof. The algorithm iteratively merges clusters, where the number of cluster is at least halved in each iteration. The merging employs the algorithm for a schedule of length $O(g(V))$ for Ψ_a.

6.5 Flow and Path Scheduling

6.5.1 Path Schedules

We consider an integrated routing and scheduling problem: Given demands $w : V \times V \to \mathbb{N}$, where $w(u, v)$ denotes the number of packets to be sent from u to v, find a minimum-length schedule S that actually delivers all packets. We denote the total demand by $W := \sum_{u,v \in V} w(u, v)$. For this problem, we use the directed distance-2 interference model with adjustable sending powers.

Let $\mathcal{P}$ be the set of routing paths that is used in the schedule S. The *dilation* $D_{\mathcal{P}}$ of $\mathcal{P}$ is the maximum number of edges in a path. The *congestion* $C_{\mathcal{P}}(e)$ of an edge e is the number of time slots in which e is not free for transmission, because there is a packet scheduled on e or any other edge that interferes with e. The congestion of $\mathcal{P}$ is defined by $C_{\mathcal{P}} := \max_{e \in E} C_{\mathcal{P}}(e)$.

Lemma 6.5.1. *[284] Let S be a schedule for path system $\mathcal{P}$. Then $T(S) \geq D_{\mathcal{P}}$ and $T(S) \geq C_{\mathcal{P}}/12$, hence $T(S) = \Omega(C_{\mathcal{P}} + D_{\mathcal{P}})$.*

Note that the congestion is some kind of aggregated interference measure, because it captures both the actual workload and the amount of interference. [284] provides an algorithm to minimize congestion that is a $O(g^2(V))$-approximation, where $g(V)$ is the diversity (6.26). Unfortunately, minimizing congestion may lead to long paths:

Lemma 6.5.2. *[284] There exist graphs for which $C_{\mathcal{P}} \cdot D_{\mathcal{P}} \geq \Omega(W)$ for all path systems $\mathcal{P}$.*

Packet collisions not only cause decreased network throughput, but also increase energy consumption. So designing interference-aware topologies and finding interference-free schedules is also believed to conserve the limited energy resources of sensor networks. We define two measures for the energy consumption of a path system $\mathcal{P}$ and schedule S. Let E' consist of all edges that are used in any time slot, and $\ell(e)$ be the number of slots in which an edge $e \in E'$ is used. Now the *unit energy* En_1 and *flow energy* En_f are

$$\text{En}_1 = \sum_{uv \in E'} d(u,v)^\alpha \quad \text{and} \quad \text{En}_f = \sum_{uv \in E'} \ell(uv)d(u,v)^\alpha \,, \tag{6.27}$$

The following lemmas assume $\alpha = 2$. Not surprising, minimizing dilation leads to high energy usage:

Lemma 6.5.3. *[284] There exist graphs for which $D_\mathcal{P} \cdot \text{En}_1 \geq \Omega(d^2)$ and $D_\mathcal{P} \cdot \text{En}_f \geq \Omega(d^2W)$, for all path systems $\mathcal{P}$, where $d = \max_{u,v \in V} d(u,v)$.*

Even worse, it turns out that small congestion and low energy consumption cannot be achieved simultaneously:

Lemma 6.5.4. *[284] There exist graphs for which the following two conditions hold for all path systems $\mathcal{P}$:*

1. *$C_\mathcal{P} = \Omega(n^{1/3}C^*)$ or $\text{En}_1 = \Omega(n^{1/3}\text{En}_1^*)$, and*
2. *$C_\mathcal{P} = \Omega(n^{1/3}C^*)$ or $\text{En}_f = \Omega(n^{1/3}\text{En}_f^*)$.*

Here, C^*, En_1^*, and En_f^* denote the optimal values for minimizing C, En_1, and En_f, resp.

6.5.2 Network Flows

After considering schedules to send some packets to their destinations once, we now turn to periodic schedules and network flows. The set $\mathcal{D} \subseteq V \times V$ of *demands* is given, and we seek *end-to-end flow rates* $f : \mathcal{D} \to [0,1]$ and a periodic schedule, such that the flows can be continuously injected into the network and are successfully delivered by the schedule. A possible objective is to maximize the total flow $\sum_{d \in \mathcal{D}} f(d)$.

Given a schedule $S = (S_1, \ldots, S_{T(S)})$, we define the flow rate $x_e(S)$ of an edge as the fraction of the time in which e is used, i.e.,

$$x_e(S) = \frac{|\{i : e \in S_i\}|}{T(S)} \,. \tag{6.28}$$

Vice versa, we say that an assignment $(x_e)_{e \in E}$ of edge rates is scheduled by S if $x_e \leq x_e(S) \, \forall e$.

Introducing schedules and interference into traditional network flows has a heavy impact on complexity: In general digraphs with directed distance-2 interference, checking feasibility of given end-to-end flow rates and of edge rates is both $\mathcal{NP}$-hard [18].

The maximum flow problem can be formulated as a linear program over the schedulable edge flow rates using any standard multi-commodity flow (MCF) formulation. For example, consider

$$\max \sum_{d \in D} f(d) \tag{6.29}$$

$$\text{s.t. } f(d) \geq \lambda f(d') \qquad\qquad \forall d, d' \in D \tag{6.30}$$

$$\sum_{su \in E} y_{su,d} - \sum_{us \in E} y_{us,d} = f(d) \qquad \forall d = (s,t) \in D \tag{6.31}$$

$$\sum_{tu \in E} y_{tu,d} - \sum_{ut \in E} y_{ut,d} = -f(d) \qquad \forall d = (s,t) \in D \tag{6.32}$$

$$\sum_{vu \in E} y_{vu,d} - \sum_{uv \in E} y_{uv,d} = 0 \qquad \begin{aligned}&\forall d = (s,t) \in D, \\ &\forall v \in V \setminus \{s,t\}\end{aligned} \tag{6.33}$$

$$x_e \geq \sum_{d \in D} y_{e,d} \qquad\qquad \forall e \in E \tag{6.34}$$

$$0 \leq y_{e,d} \leq 1 \qquad\qquad \forall e \in E, d \in D \tag{6.35}$$

$$x \text{ can be scheduled.} \tag{6.36}$$

Here, the objective (6.29) is to maximize the total flow. Inequality (6.30) introduces *fairness constraints* by bounding the unfairness

$$\frac{\min_{d \in D} f(d)}{\max_{d \in D} f(d)} \tag{6.37}$$

of the flow rates. Equations (6.31)–(6.33) are the usual flow conservation, (6.34) connects per-commodity flows and edge rates, and (6.35) defines variable bounds for the flows.

So it remains to replace (6.36) by a linear description of schedulable edge rates. Considering general conflict graphs (E, C), there is an obvious connection between schedules and independent sets and vertex colorings in the conflict graph. In fact, it can be shown that an edge rate vector is schedulable if, and only if, it is contained in the stable set polytope $\text{STAB}(E, C)$ of the conflict graph [197]. Unfortunately, there is no polynomial description of this polytope, unless $\mathcal{P} = \mathcal{NP}$.

On the other hand, this also implies that every feasible edge rate vector is a convex combination of some stable sets in the conflict graph. Assume a family $(E_1, \ldots, E_k)$, $E_j \subseteq E \; \forall j$ of k stable sets in the conflict graph is given. Then the following polytope contains schedulable edge rates only:

$$\{(x_e)_{e \in E} \mid x_e = \sum_{j : e \in E_j} \mu_j, \mathbf{1}^T \mu \leq 1, \mu \in [0,1]^k\} \tag{6.38}$$

Therefore, inserting (6.38) into (6.36) leads to a LP formulation that can be used to produce feasible solutions, albeit without quality guarantees. It can further be enhanced by embedding it into a column generation framework that repeatedly adds stable sets (i.e., increases k) [197].

We can accompany this process of generating feasible solutions (that is, lower bounds) by matching upper bounds. For that matter, we can use the same MCF formulation, but replace (6.36) by a relaxation of STAB(E, C), e.g., a linear formulation with clique-, odd-hole-, and odd-antihole-constraints [197], possibly in a repeated separation process.

This can be used to describe feasible transmission sets, which can replace the stable sets in the lower bound process. For the upper bound, the clique constraints can be adopted: Consider a maximal set $F \subseteq E$, where any two $f, f' \in F$ are mutually exclusive, i.e., $s(f, f') > 1$ or $s(f', f) > 1$, using the notation of the weighted conflict graph in Section 6.2.1. Then

$$\sum_{f \in F} x_f \leq 1 \tag{6.39}$$

is a feasible constraint, leading to a relaxation of (6.36) and hence to upper bounds on the total flow [197].

While all of this leads to computationally tractable schedulable flows with matching upper bounds, there is no guarantee on the performance. This issue is addressed now.

For the remainder of this section, we restrict ourselves to the following interference models: The directed σ-Tx model, the directed σ-Protocol model, and the bidirected distance-2 model with fixed ranges.

The schedulable edge rates obey a linear constraint: Define $I_{\geq}(uv) := \{u'v' \in I(uv) : d(u', v') \geq d(u, v)\}$, where $I(uv)$ is the set of all edges that interfere with uv. It was shown [245] that any schedule obeys

$$x_e + \sum_{e' \in I_{\geq}(e)} x_{e'} \leq c_M \quad \forall e \in E, \tag{6.40}$$

where c_M is a constant depending on the interference model M. It is known that

$$c_{\sigma-\text{Tx}} = 5 \text{ and } c_{\sigma-\text{Protocol}} = O((1 + \tfrac{1}{\sigma})^2). \tag{6.41}$$

Inequalities of type (6.40) can be added to the MCF formulation in place of (6.36), leading to a easily solvable linear program. However, it is not clear how to actually schedule the resulting edge rates. Assume that a choice of edge rates fulfills

$$x_e + \sum_{e' \in I_{\geq}(e)} x_{e'} \leq 1 \quad \forall e \in E. \tag{6.42}$$

We can turn these rates into a feasible schedule, whose length $T(S)$ is the common denominator of the edge rates, by a simple local rule for an edge e:

When all edges in $I_{\geq}(e)$ have chosen their time slots, choose any $x_e T(S)$ slots for e.

Note that this schedule may have exponential length, but there is a straightforward extension to simply use a schedule of pre-defined, constant length, with the drawback of loosing a bit of flow.

[245] shows that given edge flow rates $(x_e)_{e \in E}$ can be easily turned into a matching schedule, if they fulfill (6.40) with $c = 1$. They provide a greedy algorithm that can be implemented in a distributed way.

Note that edge rates obeying (6.40) can be transformed to fulfill (6.42) by scaling them by $\frac{1}{c_M}$, leading to the following approximation result:

Theorem 6.5.5. *[245] The integrated flow and scheduling problem with fairness constraints can be approximated within a factor of c_M.*

6.6 Chapter Notes

In this chapter, we presented various results about interference in ad hoc and sensor networks. Interference occurs due to physical effects, and it results in message loss. It marks a major difference between traditional, wired and wireless networks. From an algorithmic point of view, interference yields additional constraints (namely, avoiding it) to problems that have to be solved. Sometimes, these constraints make easy problems hard: Network flows become $\mathcal{NP}$-hard when interference is to be considered.

One goal of this chapter is to give a broad overview. We describe different models of interference, from physically motivated to highly abstract ones, as well as measures for "the interference" in a given topology. We then address different problems—minimizing interference measures, scheduling transmissions so that no interference occurs, and finding well-schedulable topologies and network flows. In parallel, we show how to address interference using different algorithmic techniques. First, interference is expressed in measures. Then we address it directly in combinatorial algorithms, and finally we model interference using linear constraints that can be used in linear and integer programming based approaches.

7

Lower Bounds

Zinaida Benenson

7.1 Introduction

This book presents efficient algorithms for a large amount of important problems in ad hoc and sensor networks. Consider a problem $\mathcal{P}$ and the most efficient algorithm $\mathcal{A}$ (known so far) solving it with running time $O(g(n))$, where n is the number of nodes in the network. $O(g(n))$ is called *upper bound* on the running time for $\mathcal{P}$. After the problem is solved, the next step is to ask: Can the problem $\mathcal{P}$ be solved more efficiently? Is the running time of the algorithm $\mathcal{A}$ *optimal*?

If somebody devises a faster algorithm for $\mathcal{P}$, then the answers to both questions are obvious. However, it can also happen that no faster algorithm is suggested for some time. Then the only way to answer these questions is to establish a *lower bound* $\Omega(f(n))$ on the running time of the algorithms for the problem $\mathcal{P}$. This task is by no means trivial, as it involves showing that even algorithms which have not been invented yet cannot be faster than $\Omega(f(n))$.

For example, many network coordination problems, such as partitioning a network into clusters (see Chapter 3), or assigning time slots in a TDMA MAC protocol (Chapter 4), can be solved using algorithms for graph coloring and maximal independent set (MIS).

Consider Algorithm 13 from Chapter 4 for finding a $(\Delta + 1)$-coloring of arbitrary graphs with running time $O(\Delta^2 + \log^* n)$[1], where Δ is the maximal node degree of the graph, and n is the number of nodes. It is also possible to color an arbitrary graph with $O(\Delta^2)$ colors in time $O(\log^* n)$. Using a c-coloring algorithm which runs in t rounds, MIS can be computed in $t + c$ rounds, and therefore, MIS can be computed in time $O(\Delta^2 + \log^* n)$ in arbitrary graphs.

---

[1] We denote the k times iterated logarithm by $\log^k x$, i.e., $\log^1 x := \log x$, and $\log^k x := \log(\log^{k-1} x)$. The least integer k for which $\log^k x \leq 2$ is denoted by $\log^* x$.

D. Wagner and R. Wattenhofer (Eds.): Algorithms for Sensor and Ad Hoc Networks, LNCS 4621, pp. 117–130, 2007.
© Springer-Verlag Berlin Heidelberg 2007

Moreover, faster algorithms are possible for unit disk graphs (UDGs). If the nodes know the distances to their neighbors, $(\Delta+1)$-coloring and MIS can be computed in time $O(\log^* n)$. Even without distance information, computing MIS for UDGs is possible in time $O(\log \Delta \cdot \log^* n)$. For graphs of bounded degree, where Δ is a constant independent of n, computing $(\Delta + 1)$-coloring and MIS is possible in time $O(\log^* n)$.

It stands out that running time of all these algorithms depends on n only slightly, as $\log^* n$ is a very slowly growing function[2]. Therefore, running time $O(\log^* n)$ can be considered constant for all practical purposes. Nevertheless, an intriguing question arises: Is it possible to design *constant time* algorithms for graph coloring and MIS?

N. Linial answered this question in negative in his seminal paper "Locality in distributed graph algorithms" [260], where he established a *lower bound* on time needed to compute MIS and coloring. He showed that, even on such simple graphs as rings, any algorithm for 3-coloring and MIS *requires* time $\Omega(\log^* n)$. This result applies to all above graph classes, as rings are certainly graphs of bounded degree, and as a UDG also can form a ring.

The above discussion exemplifies the value of the results on lower bounds:

– Lower bounds help to establish *asymptotic optimality* of an algorithm for a specific problem, e.g., one cannot color a graph faster than in $\Omega(\log^* n)$ time. This way we know when to stop looking for a better solution, as such a solution does not exist.
– Lower bounds also show *room for improvement* of the existing solutions. For example, is it possible to compute MIS on UDGs faster than $O(\log \Delta \cdot \log^* n)$?

We proceed with a formal definition of lower bound on time complexity.

Definition 7.1.1 (lower bound on time complexity). $\Omega(f(n))$ *is called a* lower bound on time complexity *of the problem* $\mathcal{P}$ *for the class of n-vertex graphs* $\mathcal{G}_n$ *if there is a constant* $c \in \mathbb{N}^+$ *such that for all n there exists a n-vertex graph* $G_n \in \mathcal{G}_n$ *such that the worst-case running time of any algorithm for* $\mathcal{P}$ *on* G_n *is at least* $cf(n)$.

The class $\mathcal{G}_n$ can consist of arbitrary graphs, or unit disk graphs (UDG), or, e.g., all graphs with diameter $\Omega(\log n)$. According to the definition, to show a lower bound of $\Omega(f(n))$ for problem $\mathcal{P}$, one has to construct for any n a graph $G_n \in \mathcal{G}_n$ such that the running time of any algorithm for $\mathcal{P}$ cannot be less than $cf(n)$ for some $c \in \mathbb{N}^+$.

Algorithms for dynamic networks with large number of nodes, such as ad hoc and sensor networks, are considered the more efficient the less their running time depends on the number of nodes in the network. *Locality* as a design principle for distributed systems means developing algorithms for solving global tasks such that the nodes of the system need to gather more

---

[2] $\log^* n \leq 4$ for n being the estimated number of atoms in the observable universe.

information from their neighborhood than from the rest of the network. Locality helps to limit state which the nodes accumulate during the protocol, and also improves fault-tolerance, as failures at some part of the network do not propagate. Thus, finding out whether a particular problem can be solved using locality-sensitive algorithms is a very important question. The above discussion on time complexity of graph coloring and MIS showed how establishing lower bounds can help here.

Following [390], we distinguish between *localized* and *local* algorithms.

Definition 7.1.2 (localized algorithms). *In a k-localized algorithm, for some parameter k, each node is allowed to communicate at most k times with its neighbors. A node can decide to retard its right to communicate; for example, a node can wait to send messages until all its neighbors having larger identifiers have reached a certain state of their execution.*

Localized algorithms can have worst-case time complexity of $\Theta(n)$, because some nodes may have to wait for other nodes before they start algorithm execution. In the worst case, this may result in only one node being active at any unit of time. On the other hand, local algorithms below do not permit linear running time.

Definition 7.1.3 (local algorithms). *In a k-local algorithm, for some parameter k, each node can communicate at most k times with its neighbors. In contrast to k-localized algorithms nodes cannot delay the algorithm execution. In particular, all nodes process k synchronized phases, and a node's operations in phase i may only depend on the information received during phases 1 to i.*

In the following, this chapter discusses fundamental limitations on design of local algorithms. In Section 7.2, lower bounds on running time of distributed algorithms for graph coloring and MIS are shown in detail. Section 7.3 presents advanced results on the locality of some problems similar to coloring and MIS. Section 7.4 discusses lower bounds for computing minimum-weight spanning tree (MST).

7.2 A Lower Bound on 3-Coloring a Ring

In this section, a lower bound $\Omega(\log^* n)$ on the time required for coloring an n-vertex ring with 3 colors and for computing MIS on an n-vertex ring is shown. This means that even in a ring, constructing a constant time algorithm for $(\Delta + 1)$-coloring is impossible. This result establishes optimality of various coloring and MIS algorithms, as discussed in the previous section.

For clarity, we repeat the definitions of c-coloring and MIS problems here.

Definition 7.2.1 (c-coloring). *A c-coloring of a graph $G = (V, E)$ for $c \in \mathbb{N}$ is an assignment of a color from the set $\{1, \ldots, c\}$ to each vertex $v \in V$ such that no two adjacent vertices have the same color.*

Definition 7.2.2 (maximal independent set (MIS)). *An* independent set *for a graph* $G = (V, E)$ *is a set of vertices* $U \subseteq V$, *no two of which are adjacent. An independent set is* maximal *if no vertex can be added to it without violating its independence. The* MIS problem *is the problem of computing a maximal independent set for a given graph.*

To show lower bounds, it is sensible to assume a powerful system model. In this case, the established lower bounds will also apply to less powerful systems. Thus, we assume synchronous and reliable communication between nodes. At each time unit, any node can send messages of *unbounded* size to all of its neighbors.

The vertices of the graph $G = (V, E)$ are labeled by unique identifiers from the set $\{1, 2, \ldots, |V|\}$. It is assumed that at time zero, i.e., before the beginning of the computation, each node knows only its own identifier.

For the graph G, $\chi(G)$ denotes the *chromatic number* of G, which is the least number of colors that suffice to color the vertices of G such that no two adjacent vertices have the same color.

Theorem 7.2.3. *Any deterministic distributed algorithm which colors the n-ring with 3 colors requires time at least $\frac{1}{2}(\log^* n - 1)$.*

This lower bound can be proven for randomized algorithms as well. However, in this section, only the proof for deterministic algorithms is presented.

This section is organized as follows. Firstly, we give the proof outline in Section 7.2.1. The proof of the main Theorem 7.2.3 is given in Section 7.2.2. Proof of Lemma 7.2.6, which is rather long, is shown in Section 7.2.3 for clarity reasons. In Section 7.2.4, two interesting additional facts are shown: the $\Omega(n)$ lower bound on 2-coloring a $2n$-ring, and the $\Omega(\log^* n)$ lower bound on computing maximal independent set (MIS) on a ring.

7.2.1 Proof Outline

The proof consists of the following steps:

1. For the n-ring R_n, we define a graph $B_{t,n}$ (Definition 7.2.4), where t is the running time for a 3-coloring algorithm, and show that every 3-coloring of R_n immediately yields a 3-coloring of $B_{t,n}$, i.e., $\chi(B_{t,n}) \leq 3$ (Lemma 7.2.5).
2. We show that $\chi(B_{t,n}) \geq \log^{(2t)} n$ (Lemma 7.2.6). This proof runs as follows:
   a) We define a family of directed graphs $D_{s,n}$ (Definition 7.2.7) and show that $\chi(B_{t,n}) \geq \chi(D_{2t+1,n})$ (Lemma 7.2.8).
   b) Then we define *dilinegraph* $DL(G)$ for each directed graph G (Definition 7.2.9), and prove Lemma 7.2.10 on the relation between $D_{s,n}$ and its dilinegraph.

c) We prove the Lemma 7.2.11: $\chi(DL(G)) \geq \log \chi(G))$ for any directed graph G.

d) Using Lemmas 7.2.10 and 7.2.11 it follows that $\chi(D_{s,n}) \geq \log^{(s-1)} n$ for all $s > 1$ (Lemma 7.2.12).

e) Then, from Lemmas 7.2.8 and 7.2.12 it follows that $\chi(B_{t,n}) \geq \log^{(2t)} n$.

3. From Lemma 7.2.6, we then derive the lower bound $\frac{1}{2}(\log^* n - 1)$ on running time t for 3-coloring of $B_{t,n}$. Then $t \geq \frac{1}{2}(\log^* n - 1)$ also applies for 3-coloring of R_n (Lemma 7.2.5).

7.2.2 Proof of Theorem 7.2.3

We now proceed to the actual proof of Theorem 7.2.3.

Consider an arbitrary deterministic algorithm for c-coloring a ring which operates in time t. At time 0, a node v of the ring R_n knows only its own identifier.

At time 1, direct neighbors of v are able to send to v all information which they possess, i.e., their own identifiers, and perhaps results of some computations. The important point is that these computations can base only upon the information known to the nodes so far, that is, on their own identifiers, as the algorithm is deterministic. It follows that, upon receiving messages from the first round of communication, v can *simulate* the computations done by its neighbors in the first round, see Fig. 7.1.

At time 2, information initially known to nodes which are at distance 2 from v, i. e., their identifiers, and, possibly, some computation results, reaches v. Again, after having received messages from the second round of communication, v can simulate all computations done by its 2-hop neighbors so far.

Thus, at time t, node v knows identifiers of all $2t$ nodes which are at distance at most t from it. No further information can reach v at time t, as all possible (deterministic) computations which its t-hop neighbors might have done in these t rounds, can be simulated by node v. Thus, knowledge of any node at time t can be denoted as a $(2t + 1)$-tuple $(x_1, \ldots, x_{2t+1})$.

It follows that, without loss of generality, we can assume that any deterministic c-coloring algorithm for an n-ring R_n works as follows:

1. For t rounds, each node collects and relays information about identifiers. Each node v ends up with a tuple $(x_1, \ldots, x_{2t+1})$ where all x_i are pairwise distinct numbers from $\{1, \ldots, n\}$. We denote the set of all such tuples $V_{t,n}$.
2. Each node v stores the color it should choose for each of $n(n-1)\ldots(n-2t)$ possible tuples of identifiers. After time t, the node chooses its color according to the gathered information.

Thus, any deterministic c-coloring algorithm defines a mapping Ψ_c from $V_{t,n}$ into the set of colors $\{1, \ldots, c\}$.

To formalize the above reasoning, we now define an auxiliary graph with the set of vertices $V_{t,n}$ where two vertices are adjacent if and only if vertices with identifiers x_t and x_{t+1} are neighbors in R_n.

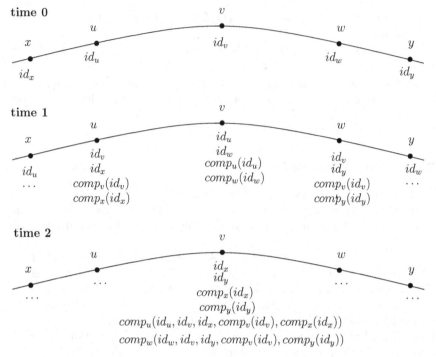

Fig. 7.1. New information known by node v and its neighbors at time 0, 1, and 2; $comp_u(\ldots)$ denotes computation done by node u.

Definition 7.2.4 (B_n). *Graph B_n with the set of vertices V_n is defined as follows: any two vertices $(_1, \ldots, _{2+1})$ and $(, _1, \ldots, _2)$ with $\neq _{2+1}$ are joined by an edge, and no other edges exist in the graph.*

Graph B_n is regular of degree $2(n-2-1)$, as any node $(_1, \ldots, _{2+1})$ is adjacent to $(n-2-1)$ nodes of form $(, _1, \ldots, _2)$ with $\neq _{2+1}$, and to $(n-2-1)$ nodes of form $(_2, \ldots, _{2+1}, z)$ with $z \neq _1$, see Figure 7.2.

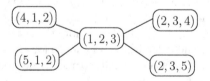

Fig. 7.2. Node $(1, 2, 3)$ in $B_{1,5}$

Lemma 7.2.5. *If R_n can be -colored in rounds, then $(B_n) \leq$.*

Proof. Let $\tau : V_n \to \{1,\dots,\}$ be a -coloring of R_n. Then τ assigns different colors to neighboring nodes. Thus, if node v gathered a tuple of indices $(_1,\dots,_{2}+1)$ ($_{+1}$ being the identifier of v), and its neighbor u with identifier gathered the tuple $(,_1,\dots,_2)$ (see Figure 7.3), then τ should satisfy: $\tau((_1,\dots,_2{+1})) \neq \tau((,_1,\dots,_2))$. Then, τ is also a -coloring for B_n, and therefore, $(B_n) \leq$. $\qquad\square$

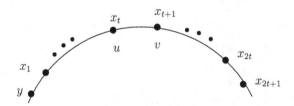

Fig. 7.3. After time t, v knows the tuple $(x_1,\dots,x_{2t+1})$, and u knows the tuple $(y,x_1,\dots,x_{2t})$.

We now consider how (B_n) depends on and n:

Lemma 7.2.6. $(B_n) \geq \log^{(2)} n$

Proof of Lemma 7.2.6 is given in Section 7.2.3.

Proof (of Theorem 7.2.3). Consider an algorithm for 3-coloring of R_n with running time . Then $(B_n) \leq 3$ (Lemma 7.2.5). Then from Lemma 7.2.6 it follows that $3 \geq \log^{(2)} n$, and therefore, $\geq \frac{1}{2}(\log^* n - 1)$. $\qquad\square$

7.2.3 Proof of Lemma 7.2.6

We first define two more auxiliary graphs and prove some of their properties.

Definition 7.2.7 ($_n$). *The vertices of directed graph $_n$ are all sequences $(_1,\dots,)$ with $1 \leq _1 \dots \leq n$. The outneighbors of $(_1,\dots,)$ are all vertices of $_n$ having the form $(_2,\dots,,b)$.*

Lemma 7.2.8. $(B_n) \geq (_2{+1} n)$

Proof. The Lemma follows immediately from the observation that graph B_n contains the underlying (undirected) graph of $_2{+1} n$ (see also Figure 7.4 for an example). $\qquad\square$

Definition 7.2.9 (dilinegraph). *Given a directed graph $G = (V,E)$, its dilinegraph (G) is a directed graph whose vertex set is E with $((u_1,u_2),(v_1,v_2))$ being an edge if $u_2 = v_1$.*

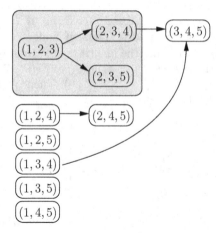

Fig. 7.4. Graph $D_{3,5}$. The underlying undirected graph is contained in $B_{1,5}$. Shaded area can be seen at Figure 7.2.

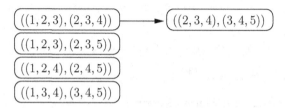

Fig. 7.5. Graph $DL(D_{3,5})$.

The dilinegraph of $D_{3,5}$ is shown in Figure 7.5.

Lemma 7.2.10. *(1)* $D_{1,n}$ *is obtained from the complete graph of order n by replacing each edge by a pair of edges, one in each direction.* *(2)* $D_{s+1,n} = DL(D_{s,n})$ *for all $s \geq 1$.*

Proof. The first claim follows immediately from the definition. For the second claim, set of edges of $D_{s,n}$ contains all pairs $((x_1, \ldots, x_s), (x_2, \ldots, x_s, y))$ with y. Then $DL(D_{s,n})$ consists of nodes of the above form. We associate each node of $DL(D_{s,n})$ with the node of $D_{s+1,n}$ of the form $(x_1, x_2, \ldots, x_s, y)$. This mapping is a bijection. It remains to show that the adjacency relations in both graphs are the same. Indeed, any edge from $DL(D_{s,n})$ has the form $(((x_1, \ldots, x_s), (x_2, \ldots, x_s, y)), ((x_2, \ldots, x_s, y), (x_3, \ldots, x_s, y, z)))$ where y, z, which maps to the edge of $D_{s+1,n}$ of the form $((x_1, \ldots, x_s, y), (x_2, \ldots, x_s, y, z))$ and vice versa. For an example, see Figure 7.6. □

Lemma 7.2.11. *For every directed graph G, $h(DL(G)) \geq \log(h(G))$.*

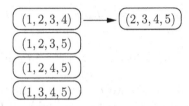

Fig. 7.6. Graph $D_{4,5}$ is isomorphic to the graph $DL(D_{3,5})$ (Figure 7.5.).

Proof. Let $k = \chi(\mathcal{C}(G))$, and consider a k-coloring $\hat{\varphi}$ of $\mathcal{C}(G)$. Then $\hat{\varphi}$ is an edge coloring of G such that if the edge e' starts in the vertex where the edge e ends, then $\hat{\varphi}(e) \neq \hat{\varphi}(e')$. We then define the coloring φ for G such that $\varphi(v) := \{\hat{\varphi}(e) \mid$ edge e ends in the vertex $v\}$.

The coloring φ uses at most 2^k colors, as each color is a subset of $\{1, \ldots, k\}$. Consider nodes u and v which are neighbors in G. We show now that $\varphi(v) \neq \varphi(v)$, i.e., that $\varphi(v)$ is a proper coloring for G.

W.l.o.g. $(u, v) = e$ is an edge of G. Then $\hat{\varphi}(e) \in \varphi(v)$. However, $\hat{\varphi}(e) \notin \varphi(u)$, as otherwise, there would have been an edge $e' = (\cdot, u)$ for some vertex $\cdot$ of G such that $\hat{\varphi}(e') = \hat{\varphi}(e)$. This is a violation of the edge coloring of H. Thus, φ is a proper coloring of G, and therefore, $\chi(G) \leq 2^k$. $\qquad\square$

Lemma 7.2.12. $\chi(\mathcal{C}_n) \geq \log^{(s-1)} n$ *for all $s > 1$.*

Proof. Proof is by induction on s.

For $s = 2$, we have to show that $\chi(\mathcal{C}_{2\,n}) \geq \log n$. From Lemma 7.2.10 it follows that $\mathcal{C}_{2\,n} = \mathcal{C}(\mathcal{C}_{1\,n})$. Then, from Lemma 7.2.11, $\chi(\mathcal{C}(\mathcal{C}_{1\,n})) \geq \log(\chi(\mathcal{C}_{1\,n})) = \log n$.

For $s > 2$, assume that $\chi(\mathcal{C}_n) \geq \log^{(s-1)} n$. We show that $\chi(\mathcal{C}_{s+1\,n}) \geq \log^{(s)} n$. From Lemma 7.2.10, $\mathcal{C}_{s+1\,n} = \mathcal{C}(\mathcal{C}_n)$, and from Lemma 7.2.11, $\chi(\mathcal{C}(\mathcal{C}_n)) \geq \log(\chi(\mathcal{C}_n))$. Then, from the induction hypothesis, $\log(\chi(\mathcal{C}_n)) \geq \log^{(s)} n$. $\qquad\square$

Finally, we prove Lemma 7.2.6.

Proof (of Lemma 7.2.6). We have to show that $\chi(B_n) \geq \log^{(2s)} n$. From Lemma 7.2.8, it follows that $\chi(B_n) \geq \chi(\mathcal{C}_{2s+1\,n})$. Then $\chi(\mathcal{C}_{2s+1\,n}) \geq \log^{(2s)} n$ (Lemma 7.2.12). $\qquad\square$

7.2.4 Additional Results

Although Theorem 7.2.3 precludes a constant-time algorithm for 3-coloring a ring, nevertheless, an algorithm which runs in time $\mathcal{O}(\log^* n)$ is very efficient, as $\log^* n$ grows very slowly with n. In contrast, for a ring with an even number of nodes, finding a 2-coloring requires time $\Omega(n)$.

Theorem 7.2.13. *Any deterministic distributed algorithm for coloring a 2n-ring with two colors requires at least $n - 1$ rounds.*

Proof. Consider an algorithm which 2-colors R_n in t rounds. Then $\chi(B_{t,2n}) \leq 2$ (Lemma 7.2.5).

If $B_{t,2n}$ can be colored with two colors, then this graph is bipartite. A graph is bipartite if and only if it contains no odd-length ring. However, for $t \leq n - 2$, $B_{t,2n}$ contains such a ring (of length $2t + 3$):

$$(1, 2, \ldots, 2t + 1), (2, \ldots, 2t + 1, 2t + 2), (3, \ldots, 2t + 3), (4, \ldots, 2t + 3, 1)$$

$$(5, \ldots, 2t + 3, 1, 2), \ldots, (2t + 3, 1, \ldots, 2t), (1, \ldots, 2t + 1) \qquad \square$$

The next theorem shows a lower bound for the MIS (maximal independent set) problem on a ring.

Theorem 7.2.14. *Any deterministic distributed MIS algorithm for the n-ring requires at least $\frac{1}{2}(\log^* n - 3)$ rounds.*

Proof. Due to Theorem 7.2.3 it suffices to show that, given a MIS for the n-ring R_n, it is possible to 3-color R_n in one round. The algorithm runs as follows:

– If the vertex v is in the MIS, then it colors itself 1 and sends message "my color is 1" to its both neighbors.
– If v is not in MIS, it sends its identifier id_v to its both neighbors. Note that as v is not in MIS, then at least one neighbor of v must be in MIS. Upon receiving the messages from round 1, v colors itself 2 if both its neighbors are in MIS. Otherwise, v must have received id_u from one of its neighbors u. If $id_v < id_u$, then v colors itself 2, otherwise it colors itself 3.

For an example execution of the algorithm, see Figure 7.7. The proof that the above algorithm properly 3-colors R_n is left as an exercise to the reader. $\square$

7.3 Locally Checkable Labelings

After realizing that neither $(\Delta + 1)$-coloring nor MIS can be computed in constant time even on rings, an intriguing question is if there are *any* non-trivial problems from graph theory which *can* be computed locally, i.e., in constant time. Naor and Stockmeyer investigate this problem in [301] for a class of problems they call *locally checkable labelings* (LCL). In an LCL problem, the vertices (or the edges) of the graph must be labeled in accordance to some rules such that the legality of the labeling can be verified locally. Vertex coloring (Def. 7.2.1) is an LCL, as any node only needs to know the colors of its neighbors to check the legitimacy of the coloring. MIS (Def. 7.2.2)

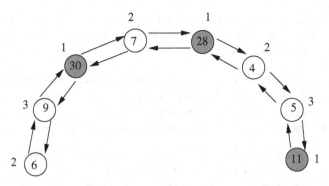

Fig. 7.7. 3-coloring a ring given an MIS (shaded nodes)

is also an LCL, as each node can check the following: If the node is in MIS, then none of its neighbors should be in MIS. If the node does not belong to MIS, then at least one of its neighbors should be in MIS.

It turns out that there is at least one LCL problem which indeed has a local algorithm for a certain non-trivial class of graphs. This problem is called *weak -coloring*.

Definition 7.3.1 (weak -coloring). *A* weak *-coloring of a graph* $G = (V, E)$ *for* $\in \mathbb{N}$ *is an assignment of a color from the set* $\{1, \ldots, \}$ *to each vertex* $v \in V$ *such that each non-isolated node has at least one neighbor colored differently.*

Weak 2-coloring exists for any graph. To find such a coloring, we can first compute the breadth-first spanning tree of the graph, and then assign one color to even levels, and a different color to odd levels. However, even weak 2-coloring cannot be computed locally for some classes of graphs, as the following theorem (given without proof) shows.

Theorem 7.3.2. *Weak coloring has the following locality properties:*

- *Consider a class of graphs of maximum degree d where every vertex has odd degree. There is a constant b such that, for every d, there is a local algorithm with running time at most* $\log^* d + b$ *that solves the weak 2-coloring problem.*
- *For graphs of even degree, there is no local algorithm which solves weak -coloring for any .*
- *If the definition of weak coloring is changed such that every vertex must have at least two neighbors colored differently, than this LCL problem cannot be solved locally.*

Sometimes, if some problem cannot be solved using a deterministic algorithm, it can be solved using a randomized algorithm having a small error probability. However, this does not apply in case of LCL problems.

Theorem 7.3.3. *Randomization cannot make an LCL problem local. That is, if an LCL problem has a randomized local algorithm, then it also has a deterministic local algorithm.*

It would be very convenient to have an algorithm which, given any LCL problem, decides if this problem has a local algorithm. Unfortunately, such an algorithm does not exist. More precisely, it can be shown that, if such an algorithm was possible, it could be used to solve the Halting Problem.

Theorem 7.3.4. *It is undecidable whether a given LCL problem has a local algorithm. However, there exists an algorithm which, given any LCL problem, decides if this problem has an algorithm which operates in time t.*

7.4 Minimum-Weight Spanning Trees

Algorithms for LCL problems such as coloring and MIS are essentially local, as their running time depends on n only "slightly". This fits well with the local nature of these problems: The decision if a node should join the MIS, or which color should the node take, is largely independent on the nodes at the far end of the network.

On the other hand, some problems seem to require global knowledge about the network. For these problems, an algorithm may be considered local if it runs in polylogarithmic time. One of such problems is computing minimum-weight spanning tree (MST) of a given weighted graph. The MST can be used for efficient broadcast, see Chapter 12.

Definition 7.4.1 (minimum-weight spanning tree (MST)). *For a weighted graph $G = (V, E, w)$ with the weight function $w : E \to \mathbb{R}^+$, minimum-weight spanning tree is a spanning tree with the minimal total weight (sum of the weights of all its edges).*

Recall that in Section 7.2, lower bound on 3-coloring a ring was proven for synchronous and reliable communication with unbounded messages. That is, in each round each node can send messages of unbounded size to all its neighbors. In this model, MST can be constructed in time $O(D)$, where D is the graph diameter (the maximum distance in hops between any two vertices). This can be done, e. g., by collecting the entire graph topology at some vertex, computing the MST locally, and then broadcasting the result to all nodes. Moreover, it can be shown that in this model, MST construction requires $\Omega(D)$ time.

However, solving the MST problem becomes much more involved in the *bounded message model*, where each node can send messages of size at most B bits to its neighbors. Usually, $B = \mathcal{O}(\log n)$ is considered. Of course, the lower bound $\Omega(D)$ also applies in this stronger model. However, more precise lower bounds can be shown here.

In the bounded message model, one of the most efficient protocols for distributed MST construction has the time complexity $O(D + \sqrt{n} \log^* n)$. Does a faster (e.g., polylogarithmic time) algorithm exist for MST computation? The following theorem shows the asymptotical near-optimality of the above MST construction algorithm.

Theorem 7.4.2. *Any distributed algorithm for the distributed MST construction requires* $\Omega(D + \sqrt{n}/\log n)$ *time on graphs of diameter* $D = \Omega(\log n)$.

This result is shown by construction for each n a special n-vertex graph where the MST problem cannot be solved faster than $\Omega(D + \sqrt{n}/\log n)$. Interestingly, even for graphs with very low constant diameter, there is no polylogarithmic MST algorithm. The following theorem presents more exact lower bounds on distributed MST construction.

Theorem 7.4.3. *Any distributed algorithm for the distributed MST construction requires:*

- $\Omega(\sqrt{n/\log n})$ *for graphs of diameter* n^δ, $0 < \delta < 1/2$,
- $\Omega(\sqrt{n}/\log n)$ *for graphs of diameter* $\Theta(\log n)$,
- $\Omega(\sqrt[3]{n}/\log n)$ *for graphs of diameter 4,*
- *and* $\Omega(\sqrt[4]{n/\log n})$ *for graphs of diameter 3.*

The above results apply not only for deterministic, but also for randomized algorithms. On the other hand, MST for graphs of diameter 2 can be computed efficiently in time $O(\log n)$.

Sometimes, if the desired efficiency results cannot be achieved with exact algorithms, there are approximation algorithms which solve problems less exact, but with better efficiency results. For the MST problem, this means that the resulting *H-approximate MST* has the weight at most H times greater than the exact MST. H is called *approximation ratio*. Thus, even if one cannot compute exact MST in polylogarithmic time, perhaps it is possible to compute its H-approximation efficiently? Unfortunately, as the following theorem shows, the MST problem is even hard to approximate.

Theorem 7.4.4. *Approximating the MST within an approximation ratio* H *requires time* $\Omega\left(\sqrt{\frac{n}{H \cdot \log n}}\right)$.

In particular, this means that approximating the MST problem within any constant factor ($H = O(1)$) requires $\Omega(\sqrt{n/\log n})$ time, which is the same lower bound as for the exact MST construction, cf. Theorem 7.4.3. Moreover, for any $0 < \epsilon < 1$, an $\left(\frac{n}{\log n}\right)^{1-\epsilon}$-approximation of the MST problem requires $\Omega\left(\left(\frac{n}{\log n}\right)^{\epsilon/2}\right)$ time, i. e., it cannot be solved locally (in polylogarithmic time).

7.5 Chapter Notes

In this chapter, we discussed lower bounds on running time of distributed algorithms for such important problems as graph coloring, MIS, and MST. These results help to establish optimality of existing algorithms, and pose a challenge to find algorithms which meet lower bounds.

The distinction between localized and local algorithms originates from [390].

Lower bounds of $\Omega(\log^* n)$ on computing 3-coloring and MIS in a ring are shown in [260]. The LCL problems are investigated in [301]. For further results on lower bounds for approximability of MIS and minimum dominating set problems, see also Section 3.4.4. Most recent results on lower bounds for graph coloring, MIS, and related problems can be found in [233] and [239].

Lower bounds on the exact MST construction for the unbounded message model, and for bounded message model for graphs of diameter $O(\log n)$ originate from [314]. Results for graphs of constant diameter are shown in [264]. In [110], refined lower bounds for the bounded message model are presented. Lower bounds on approximation algorithms for MST are also shown in [110]. An interesting open question is which lower bounds on MST construction apply for UDGs.

As stated in corresponding sections, the above results apply not only for deterministic algorithms, but also for randomized ones.

In this chapter, we also used for comparison results on upper bounds for different problems. For origins of Algorithm 13, see Chapter Notes to Chapter 4. The algorithm for $O(\Delta^2)$-coloring of arbitrary graphs with running time $O(\log^* n)$ is presented in [260]. On the relation of coloring and MIS see, e.g., [313, page 93].

Algorithms with running time $O(\log^* n)$ for $(\Delta + 1)$-coloring and MIS in UDGs with known distances are obtained in [234]. An $O(\log \Delta \cdot \log^* n)$ MIS algorithm for UDGs without distance information originates from [229].

The protocol for distributed MST construction with running time $O(D + \sqrt{n} \log^* n)$, where D is the diameter of the graph, is given in [246]. Most efficient so far MST algorithm is due to Elkin [109]. A method to compute MST in graph of diameter 2 in time $O(\log n)$ is shown in [264].

8

Facility Location

Christian Frank

8.1 Introduction

The facility location problem has occupied a central place in operations research since the early 1960's, as it models design decisions on the placement of factories, warehouses, schools, or hospitals to serve a set of customers efficiently. For an overview of previous work see the survey chapter by Cornuejols, Nemhauser & Wolsey in [289]. Recently, it has received renewed research interest because it can be used to place servers or proxies in communication networks [380].

We apply the facility location problem to configuration and deployment issues in wireless sensor networks. More specifically, we discuss how algorithms for the facility location problem can be used to assign functionality to certain nodes in a wireless sensor network, i.e., choose a subset of designated nodes as *service providers* of a particular service.

Such assignment of functionality to designated nodes is a key element in the often-cited ability for *self-configuration* [64] that is required in many sensor-network applications. Consider clustering, which is a prominent sensor network configuration problem (see Chapter 3), where a subset of cluster leaders is selected to provide a service to their neighboring nodes, while the latter (slave nodes) can shut down some functionality and thus save power. Clustering can for instance be used to improve the efficiency of data delivery (e.g., flooding, routing). In this setting, cluster leaders act as sole direct communication partners for the associated slave nodes, which allows slave nodes to synchronize to their cluster leader's transmission schedule and shut down their transceivers based on this schedule.

As slave nodes save power, it is desirable to have only few cluster leaders (as long as each slave node has a cluster leader as a neighbor) to enable energy savings at many slave nodes. In this sense an optimal configuration is modeled by the minimum dominating set problem discussed in Chapter 3. An example of such a configuration is shown in Figure 1(a).

D. Wagner and R. Wattenhofer (Eds.): Algorithms for Sensor and Ad Hoc Networks, LNCS 4621, pp. 131–159, 2007.

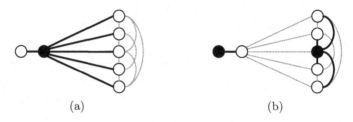

(a)   (b)

Fig. 8.1. Different clustering configurations

However, such a strategy increases the distance between cluster leaders and their associated nodes. If network nodes can vary their transmission power, such a configuration is not energy optimal. In particular, an energy-optimal configuration would require placing cluster leaders as close as possible to their associated nodes. An example of this strategy is shown in Figure 1(b). Here an optimal balance between transmission energy expenditure and the effort involved in operating cluster leaders must be found.

Apart from clustering, solving the facility location problem is useful for deploying services to certain nodes in the network. This will be increasingly relevant as wireless sensor networks pursue the path of other distributed systems platforms towards re-usable component-based systems that are supported by adequate deployment middleware (e.g., [106, 178]). While present research prototypes (see Chapter 1) are still based on *application-specific* design and implementation of major parts of the system, a key for the wide-spread adoption of sensor-networks is to provide a large pool of re-usable and pluggable components that can be used to build a new application. Such components could provide a service for their network neighborhood and thus need not be executed on *every* single node. *Where* such components will be executed in the network will be addressed by distributed algorithms solving the facility location problem.

In the following, we formally introduce the facility location problem in Section 8.2, discuss centralized approximations of the problem in Section 8.3, and describe distributed approximation algorithms in detail in Sections 8.4 and 8.5. We will conclude the chapter with an outlook on how the presented approaches can provide a basis for designing *local* facility location algorithms that are particularly suited for wireless sensor networks in Section 8.6.

8.2 Problem Definition

The facility location problem formulates the geographical placement of *facilities* to serve a set of *clients* or customers, sometimes also called cities. We will define the traditional facility location problem in Section 8.2.1. In wireless sensor networks *facilities* can represent any kind of service provider (like cluster heads, time synchronization hubs, or network sinks) that can be used by *client* nodes nearby. We will discuss in more detail how the facility location problem can be applied to the multi-hop wireless sensor network model in the subsequent Section 8.2.2.

8.2.1 Problem Formulations

A number of problem formulations exist in the domain of the facility location problem. Generally, these are based on connecting clients to a set of open facilities at minimum cost.

Definition 8.2.1 (Uncapacitated Facility Location Problem – UFL). *Given a bipartite graph G with bipartition $(C,)$ where C is a set of clients and is a set of potential facilities, a fixed cost $_i \in \mathbb{R}_+$ for opening each facility $i \in$ and connecting costs $_{ij} \in \mathbb{R}_+$ for connecting client $\in C$ to facility $i \in$: Return a subset $S \subseteq$ of open facilities and a mapping $: C \to S$ of clients to open facilities that minimize the total costs (fixed costs of open facilities and connection costs),*

$$\sum_{i \in} {}_i + \sum_{j \in} {}_{(j)j} \cdot$$

Figure 8.2 sketches the elements introduced above. We will use $n = |C|$ and $= |\ |$ to refer to the number of clients and facilities, respectively.

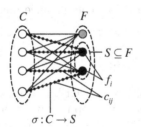

Fig. 8.2. Elements of the facility location problem

In the above UFL formulation, the fixed costs $_i$ enable the trade-off to open additional facilities at additional costs if these can reduce connecting costs. Alternatively, in the following problem formulation, opening costs $_i$ are zero. Instead, a fixed number of k facilities will be chosen from such that the sum of the distances from each client to its next facility is minimized. This models the problem of finding a minimum-cost clustering of k clusters [375, chap. 25].

Definition 8.2.2 (The k-Median Problem – k-MN). *Given a bipartite graph G with bipartition $(C,)$ where C is a set of clients and is a set of potential facilities, an integer k limiting the number of facilities that can be* *opened, and connecting costs* $c_{ij} \in \mathbb{R}^+$ *for each* $i \in F$ *and* $j \in C$: *Return a subset* $S \subseteq F$ *of* $|S| \leq k$ *open facilities and a mapping* $\sigma : C \to S$ *of clients to open facilities that minimize the total connection costs,*

$$\sum_{j \in C} c_{\sigma(j)j} \, .$$

Notice that the facility location problem is a rather generic problem formulation. For example, it is a generalized version of the minimum dominating set (MDS) problem. Consider a graph $G = (V, E)$ which is an instance of MDS. It can be reduced to UFL if F and C are set to be a copy of the same vertices V, $c_{ij} = 0$ for $(i, j) \in E$ and ∞ otherwise, and the opening costs f_i are set to a constant f_c for all i.

Due to its wide range of applications, many variants of the facility location problem exist. A common generalization of UFL and k-MN is $k-$UFL that includes non-zero opening costs and the additional constraint that at most k facilities can be opened. The k-UFL problem corresponds to k-MN if opening costs f are zero and to UFL if k is set to $|F|$. The *capacitated* facility location problem refers to an additionally constrained version of UFL, in which each potential facility $i \in F$ can serve at most u_i clients. Last but not least, an instance of UFL or k-MN is called *metric* if the connection costs c_{ij} satisfy the triangle inequality, which requires that, given any three nodes i, j, k the cost of the direct edge (i, j) is smaller than the path over an additional hop ($c_{ij} \leq c_{ik} + c_{kj}$). Note that this implies that the edge (i, j) must exist if (i, k) and (k, j) exist and therefore a metric instance must be complete or at least consist only of cliques.

8.2.2 Sensor Network Models

The above "traditional" definition of UFL and k-MN problems does not immediately accommodate the multi-hop network model used in wireless sensor networks. In multi-hop sensor networks, the network communication graph $G = (V, E)$ will be "unfolded" into a UFL instance in various ways, depending on the given application.

In our introductory examples of finding an energy-efficient clustering or deploying services onto network nodes, we have regarded special cases of the facility location problem, in which all nodes V of the network communication graph simultaneously take on the role of a client and of a potential facility, thus $F = V$ and $C = V$. Specifically, for an energy efficient clustering it is required that clients connect to facilities which are their direct neighbors (Figure 3(a)). When deploying services to network nodes, it is sometimes sufficient for clients to connect to service provider nodes that are multiple network hops away (Figure 3(b)), which allows for a smaller number of service nodes.

In other settings, only some nodes have the necessary capabilities (e.g., remaining energy, available sensors, communication bandwidth, or processing

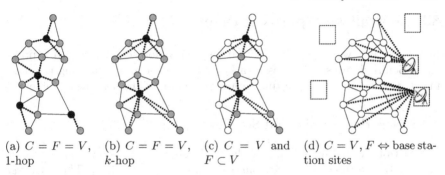

(a) $C = F = V$, (b) $C = F = V$, (c) $C = V$ and (d) $C = V, F \Leftrightarrow$ base sta-
1-hop k-hop $F \subset V$ tion sites

Fig. 8.3. Possible applications of the facility location problem in wireless sensor networks

power) to execute a service. In these cases, a subset of eligible nodes can be selected as facilities based on their capabilities beforehand [135], which results in $C = V$ but $F \subset C$ (Figure 3(c)).

Another application of UFL is to obtain an optimal deployment of base stations (sometimes referred to as *sinks* or *gateways*) to connect the sensor network to a background infrastructure. Here F contains a set of potential basestation sites, while installed base stations then provide back-end connectivity to all network nodes ($C = V$). In this case, S represents the minimum-cost deployment of an optimally placed number of base stations (Figure 3(d)).

In all cases, c_{ij} denote some kind of communication cost between a facility node i and a client node j. In a communication graph $G = (V, E)$ of a multi-hop network, we can assume that edge costs are given for all $(i, j) \in E$, which represent the quality of the link, e.g., based on dissipated energy, latency, or interference. Depending on the indicator used to represent link quality, the resulting facility location instance may not be *metric*. However, we can obtain a metric instance if one sets c_{ij} to the costs of the shortest i-j-path in G, which additionally allows to address scenarios in which clients and facilities are allowed to be an arbitrary number of network hops apart.

In the remaining chapter, algorithms will be presented in terms of the standard bipartite graph model as given in Definition 8.2.1. This model can directly be applied to placing base stations (Figure 3(d)), and may further be used, after duplicating each network node into a facility and client node interconnected with zero costs, to compute an energy efficient clustering (Figure 3(a)). However, when facilities can be multiple hops away (Figures 3(b) and 3(c)), additional adaptations of the distributed algorithms we present in this chapter are required. We will preview such adaptations in Section 8.6.

8.3 Centralized Approximations

In this section, we discuss centralized approximation schemes for UFL problems. Before doing so, we present a lower bound for approximating the non-metric version of UFL, resulting from a reduction of the set cover problem.

8.3.1 Negative Results and Set Cover

It turns out that an instance of the well-known *set cover problem* can be directly transformed into a *non-metric* instance of UFL [380]. The *set cover problem* is defined as follows:

Definition 8.3.1 (Set Cover Problem). *Given a pair $(U, \mathbf{S})$, where U is a finite set and $\mathbf{S}$ is a family of subsets of U with $\bigcup_{S \in \mathbf{S}} S = U$, and weights $c : \mathbf{S} \to \mathbb{R}^+$ for each such subset: Find a set $\mathcal{R} \subseteq \mathbf{S}$ with $\bigcup_{S \in \mathcal{R}} S = U$ that minimizes the total weight $\sum_{S \in \mathcal{R}} c(S)$.*

There are well-known lower bounds for the approximation of the *set cover problem*. Raz and Safra [330] showed that there exists a constant $\mathcal{X} > 0$ such that, unless P=NP, there is no polynomial-time approximation algorithm that produces, in the worst-case, a solution whose cost is at most $\mathcal{X} \ln |U|$ times the optimum. Feige [123] showed that such an algorithm does not even exist for any $\mathcal{X} < 1$ unless every problem in NP can be solved in $O(n^{O(\log \log n)})$ time.

Reducing the *set cover problem* to UFL is straightforward: Let $(U, \mathbf{S}, c)$ be an instance of the *set cover problem*. We construct an instance of UFL, in which $C := U$, $F := \mathbf{S}$, and opening costs $f_s := c(S)$ for each $S \subseteq \mathbf{S}$. Further, for each $j \in C$ and $S \in F$, we set connection costs c_{Sj} to zero if $j \in S \in \mathbf{S}$ and to ∞ if $j \in U \setminus S$.

As a consequence, the above negative results directly apply to the *non-metric* version UFL and the best-possible approximation factor of a polynomial algorithm is logarithmic in the number of clients.

8.3.2 A Simple Greedy Algorithm

Conversely, Chvátal's algorithm [74] already provides a $1 + \ln |U|$ approximation of the *set cover* problem, and thus, a reduction of UFL to *set cover* will already provide a good approximation for UFL. For this reduction, we need the notion of a *star* (i, B) consisting of a facility i and an arbitrary choice of clients $B \subseteq C$. The *cost* of a star (i, B) are $f_i + \sum_{j \in B} c_{ij}$ while its *cost efficiency* is defined as the star's *cost* per connected client in B:

$$c(i, B) = \frac{f_i + \sum_{j \in B} c_{ij}}{|B|} \tag{8.1}$$

Given an instance (C, F, c, f) of UFL, we can generate a *set cover* instance by setting $U := C$ and $\mathbf{S} := 2^C$, where $c(B)$ of $B \in 2^C$ are set to the minimum cost of a star (i, B) among the possible stars built using different facilities i, namely to $c(B) = \min_{i \in F} f_i + \sum_{j \in B} c_{ij}$.

Because the resulting *set cover* instance has exponential size, it cannot be generated explicitly. Nonetheless, one can apply Chvátal's algorithm [74] by considering only the most relevant star in each step as proposed by Hochbaum [185]. This method (Algorithm 20) provides the first algorithm approximating the *non-metric* version of UFL.

Algorithm 20: Chvátal's algorithm applied to UFL

---

1   set $U = C$
2   **while** $U \neq \emptyset$ **do**
3   find most cost-efficient star (i, B) with $B \subseteq U$
4   open facility i (if not already open)
5   set $\sigma(j) = i$ for all $j \in B$
6   set $U = U \setminus B$

The algorithm greedily finds the most cost-efficient star (i, B), opens the facility i, connects all clients $j \in B$ to i, and from this point on disregards all (now connected) clients in B. The algorithm terminates when all clients are connected. Note that in spite of there being exponentially many sets $B \subseteq U$, the most efficient star can be found in polynomial time: For each facility i, clients j can be stored pre-sorted by ascending connection costs to i. Given some number k, the most cost-efficient star with k clients will consist of a facility i and the first k clients that have lowest connection costs to i – all other subsets of k clients can be disregarded, as these cannot be more efficient. Hence, at most n different sets must be considered (instead of 2^n).

The results for *set cover* [74] therefore imply that Algorithm 20 provides an approximation ratio of $1 + \ln |C|$ for the non-metric version of UFL, which is very close to its lower bound. For details on Chvátal's set cover algorithm [74] see also Theorem 16.3 in [224].

8.3.3 Improvements for Metric UFL

Jain et al. [196] showed that in the metric case, Algorithm 20 does not perform much better, namely its approximation is still at least $\log n / \log \log n$. However, a slight revision [196], which is shown in Algorithm 21, can guarantee a much better approximation for the metric case of UFL.

Interestingly enough, this algorithm – discovered and analyzed 21 years later – only adds one line to the end of the loop: Once a facility i has been opened, its opening cost f_i is set to zero. This allows facility i to be chosen again to connect additional clients in later iterations, based on a cost-

---

Algorithm 21: Centralized *metric* UFL algorithm by Jain et al.

---
1   set $U = C$
2   **while** $U \neq \emptyset$ **do**
3 | find most cost-efficient star (i, B) with $B \subseteq U$
4 | open facility i (if not already open)
5 | set $\sigma(j) = i$ for all $j \in B$
6 | set $U = U \setminus B$
7 | **set $f_i = 0$**

---

efficiency that disregards i's opening costs f_i – as the facility i has already been opened before in order to serve other clients. This slight change – together with an analysis we will outline in the next section – provides a constant 1.861-approximation of the *metric* UFL problem. Note that this algorithm can be executed in a distributed manner, albeit in a linear number of rounds. See Section 8.4 for details.

8.3.4 LP-Duality Based Analysis

Algorithm 21 provides a strikingly improved approximation guarantee for the metric case. Its analysis is based on interpreting its basic steps as updates of variables in a primal-dual pair of linear programs that formulate the facility location problem.

Specifically, UFL is formulated in terms of the integer linear program (ILP) shown below. Based on this formulation [380], various approximation schemes have been developed – including the distributed approximations we will present in this chapter. The ILP contains two different sets of binary variables y_i and x_{ij}, where y_i indicate that facility i is opened and x_{ij} that client j is connected to facility i.

$$\text{minimize} \sum_{i \in F} f_i y_i + \sum_{i \in F} \sum_{j \in C} c_{ij} x_{ij}$$

$$\text{subject to} \sum_{i \in F} x_{ij} \geq 1 \qquad\qquad \forall j \in C \ (1) \qquad (8.2)$$

$$x_{ij} \leq y_i \qquad\qquad \forall i \in F, \forall j \in C \ (2)$$
$$x_{ij}, y_i \in \{0, 1\} \qquad\qquad \forall i \in F, \forall j \in C \ (3)$$

Constraint (1) ensures that each client is connected to at least one facility (the cost function and positive c_{ij} ensure that no more than one facility is connected) while (2) demands that a client can only be connected to a facility i if the facility is open. The corresponding LP relaxation of UFL (we will hence call it UFL$_{frac}$) changes the integer constraints (3) to $x_{ij}, y_i \geq 0$.

The dual linear program of *UFL*$_{frac}$ is

$$\text{maximize} \quad \sum_{j \in C} \alpha_j$$

$$\text{subject to} \quad \sum_{j \in C} \beta_{ij} \le f_i \qquad \forall i \in F \qquad (8.3)$$

$$\alpha_j - \beta_{ij} \le c_{ij} \ \forall i \in F, \forall j \in C$$
$$\beta_{ij}, \alpha_j \ge 0 \quad \forall i \in F, \forall j \in C .$$

In LP-duality based schemes, each iteration of a combinatorial algorithm (e.g., Algorithm 21 above) is interpreted as an update of the primal variables y_i and x_{ij}, and of the corresponding dual variables α_j and β_{ij}. The primal updates simply reflect the corresponding steps taken in the algorithm, e.g., if the algorithm opens facility i, the respective y_i is set to 1, while the dual updates are used to provide a certain approximation guarantee.

Recall the weak duality theorem: If $\mathbf{p} = (x_{ij}, y_i)$ and $\mathbf{d} = (\alpha_j, \beta_{ij})$ are feasible solutions for the primal and dual program, respectively, then the dual objective function bounds the primal objective function (for an introduction to LP-duality see [375, chap. 12]). That is, for a minimization problem like UFL, the costs of the optimal primal solution $\mathbf{p}_{\text{opt}}$ lie between

$$g(\mathbf{d}) \le f(\mathbf{p}_{\text{opt}}) \le f(\mathbf{p}) \qquad (8.4)$$

where $f(\mathbf{p})$ and $g(\mathbf{d})$ denote the primal and the dual objective functions, respectively. Moreover, $g(\mathbf{d}) = f(\mathbf{p}_{\text{opt}}) = f(\mathbf{p})$, iff all of the so-called *complementary slackness conditions* are satisfied. For the programs in (8.2) and (8.3), the complementary slackness conditions are

$$\forall i \in F, j \in C : x_{ij} > 0 \Rightarrow \alpha_j - \beta_{ij} = c_{ij} \qquad (8.5)$$

$$\forall i \in F : y_i > 0 \Rightarrow \sum_{j \in C} \beta_{ij} = f_i \qquad (8.6)$$

$$\forall j \in C : \alpha_j > 0 \Rightarrow \sum_{i \in F} x_{ij} = 1 \qquad (8.7)$$

$$\forall i \in F, j \in C : \beta_{ij} > 0 \Rightarrow y_i = x_{ij}. \qquad (8.8)$$

These laws of LP-duality allow designing approximation algorithms in two specific ways. The first class of algorithms is based on the so-called *primal-dual* scheme: The algorithm produces a pair of *feasible* solutions $\mathbf{p}$ and $\mathbf{d}$ and in its iterations tries to satisfy more and more of the conditions (8.5-8.8). In particular, the approximation algorithm [198], which we will detail in the next section, satisfies all conditions but condition (8.5). Instead, it satisfies a relaxed version of (8.5), namely, $x_{ij} > 0 \Rightarrow (1/3)c_{ij} \le \alpha_j - \beta_{ij} \le c_{ij}$. As all other conditions are satisfied, it can be shown that for each result of the algorithm consisting of a primal solution $\mathbf{p}$ and a dual solution $\mathbf{d}$, it holds that

$g(\mathbf{d}) \leq f(\mathbf{p}) \leq 3 \times g(\mathbf{d})$. This implies that a primal solution $\mathbf{p}$ is provided with an approximation guarantee of 3 for the metric version of the facility location problem. We describe this algorithm in detail in the next Section 8.3.5.

A second class of algorithms, instead, employs the scheme of *dual fitting*, which provides a basis for both the analysis of Algorithm 21 and the analysis of a fast distributed algorithm we will present in Section 8.5. *Dual fitting* based algorithms perform the updates of the primal and dual variables such that these obey *all* complementary slackness conditions, i.e., the two objective functions are equal throughout the algorithm (in the general case, the primal is at least bounded by the dual).

However, while the primal variables represent a feasible solution of the primal program, the dual variables are allowed to become *infeasible*. Nevertheless, the authors show that the degree of infeasibility of α_j and β_{ij} is bounded and that – after dividing all dual variables by a suitable factor γ – the dual is feasible and does *fit* into the dual LP, that is, satisfies all dual constraints. In [196], the authors then formulate an additional LP (the so-called *factor-revealing* LP) to find the minimal γ that suffices for all problem instances. For Algorithm 21 they give $\gamma = 1.861$. By the laws of LP duality (8.4) and considering the objective functions for UFL, this implies that Algorithm 21 provides a 1.861-approximation for metric UFL instances.

8.3.5 A 3-Approximation Algorithm

The following algorithm demonstrates how an approximation factor (in this case a factor of 3 for metric instances) can be obtained based on LP-duality. We include its analysis as it is rather simple, compared to the one stated for instance for Algorithm 21.

Consider the following algorithm that consists of two phases. In the following, we give the algorithm as described by Vazirani [375, chap. 24]. The original version is found in [198].

Phase 1. Phase 1 uses a notion of time that allows associating events with the time at which these happened. Time starts at 0. Motivated by the desire to find a dual solution that is as large as possible, each client uniformly raises its dual variable α_j by 1 in unit time. Initially, all clients start at $\alpha_j = 0$ and all facilities are closed. There are two events that may happen:

Tight edge: When $\alpha_j = c_{ij}$ for some edge (i, j), the algorithm will declare this edge to be *tight*. After this event, if the facility i is not yet *temporarily open* (the event in which facilities are opened is described below), the dual variable β_{ij} corresponding to the tight edge will be raised at the same rate as α_j ensuring that the first dual constraint ($\alpha_j - \beta_{ij} \leq c_{ij}$) is not violated. For $\beta_{ij} > 0$ the edge (i, j) is called *special*. If facility i is already temporarily open, the client j is simply connected to i and stops raising its dual variables.

Facility paid for: Facility i is said to be *paid for* if $\sum_{j \in C} \beta_{ij} = f_i$. When this happens, the algorithm declares this facility *temporarily open*. Furthermore, all unconnected clients having tight edges to i are declared *connected*. After this, facility i is called the *connecting witness* of the just connected clients $\{j\}$ and the respective dual variables $\{\alpha_j\}$ and $\{\beta_{ij}\}$ are not raised any further.

Phase 1 completes after all clients are connected to a temporarily open facility. If several events happen simultaneously, they are processed in arbitrary order.

Notice how the events in Phase 1 are designed to satisfy complementary slackness conditions (8.5) and (8.6), respectively. The problem at the end of Phase 1, however, is that the client may have contributed to temporarily opening more than one facility (and thus too many facilities are temporarily open)[3].

Phase 2. Let F_t denote the set of temporarily open facilities and T denote the subgraph of G consisting of edges that were special at the end of Phase 1. Let T^2 denote the graph that has edge (u, v) iff there is a path of length at most 2 between u and v in T. Further let H be the subgraph of T^2 induced on F_t. Compute a maximal independent subset, I, of these vertices w.r.t. to the graph H. All facilities in set I are declared *open*.

Finally, each client j is connected to a facility in I as follows: If there is a tight edge (i, j) and facility i is open (i.e., $i \in I$), j is declared *directly connected* to i and the algorithm sets $\sigma(j) := i$ and $\beta_{i'j} := 0$ for all $i' \neq i$. Otherwise, consider the tight edge (i, j) such that i was the connecting witness for j. Since $i \notin I$, its closing witness, say i' must be in I. Thus, j is declared *indirectly connected* to i' and $\sigma(j)$ is set to i'.

The resulting *open* facilities and *connected* clients define a primal integral solution: That is, $y_i = 1$ iff $i \in I$ and $x_{ij} = 1$ iff client j is connected to i. Further, α_j and β_{ij} constitute a dual solution.

Please note how the algorithm is formulated in terms of the dual variables which then drive the execution and determine which facilities are opened and which clients are connected. Similar re-formulations in terms of the duals exist for example for Algorithm 21, which then opens up the way for its analysis [196]. Let us proceed to analyze the approximation ratio of this algorithm.

Analysis. The analysis relies on the following three points, of which the first two can be derived from the respective updates of the algorithm:

---

[3] Consider the complementary slackness condition (8.6) that each facility, which is open, should be fully paid for $\sum_{j \in C} \beta_{ij} = f_i$. Consider a facility i that has become temporarily open (because client j contributed β_{ij} to the above sum) *after* client j has been connected to a different facility (say i'). Thus, $y_i \neq x_{ij}$ but $\beta_{ij} > 0$. This violates condition (8.8), namely that $\beta_{ij} > 0 \Rightarrow y_i = x_{ij}$, and implies that in an optimal solution only one β_{ij} can be non-zero for a client j. These violations are repaired by closing some temporarily open facilities in Phase 2.

1. As a result of Phase 2, every client is connected to a facility and is only connected if this facility is also open. Therefore, y_i and x_{ij} are a feasible (and integral) solution of the primal ILP of eq.(8.2).
2. Similarly, the updates of the dual variables in the algorithm, never violate the dual constraints of eq. (8.3), and the resulting α_j and β_{ij} are a feasible solution for the dual LP.
3. The dual objective function bounds the primal objective function from above by a factor of 3. Note that the authors even show a stronger inequality (Theorem 8.3.5) which allows using this algorithm as a subroutine in an approximation of the k-Median problem.

Note that the algorithm satisfies all complementary slackness conditions for directly connected clients. For a client j *indirectly connected* to facility i (for which $\beta_{ij} = 0$), the respective complementary slackness condition $\alpha_j = c_{ij} + \beta_{ij}$ is relaxed to $\frac{1}{3}c_{ij} \leq \alpha_j \leq c_{ij}$.

For easier analysis, the authors split up the dual variables α_j into two sub-terms $\alpha_j = \alpha_j^e + \alpha_j^f$. By α_j^e they refer to the amount in α_j that is used towards paying for the connection costs to the facility $\sigma(j)$. By α_j^f they denote the amount in α_j that is used by client j to pay for opening a facility in I (j's part in the sum $\sum_{j \in C} \beta_{ij} = f_i$).

If i is indirectly connected, then $\alpha_j^f = 0$ and $\alpha_j^e = \alpha_j$. If j is directly connected to i, by definition of the algorithm, $\alpha_j = c_{ij} + \beta_{ij}$, that is, $\alpha_j^f = \beta_{ij}$ and $\alpha_j^e = c_{ij}$.

In the following, we give the proof as stated in [375, chap. 24]:

Lemma 8.3.2. *Let* $i \in I$. *Then,*

$$\sum_{j:\sigma(j)=i} \alpha_j^f = f_i.$$

Proof. The condition for declaring facility i temporarily open at the end of Phase 1 was that it is completely paid for, i.e,

$$\sum_{j:(i,j) \text{ is special}} \beta_{ij} = f_i.$$

In Phase 2, as $i \in I$, all clients that have contributed to i in Phase 1 will be directly connected to i. For each such client, $\alpha_j^f = \beta_{ij}$. Any other client j' that is connected to facility i must satisfy $\alpha_{j'}^f = 0$. The lemma follows.

Corollary 8.3.3. $\sum_{i \in I} f_i = \sum_{j \in C} \alpha_j^f.$

Because $\alpha_j^f = 0$ for indirectly connected clients, only the directly connected clients pay for the cost of opening facilities. The sum of all these clients pays to open all facilities.

Lemma 8.3.4. *For a client indirectly connected to a facility i, $c_{ij} \leq 3\alpha_j^e$.*

Proof. Let i' be the connecting witness for a client j. Since j is indirectly connected to i, (i, i') must be an edge in H. In turn, there must be a client, say j' such that (i, j') and (i', j') were both special edges at the end of phase 1. Let t_1 and t_2 be the times at which i and i' were declared temporarily open during Phase 1.

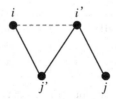

Since edge (i', j) is tight, $\alpha_j \geq c_{i'j}$. Below we will show that $\alpha_j \geq c_{ij'}$ and $\alpha_j \geq c_{i'j'}$. Then, the lemma will follow by using the triangle inequality[4]

$$c_{ij} \leq c_{i'j} + c_{i'j'} + c_{ij'} \leq 3\alpha_j.$$

Since edges (i', j') and (i, j') are tight, $\alpha_{j'} \geq c_{ij'}$ and $\alpha_{j'} \geq c_{i'j'}$. Because both these edges are special, they must both have gone tight before either i or i' is declared temporarily open. Consider the time $\min(t_1, t_2)$. Clearly, $\alpha_{j'}$ cannot be growing after this time, that is, $\alpha_{j'} \leq \min(t_1, t_2)$. Finally, since i' is the connecting witness for j, $\alpha_j \geq t_2$. Therefore, $\alpha_j \geq \alpha_{j'}$, and the required inequalities follow.

Theorem 8.3.5. *The primal and dual solutions constructed by the algorithm satisfy*

$$\sum_{i \in F} \sum_{j \in C} c_{ij} x_{ij} + 3 \sum_{i \in F} f_i y_i \leq 3 \sum_{j \in C} \alpha_j.$$

Proof. For a client j directly connected to i, $c_{ij} = \alpha_j^e \leq 3\alpha_j^e$. Combining this with Lemma 8.3.4 one obtains

$$\sum_{i \in F} \sum_{j \in C} c_{ij} x_{ij} \leq 3 \sum_{j \in C} \alpha_j^e. \tag{8.9}$$

The theorem results from adding the equation of Corollary 8.3.3 multiplied by 3 to the above (8.9).

8.3.6 Summary

We conclude this section with a summary of known lower bounds and centralized approximation algorithms in Figure 8.4.

---

[4] This is where the analysis assumes a *metric* instance of UFL.

Variant	Lower bound	Approx. factor	Time compl.	Reference (first author)
Set cover	$\Omega(\ln(|U|))$	$1 + \ln(|U|)$	$|U|^3$	Chvátal [74]
General UFL	$\Omega(\ln(n))$	$1 + \ln(n)$	$(n+m)^3$	Hochbaum [185]
Metric UFL	1.463 in [173]	3.16	LP	Shmoys [347]
		3	$nm\log(nm)$	Jain [198]
		2.41	LP	Guha [173]
		1.861	$nm\log(nm)$	Jain [196]
		1.61	$(n+m)^3$	Jain [196]
		1.52	$(n+m)^3$	Mahdian [272]

Fig. 8.4. Approximation algorithms for UFL. In the above notation, n denotes the number of clients and m denotes the number of facilities.

In the table, we noted Algorithm 20 by Hochbaum [185], the respective equivalent set-cover algorithm by Chvátal [74], the factor 3-approximation algorithm [198, 375] presented in Section 8.3.5, and Algorithm 21 by Jain et al. [196]. We also want to mention a second improved algorithm in the same paper [196] for which the authors prove an approximation ratio of 1.61.

The first known constant-factor approximation by Shmoys et al. [347] is also included in Figure 8.4 together with an LP-based algorithm by Guha and Khuller [173], who proved a lower bound of 1.463 for *metric* UFL. Moreover [173] and [196] were combined by Mahdian et al. [272] to obtain the currently best known approximation factor of 1.52 for this problem (also based on dual fitting). Note that the above list of centralized approximations and variants is not exhaustive, for further references refer to [272].

In the remainder of this chapter, we present distributed facility location algorithms which are built around the elements introduced in this section, most prominently the concept of opening facilities which have a good *cost-efficiency* – and the parallelization of this step.

8.4 Simple Distributed Approximation

In this section we discuss how Algorithm 21 can be executed in a distributed manner. Consider the distributed re-formulation we give in Algorithm 22 (facilities) and 23 (clients).

Please be aware that the distributed re-formulation still requires a *metric* instance which demands that the underlying bipartite graph is complete (or at least consists of cliques). In a graph derived from a multi-hop network, this implies global communication among nodes of a connected component. These algorithms are therefore not efficient for large multi-hop networks. Note that this restriction does not apply to the faster algorithm presented in Section 8.5 as it does not require a metric instance.

Assume that after a neighbor discovery phase, each client j has learned the set of neighboring facilities, which it stores in the local variable F, and vice versa each facility i has learned the set of neighboring clients, which it stores in the local variable C. Moreover, after initialization, each client j knows its respective c_{ij} for all $i \in F$ and equally each facility i knows its respective c_{ij} for all $j \in C$.

Algorithm 22: Distributed Formulation of Algorithm 21 for Facility i

---
1 set $U = C$
2 **while** $U \neq \emptyset$ **do**
3 | find most cost-efficient star (i, B) with $B \subseteq U$
4 | **send** $c(i, B)$ to all $j \in U$
5 | **receive** "connect-requests" from set $B^* \subseteq U$.
6 | **if** $B^* = B$ **then**
7 | | open facility i (if not already open)
8 | | **send** "open" to all $j \in B$
9 | | set $\sigma(j) = i$ for all $j \in B$
10 | | set $U = U \setminus B$
11 | | set $f_i = 0$
12 | **else**
13 | | **send** "not-open" to all $j \in B^*$
14 | **receive** "connected" from set B'
15 | set $U = U \setminus B'$

Algorithm 23: Distributed Formulation of Algorithm 21 for a Client j

---
1 **while** *not connected* **do**
2 | **receive** $c(i, B)$ from all $i \in F$
3 | $i^* = \mathrm{argmin}_{i \in F}\, c(i, B)$ // *use node ids to break ties among equal* $c(i, B)$
4 | **send** "connect-request" to i^*
5 | **if** *received* "open" *from* i^* **then**
6 | | set $\sigma(j) = i^*$
7 | | **send** "connected" to all $i \in F$

Note that for ease of exposition, we describe the two algorithms in terms of a synchronous message passing model [313] although the algorithms do not depend on it. In particular, in each round, facilities and clients perform one iteration of the while loop.

Each facility i maintains a set of uncovered clients U which is initially equal to C (line 1). Facilities start a round by finding the most cost-efficient star (i, B) with respect to U and sending the respective cost efficiency $c(i, B)$ to all clients in B (lines 3-4). In turn, the clients receive cost-efficiency numbers $c(i, B)$ from many facilities (line 2). In order to connect the most cost-efficient star among the many existing ones, clients reply to the facility i^* that has sent the lowest cost-efficiency value with a "connect request" (lines 3-4). Facilities in their turn collect a set of clients B^* which have sent these "connect requests" (line 5). Intuitively, a facility should only be opened if $B = B^*$, that is, if it has connect requests from all clients B in its most efficient star (line 6). This is necessary, as it could happen that some clients in B have decided to connect to a different facility than i as this facility spans a more cost-efficient star. So, if all clients in B are ready to connect, facility i opens, notifies all clients in B about this, removes the connected clients from B, and sets its connection costs to 0 (lines 7-11).

If a client receives such an "open" message from the same facility i^*, which it had previously selected as the most cost-efficient (line 5), it can connect to i^*. Further it notifies all other facilities $i \neq i^*$ that it is now connected, such that these facilities can update their sets of uncovered clients U (line 15) and start the next round.

Note that the distributed algorithms correctly implement Algorithm 21: Consider the globally most efficient star (i^*, B^*) selected by the centralized algorithm. In the distributed version, the clients in B^* will all receive the respective cost-efficiency from i^* and select i^* as the most cost-efficient facility – if more than one most cost-efficient facility exists, the clients agree on one of them, which has the smallest id. Therefore all clients in B^* will reply a "connect-request" to i^*, which will open facility i^*.

Thus, the distributed algorithms connect the most efficient star in each communication round. However, additionally, in the same round other facilities and associated stars may be connected. This is particularly the case if the instance graph is not complete, which often occurs in instances derived from ad hoc and sensor networks, and results in these stars being connected *earlier* than in the centralized algorithm. Regarding such stars, we show below that, given any star (i, B) connected in round k and any client $j \in B$, no star (i^+, B^+) connected in a later round $k^+ > k$ exists such that $j \in B^+$ and $c(i^+, B^+) < c(i, B)$. This implies that for each (i, B), the centralized algorithm would either also connect (i, B) before it will connect any (i^+, B^+) or, if $c(i, B) > c(i^+, B^+)$, that (i, B) is completely disjoint from any (i^+, B^+) that has fallen behind compared to the centralized algorithm.

Assume the contrary, namely that such stars (i, B) and (i^+, B^+) exist with $j \in B$ and $j \in B^+$. Observe that the star (i^+, B^+) could have equally been connected in the current round k (by assumption, clients in B^+ are uncovered and facility i^+ is not open in round k^+; this also holds for round k). In round k, however, client j has sent its "connect-request" message to i as this is required for connecting (i, B) – which contradicts the assumption that $c(i^+, B^+) < c(i, B)$.

Note that similarly to the above re-formulation of Algorithm 20, also the improved version presented in the same paper giving a 1.61 approximation guarantee and the 3-approximation algorithm of Section 8.4 can be re-formulated for distributed execution.

However, analogously to the maximal independent set algorithm presented Section 3.3.2, the worst-case number of rounds required by Algorithms 22 and 23 is $\Omega(m)$, because there can be a unique point of activity around the globally most cost-efficient facility i^* in each round: Consider for instance a chain of m facilities located on a line, where each pair of facilities is interconnected by at least one client, and assume that facilities in the chain have monotonously decreasing cost-efficiencies. Each client situated between two facilities will send a "connect-request" to only one of them (the more cost-efficient) and therefore the second cannot open. In this example, only the facility at the end of the chain can be opened in one round.

The linear number of rounds required in the worst-case is impracticable in many cases. Nevertheless, the constant approximation guarantees of 1.861 – or even 1.61 for a similar algorithm – inherited from the centralized versions are intriguing and the *average* number of rounds required for typical instances in sensor networks remains to be explored. A much faster distributed algorithm, which runs in a constant number of rounds and can additionally handle arbitrary (non-metric) instances, is presented in the following Section 8.5.

8.5 Fast Distributed Approximation

Moscibroda and Wattenhofer [294] present the first distributed algorithm that provides a non-trivial approximation of UFL in a number of rounds that is *independent of the problem instance*. The approach follows the line of a number of recent approximation algorithms, e.g. [238], that employ an integer linear program (ILP) formulation of the addressed optimization problem, solve its linear relaxation in a distributed manner, and finally employ schemes to convert the fractional results to an integer solution with approximation guarantees.

The algorithm consists of two phases. In the first phase, given an arbitrary integer k, the algorithm computes an $O(\sqrt{k}(m\rho)^{1/\sqrt{k}})$ approximation of the LP relaxation in $O(k)$ communication rounds. In this bound, ρ is a factor depending on the coefficients c_{ij} and f_i and further it is required that $c_{ij} \geq 1$ and $f_i \geq 1$. To dispose of the dependency on ρ in the approximation ratio and to allow arbitrary coefficients, the authors give a transformation algorithm that runs in $O(k^2)$ rounds and yields an approximation ratio of $O(k(mn)^{1/k})$ [294]. Then in a second phase, the fractional result is rounded into a feasible integer solution of the ILP, worsening the approximation ratio only by a factor of $\log(m + n)$. The authors provide distributed algorithms that can be executed on facility and client nodes, respectively, for both phases of the algorithm. Because the algorithms – especially of the first phase – are quite intricate, in the following we will omit details on how the dual variables are updated. These updates, based on *dual fitting* (see Section 8.3.5), are designed to implement the required complementary slackness conditions, while allowing the dual to become infeasible – but only slightly, such that the dual variables divided by a certain factor yield a feasible dual solution.

Note that the presented approach is, in the context of distributed LP approximation [32, 267], the first to approximate a linear program (namely equation (8.2)) which is not a *covering or packing* LP, i.e., a linear program in which negative coefficients are allowed.

8.5.1 LP Approximation

In order to obtain an approximate solution of the relaxed linear program UFL_{frac}, facilities and clients execute Algorithms 24 and 25, respectively. The algorithms perform stepwise parallel increases of the variables x_{ij} and y_i, which start at zero. Each facility i will increase the variable y_i determining whether facility i should be open, while each client j will increase its connection variables x_{ij} determining to which facilities i it is connected. These increases are made in k synchronous parallel steps, determined by an outer loop (line 4) and a nested inner loop (line 5) which are each executed $\lceil \sqrt{k} \rceil$ times. Note that a synchronous message passing model as in [313] is assumed; the authors state that a synchronizer [24] can be used to make the algorithm work in asynchronous settings.

Algorithm 24: LP Approximation - Facility i

---

1 $h := \lceil \sqrt{k} \rceil$
2 $\rho := \max_{j \in C} \min_{i \in F}(c_{ij} + f_i)$
3 $y_i := 0;\ U := C$
4 **for** $s := 1$ *to* h *by* 1 **do**
5   **for** $t := h - 1$ *to* 0 *by* -1 **do**
6   **receive** "covered" from subset $B^- \subseteq C$
7          $U := U \setminus B^-$
8   **if** $\exists$ *star* (i, B) *with* $B \subseteq U$ *and* $c(i, B) \leq \rho^{s/h}$ **then**
9   find largest star (i, B_i') with $B_i' \subseteq U$ and $c(i, B_i') \leq \rho^{s/h}$
10   **if** $y_i < m^{-t/h}$ **then**
11   raise y_i to $m^{-t/h}$
12   **send** y_i to all $j \in B_i'$
13   *update duals (omitted)*

Algorithm 25: LP Approximation - Client j

---

1 $h := \lceil \sqrt{k} \rceil$
2 **send** c_{ij} to all $i \in F$
3 $\forall i \in F : x_{ij} := 0$
4 **for** $s := 1$ *to* h *by* 1 **do**
5   **for** $t := h - 1$ *to* 0 *by* -1 **do**
6   **if** $\sum_{i \in F} x_{ij} \geq 1$ **then send** "covered" to all $i \in F$
7   **receive** y_i from all $i \in F$ with $j \in B_i'$
8   **forall** $i \in F$ *with* $j \in B_i'$ **do**
9   **if** $\sum_{l \in F} x_{lj} < 1$ **then**
10                 $x_{ij} := y_i$
11   *update duals (omitted)*

In each round, the *clients* compute whether they are *covered*, i.e., whether they are fractionally connected to a facility ($\sum_{i \in F} x_{ij} \geq 1$), and send this information to all facilities in line 6. Each *facility* receives this information and computes the set U of *uncovered* clients in lines 6 and 7.

The *facility* algorithm will then perform parallel increases of the y_i variables. The choice of which of the y_i to increase is based – as in the centralized Algorithms 20 and 21 presented above – on how beneficial a facility is to the set U of yet *uncovered* clients: A *facility* node i will increase the respective y_i if there exists a star (i, B) with $B \subseteq U$ that has a cost efficiency[5] smaller than a step-wise increasing threshold $\rho^{s/h}$ (line 8). Note that the threshold $\rho^{s/h}$ depends on the outer s-loop (an exemplary increase of $\rho^{s/h}$ with s is shown in Figure 6(a), we will discuss its function later). If such a star exists, a facility computes the *largest* such star (i, B_i') which still has a cost efficiency that is less than $\rho^{s/h}$ (line 9). Note that the set B_i' can simply be constructed as $B_i' := \{j | c_{ij} \leq \rho^{s/h}\}$, see the original paper for details and proof [294]. Further, a facility i increases its y_i to a new level $m^{-t/h}$ in line 11. This *raise level* $m^{-t/h}$ depends on the inner t-loop and increases up to 1 in the last iteration of the loop (see Figure 6(a) for an example). Facility i then notifies all clients in its largest star B_i' of its increase of y_i in line 12.

The corresponding *client* algorithms then receive the respective y_i from such facilities i in line 7. Clients then increase their x_{ij} to y_i in line 10 unless they have become covered earlier. This means that a client j connects itself (fractionally) to all facilities i which – in this step – increased y_i and thus have become "more open" in the fractional sense.

---

[5] We refer to the cost efficiency as defined in eq. (8.1).

8.5.2 Example

In the following, we sketch an exemplary execution of Algorithms 24 and 25 using an instance as shown in Figure 8.5. It shows a network of three facilities and four clients with respective opening and connection costs.

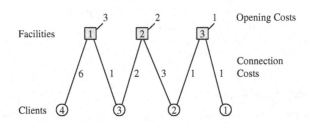

Fig. 8.5. An example instance with three facilities and four clients

The actual execution of the algorithms is shown in Figure 8.6. In this example, we let the algorithms perform a total of $k = 16$ steps, resulting in $= \sqrt{k} = 4$ steps in each loop. We show the two values that drive the algorithm's execution, the *efficiency threshold* and the *raise level* ¯ , in Figure 6(a).

The *efficiency threshold* is important for parallelization of the central greedy step of the centralized approximation algorithms in Section 8.3 above (opening the facility that constitutes the most cost-efficient star). Figure 6(a) shows an example of how the threshold is increased from ¹ up to . The idea of the outer s-loop is to increase the $_i$ of facilities with comparably good cost efficiency and further connect the respective clients. In its last step, an efficiency threshold of is high enough to include all not yet connected clients in the last step (see also Lemma 8.5.1). In the chosen instance of Figure 8.5, $= 9$. That is, in the last step, client 4 can be connected by facility 1 forming a single-client star with a cost-efficiency of 9, while all other clients can be connected (possibly earlier) via stars that are more cost-efficient.

In Figure 6(b), we give the respective variables $_i$ and $_{ij}$. By $_j$ we refer to the sum $\sum_i {}_{ij}$ of all connections of client . We use $_j$ to emphasize whether the client is already *covered* or still active.

Consider the first step of the s-loop ($s = 1$). At this time, only facility 3 can form a star with cost efficiency smaller than ¹ ⁴ 1.73. In this case, clients 1 and 2 are part of this star, which is also facility 3's largest star B'_3 with *cost efficiency* less than ¹ ⁴ in line 9 of the *facility* algorithm. The inner -loop then raises $_3$ and sends this update to all $\in B'$. Clients 1 and 2 thus set $_{31}$ and $_{32}$ to the same amount in line 10.

The *raise level* ¯ , also shown in Figure 6(a), denotes the value to which the variables $_i$ are raised in each step of the inner -loop (the respective $_{ij}$ are raised to the same level unless already covered). In this example the

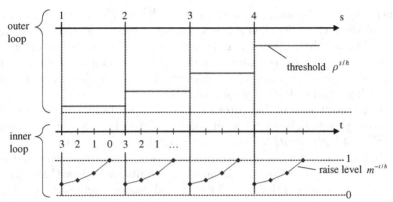

(a) Efficiency threshold $\rho^{s/h}$ and raise level $m^{-t/h}$. The outer s-loop and the nested t-loop each perform $h = \sqrt{k} = 4$ steps.

Iterations s	1				2				3				4				Final
t	3	2	1	0	3	2	1	0	3	2	1	0	3	2	1	0	
Facilities y_1												▪				▪	▪
y_2												▪					▪
y_3				▪													▪
Client 1 x_{11}																	
x_{21}																	
x_{31}				▪													▪
X_1				▪													▪
Client 2 x_{12}																	
x_{22}																	
x_{32}				▪													▪
X_2				▪													▪
Client 3 x_{13}												▪					▪
x_{23}												▪					▪
x_{33}																	
X_3												▪					▪
Client 4 x_{14}												▪				▪	▪
x_{24}																	
x_{34}																	
X_4												▪				▪	▪

(b) Facility and client variables during execution. Full/missing bars indicate 1 or 0, respectively. Bars are omitted if no raise occurs.

Fig. 8.6. Execution of Algorithms 24 and 25 on instance from Fig. 8.5.

inner loop is performing four steps from $m^{-3/4}$ to $m^{0/4} = 1$. Thus, in the final step of the t-loop, the clients 1 and 2 become *covered* as they are connected to facility 3. Thus, clients 1 and 2 are removed from the set of uncovered clients U and are not considered further in the computation of stars at later steps. Because of this, the cost efficiency of the most efficient star is always increasing at all facilities (proof omitted).

In the next step of the outer loop ($s = 2$), no facility is efficient enough in connecting the clients remaining in U (i.e., can form a star with efficiency $\leq \rho^{2/4} = 3$ with clients 3 and/or 4), therefore nothing happens.

In the next iteration of the outer loop ($s = 3$), the *efficiency threshold* is increased again to $\rho^{3/4} \approx 5.2$. As facilities execute their algorithms in parallel, at this time both facilities 1 and 2 have found stars with efficiency levels $c(i, B)$ that are below the threshold. In this example, facility 1 can connect the client set $\{3, 4\}$ with cost efficiency $c(1, \{3, 4\}) = 5$ and facility 2 can connect the client 3 with efficiency $c(2, \{3\}) = 4$. Note that both facilities have the client 3 in their respective *largest stars*, B_1' and B_2'. In this setting, the role of the inner t-loop becomes evident, namely to ensure that the sum $\sum_{i \in F} x_{ij}$ is not increased too much if there are many facilities that simultaneously increase their y_i and thus cause an increase of many x_{ij} of a given client j.

In the given example, in each step of the inner t-loop, both facilities increase their respective y_i, and each facility then causes an update of the respective x_{ij}, that is, facilities 1 and 2 contribute to increases of x_{13} and x_{23}, respectively, and facility 1 additionally contributes to increases of x_{14}. Therefore, as X_3 is increased twice in each step, the client already becomes (fractionally) covered by facilities 1 and 2 in the second-to-last step of the inner loop. In the last t-step, client 3 is already covered, and client 4 is the only client remaining in U, with which neither facility can form an efficient-enough star, therefore client 4 is not covered any further.

In the final step of the outer loop ($s = 4$), all yet (fractionally) uncovered clients (here client 4) will be connected, as the *efficiency threshold* is set to ρ. In this example, the increases of the inner loop only have an effect after the *raise level* is above the actual y_1, which is in the very last step. In this last step of the inner t-loop, y_1 and x_{41} are both raised to 1.

8.5.3 Analysis

In the following, we make a few observations about Algorithms 24 and 25. For brevity, we omit large parts of the proof. These can be readily found in [294].

Lemma 8.5.1. *Algorithms 24 and 25 produce a feasible solution for* UFL_{frac}.

Proof. The feasibility of the second LP constraint, $x_{ij} \leq y_i$ for all $i \in F$ and $j \in C$, directly follows from the algorithm definition in lines 11, 12 and 7, 10 of the facility and client algorithms, respectively. The increase of a connection variable Δx_{ij} never exceeds the previous increase of the corresponding Δy_i.

As for the first LP constraint, $\sum_i x_{ij} \geq 1$ for all $j \in C$, assume for contradiction that j is a client which is still *uncovered* at the end of the algorithm, i.e., $\sum_i x_{ij} < 1$ and $j \in U$. Now, consider the very last iteration of the inner loop ($s = h, t = 0$). By the definition of ρ there exists at least one facility i forming a star $(i, \{j\})$ with cost efficiency $c(i, \{j\}) \leq \rho^6$. Because $s = h$, facility i will increase its variable to $m^{-t/h} = 1$ in line 11. Subsequently, j will set $x_{ij} = 1$ in line 10, which contradicts the assumption that $\sum_i x_{ij} < 1$ at the end of the algorithm.

Further, the algorithm achieves approximation guarantees through a careful handling of the dual variables α_j and β_{ij} of the corresponding dual LP (eq. (8.3)). Essentially, the method of *dual fitting* is applied in a distributed setting. While we omit the details of the dual updates in line 11 of Algorithm 24 and line 13 of Algorithm 25, we highlight the following property of these algorithms:

Lemma 8.5.2. *At the end of each t-loop, the primal objective function $\sum_{j \in C, i \in F} c_{ij} x_{ij} + \sum_{i \in F} f_i y_i$ is equal to the dual objective function $\sum_{j \in C} \alpha_j$.*

Furthermore, the algorithm ensures that dual infeasibility is bounded by a factor γ, and thus in a final step, the dual variables can be divided by γ to yield a feasible dual solution. For details we refer to the original paper [294]. Because of basic laws of LP-duality, the primal solution is then a γ-approximation of UFL_{frac}.

This leads to the approximation guarantee of Algorithms 24 and 25.

Theorem 8.5.3. *For an arbitrary integer $k > 0$ algorithms 24 and 25 compute a $O(\sqrt{k}(m\rho)^{(1/\sqrt{k})})$ approximation of UFL_{frac}.*

For this bound it is required that $c_{ij} \geq 1$ and $f_i \geq 1$. In a re-formulation that is independent of ρ and allows arbitrary coefficients, the algorithm uses a scaling technique that requires $O(k^2)$ steps. In this case the approximation factor becomes $O(k(mn)^{1/k})$. We will discuss implications of the obtained approximation in Section 8.5.5.

8.5.4 Rounding

So far, the solution computed in the first phase of the algorithm is fractional, i.e., it solves UFL_{frac} without requiring the variables y_i and x_{ij} to be integers. A second rounding phase, shown in Algorithm 26 and Algorithm 27, is therefore used to construct a feasible integer solution for a UFL instance. For this, each *facility i* sets its y_i to 1 with probability p_i and to 0 otherwise, and sends y_i

---

[6] Consider the definition of $\rho = \max_{j \in C} \min_{i \in F}(c_{ij} + f_i)$ in line 2 and let i be the facility for which the term $\min_{i \in F}(c_{ij} + f_i)$ is minimal for the uncovered client j. The cost efficiency of the star $(i, \{j\})$ is at most $c_{ij} + f_i$. By the definition of $\rho = \max_{j \in C}(\ldots)$, the cost efficiency of this star is less or equal to ρ.

to all clients. Later, if they receive a JOIN message from any client, they set their y_i to 1.

Each *client* computes its current contribution to the fractional service costs C_j^* (line 1) and the set V_j of facilities within a $\ln(n+m)$ factor of C_j^* (line 2). Then each j receives y_i from all facilities i and defines its neighborhood N_j as the open facilities in V_j. If at least one such neighboring facility is available, the open facility with least service cost is chosen (line 7) or, otherwise, the closest facility is sent a JOIN message in line 11.

Algorithm 26: Rounding - Facility i

1 $p_i := \min\{1, \hat{y}_i \cdot \ln(n+m)\}$

2 $y_i := \begin{cases} 1 \text{ with probability } p_i \\ 0 \text{ otherwise} \end{cases}$

3 **send** y_i to all clients $j \in C$

4 **if receive** JOIN *message* **then** $y_i := 1$.

Algorithm 27: Rounding - Client j

1 $C_j^* := \sum_{i \in F} c_{ij} \hat{x}_{ij}$

2 $V_j := \{i \in F | c_{ij} \le \ln(n+m) \cdot C_j^*\}$

3 **receive** y_i from all $i \in F$

4 $N_j := V_j \cap \{i \in F | y_i = 1\}$

5 **if** $N_j \ne \emptyset$ **then**

6 $\quad i' := \operatorname{argmin}_{i \in F} c_{ij}$

7 $\quad x_{i'j} := 1$

8 **else**

9 $\quad i' := \operatorname{argmin}_{i \in F} (c_{ij} + f_i)$

10 $\quad x_{i'j} := 1$

11 $\quad$ **send** JOIN message to facility i'

Theorem 8.5.4. *Let $\hat{x}_{ij}, \hat{y}_i$ for each $j \in C$ and $i \in F$ denote the fractional solution with cost at most $\gamma \times OPT$ obtained by Algorithms 24 and 25. In two rounds of communication, Algorithms 26 and 27 produce an integer solution x_{ij}, y_i with cost at most $O(\log(m+n))\gamma OPT$ in expectation.*

Proof. It follows from the definition of C_j^* and V_j that $\sum_{j \in F \setminus V_j} \hat{x}_{ij} \le 1/\log(n+m)$, because if not, C_j^* would be larger. The design of Algorithms 24 and 25 guarantees the invariant $\sum_{i \in V_j} \hat{x}_{ij} = \sum_{i \in V_j} \hat{y}_i$ and $\sum_{i \in F} \hat{x}_{ij} \ge 1$, thus it holds that

$$\sum_{i \in V_j} \hat{y}_i = \sum_{i \in V_j} \hat{x}_{ij} \geq 1 - \frac{1}{\log(n+m)}. \tag{8.10}$$

For each client j having $N_j \neq \emptyset$, the connection costs c_{ij} are at most $\log(n + m) \cdot C_j^*$ by the definition of the neighborhood V_j. It follows that these clients account for total connection costs of at most $\log(n+m) \sum_{j \in C, i \in F} c_{ij} \hat{x}_{ij}$. A facility declares itself *open* in line 2 of Algorithm 26 with probability $\min\{1, \hat{y}_i \cdot \log(n+m)\}$. The expected opening costs of facilities opened in line 2 are thus bounded by the value $\log(n+m) \sum_{i \in F} \hat{y}_i f_i$.

It remains to bound the costs incurred by clients that are *not* covered, i.e., $N_j = \emptyset$, and facilities that are opened by a JOIN message. The probability q_j that a client j does *not* have an open facility in its neighborhood is at most

$$
q_j = \prod_{i \in V_j} (1 - p_i) = \left(\sqrt[n+m]{\prod_{i \in V_j} (1 - p_i)} \right)^{n+m}
$$

$$
\leq \left(\frac{\sum_{i \in V_j} (1 - p_i)}{n + m} \right)^{n+m}
$$

$$
\stackrel{|V_j| \leq m}{\leq} \left(1 - \frac{\ln(n+m) \sum_{i \in V_j} \hat{y}_i}{n + m} \right)^{n+m}
$$

$$
\stackrel{\text{Eq.8.10}}{\leq} \left(1 - \frac{\ln(n+m)}{n + m} \left(1 - \frac{1}{\ln(n+m)} \right) \right)^{n+m}
$$

$$
= \left(1 - \frac{\ln(n+m) - 1}{n + m} \right)^{n+m}
$$

$$
\leq e^{-\ln(n+m)-1} \leq \frac{1}{e(n+m)}.
$$

The first inequality follows from the fact that for every sequence of positive numbers the geometric mean is smaller than or equal to the arithmetic mean of these numbers. An uncovered client sends a JOIN message to the facility $i \in F$ that minimizes $c_{ij} + f_i$. Each of these costs is at most $\sum_{j \in C, i \in F} c_{ij} \hat{x}_{ij} + \sum_{i \in F} \hat{y}_i f_i$ because $\hat{x}$ and $\hat{y}$ would not represent a feasible solution otherwise. Combining this with the above results, the total expected cost $\mu = E[ALG]$ is

$$
\mu \leq \ln(n+m) \left(\sum_{j \in C, i \in F} c_{ij} \hat{x}_{ij} + \sum_{i \in F} \hat{y}_i f_i \right) +
$$

$$
\frac{n}{e(n+m)} \left(\sum_{j \in C, i \in F} c_{ij} \hat{x}_{ij} + \sum_{i \in F} \hat{y}_i f_i \right)
$$

$$
\leq (\ln(n+m) + O(1)) \gamma \, OPT.
$$

This concludes the proof of Theorem 8.5.4.

The above rounding algorithm obtains a solution that holds in expectation. If one is interested in results that hold with high probability, it is possible to change Algorithm 26 to obtain the above approximation with probability $1 - n^{-1}$. However, this requires additional coordination of the opening of facilities by a single leader node.

8.5.5 Obtainable Approximation

Note that the resulting approximation ratio of $O(\log(m+n) \times \sqrt{k}(m\rho)^{1/\sqrt{k}})$ deserves further discussion. Generally, the algorithm permits an arbitrary number of steps k. Against intuition however, the approximation factor cannot be improved arbitrarily with increasing k. In contrast, for $k \to \infty$ the approximation ratio even tends towards ∞.

That is, for any given instance size, there exists an optimal number of rounds k_{opt} for which the approximation factor is minimal. After more than k_{opt} steps, the approximation ratio degrades again, consequently it only makes sense to chose $k \in [1, k_{\mathrm{opt}}]$. The best approximation guarantee is obtained for $k_{\mathrm{opt}} = \log^2(m\rho)$ or, respectively in the formulation independent of ρ, $k_{\mathrm{opt}} = \log^2(mn)$. This "point of operation" is interesting when comparing the algorithm to previous centralized algorithms and lower bounds shown in Figure 8.4.

Note that at k_{opt} rounds of execution, the last term of the approximation ratio, $(m\rho)^{1/\sqrt{k}}$, becomes a constant[7] which may be neglected in O-notation. The algorithm therefore obtains a $O(\log(m + n) \log(m\rho))$ approximation in $O(k_{\mathrm{opt}} = \log^2(m\rho))$ steps.

Similarly, the best approximation ratio that is independent of ρ can be computed. In Figure 8.7, we give an overview of the best achievable approximation factors using an optimal number of rounds k_{opt}, together with the slow but simple distributed algorithm for metric instances given in Section 8.4.

Variant	Approximation Factor	Comm. Rounds
Distributed metric UFL	1.861 (or 1.61)	m
Distributed general UFL		
– dependent on ρ	$\log(m+n)\sqrt{k}(m\rho)^{1/\sqrt{k}}$	k
	$\log(m+n)\log(m\rho)$	$k_{\mathrm{opt}} = \log^2(m\rho)$
– independent of ρ	$\log(m+n)\sqrt{k}(mn)^{1/\sqrt{k}}$	k
	$\log(m+n)\log(mn)$	$k_{\mathrm{opt}} = \log^2(mn)$

Fig. 8.7. Trade-off between approximation guarantee, number of communication rounds, and dependency on ρ

---

[7] The equation $c = (m\rho)^{1/\log_a(m\rho)}$ can be solved by applying $\ln(\dots)$ to $c = (m\rho)^{\ln a / \ln(m\rho)}$ which yields $\ln c = \ln a$, that is, c is the basis of the logarithm.

Summing up, the distributed algorithm for *general* UFL instances described in this section provides the first distributed approximation whose running time is independent of the instance size. While the algorithm is not as flexible as to allow to come arbitrarily close to the centralized lower bound, when it is executed for a logarithmic number of rounds, its approximation ratio is within a logarithmic factor of the centralized lower bound.

In contrast, for the *metric* case, we have presented a (some will argue prohibitively slow) distributed version of the centralized greedy Algorithm 21, which in exchange provides a much better approximation ratio (Section 8.4). Note that such a distributed algorithm can be parallelized further, for instance by employing Luby's classic algorithm for computing a maximal independent set [266] (also described in Section 3.3.3). This technique would be particularly suitable to implement a fast distributed version of the centralized 3-approximation algorithm described in Section 8.3.5.

8.6 Discussion and Outlook

We presented a simple distributed algorithm for metric instances of the facility location problem in Section 8.4. Further, we described a fast distributed algorithm [294] for general instances in the previous Section 8.5. Both were not particularly designed for the models applicable to multi-hop sensor networks. The algorithms operate on bipartite graphs, in which clients communicate with all eligible facilities in each round. The first algorithm, although distributed, preserves its approximation factor only with global communication among all nodes – as it requires a metric instance which implies that the underlying bipartite graph is complete. The second algorithm can be used for settings in which facilities and clients are direct neighbors. For example, it could be used to compute an energy-efficient clustering (Figures 1(b) and 3(a)) as such scenarios can be unfolded into bipartite graphs in which some edges are missing.

Notice that each of the algorithms has additional disadvantages even in this one-hop setting. The first algorithm for metric instances can be prohibitively slow, although we have noted that there is potential for accelerating a similar algorithm. The second algorithm, while in turn very fast and even suitable for general instances, requires that the coefficient ρ – computed from the costs of the respective instance – is known at all nodes. Note that computing and distributing ρ is inherently non-local. Nevertheless, in some settings, ρ can be determined in advance, e.g., prior to deployment. Also, adaptations could be provided to make the presented algorithm independent of ρ.

Further, for applications in which clients may connect to facilities that lie more than one hop away, both algorithms are not efficient, as each round requires "global" communication between eligible clients and facilities.

However, the presented approaches can well be adapted to these multi-hop settings: Notice that the main idea of all presented algorithms is to open –

either sequentially or in parallel – the facilities $\{i\}$ that span the most *cost-efficient* stars $\{(i, B)\}$. We noted above that in spite of an exponential number of sets of uncovered clients $B \subseteq U$, a facility can find such a star by considering the clients ordered by the respective connection costs $_{ij}$ to facility i. In most multi-hop models, the clients are already ordered by $_{ij}$, as $_{ij}$ will reflect a measure of network proximity between i and . That is, a facility can consider clients in the order of their network proximity to itself. As a consequence, the clients in a least-cost star of a given size, will always be in a local network neighborhood around a potential facility i, as sketched in Figure 8.8.

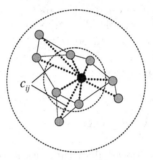

Fig. 8.8. A star in the multi-hop model

Similarly to the *efficiency threshold* – which determines which stars are connected in a given step of the fast distributed algorithm – the *scope* around each potential facility i could be increased, thus raising the connection variables $_{ij}$ at close clients at first and at more remote clients later.

This should enable approximating different variants of the facility location problem even in the multi-hop model. For example, one could stop the algorithm after a given step k and solve facility location for the case that clients and facilities can be at most $scope(k)$ apart. Moreover, if the scope is doubled at each step, one may obtain an approximation for arbitrary distances between facilities and clients.

Furthermore, some wireless sensor network applications may only require a special case of UFL, for which better guarantees can be given. We have mentioned that most instances in multi-hop networks will be metric. Further, the costs of executing a service at a given node – constituting the opening costs of a facility – could be assumed to be known and equal at all nodes. Similarly, a fault tolerant version of UFL could be applied, in which each client   must be connected to at least r_j facilities. The fault-tolerant case is also made easier, if the redundancy requirement r_j is the same at all nodes.

8.7 Chapter Notes

We close with a few notes on further reading. Thorup [360] describes a method for improving the centralized runtime both of the presented Algorithm 21 and of an advanced version [196] that provides a 1.61 approximation factor. Chudak et al. [73] have discussed how the centralized algorithm of Section 8.3.5 [198] can be adapted to obtain a fast distributed algorithm which solves a constrained version of the problem requiring facilities and associated clients to be at most three network hops apart. Recent work [158] discusses a constant-time distributed algorithm for a constrained version of the problem in which opening costs at all nodes are equal. A distributed algorithm based on hill-climbing [228] addresses a version of the problem in which exactly k facilities are opened. Although in [228] the worst-case time complexity and the obtained approximation factor are not discussed explicitly, the approach is expected to adapt well to instances which change over time. Finally, in current work [136], we give more involved versions of Algorithms 22 and 23, which can be implemented in multi-hop networks and inherit the approximation factor of 1.61 from [196]. Although in theory requiring a high worst-case runtime, in practice these re-formulations terminate in few communication rounds and require only local communication confined to small network neighborhoods.

9

Geographic Routing

Aaron Zollinger

9.1 Introduction

The one type of routing in ad hoc and sensor networks that currently appears to be most amenable to algorithmic analysis is geographic routing. This chapter contains an introduction to the problem field of geographic routing, presents a specific routing algorithm based on a synthesis of the greedy forwarding and face routing approaches, and provides an algorithmic analysis of the presented algorithm from both a worst-case and an average-case perspective.

Routing in a communication network is the process of forwarding a message from a source host to a destination host via intermediate nodes. In wired networks, routing is commonly a task performed by routers, special fail-safe network hosts particularly designed for the purpose of forwarding messages with high performance. In ideal wireless ad hoc networks, in contrast, every network node may act as a router, as a relay node forwarding a message on its way from its source node to its destination node. This process is particularly important in ad hoc networks, as network nodes are assumed to have restricted power resources and therefore try to transmit messages at low transmission power, leading to the effect that the destination of a message can typically not be reached directly from the source. The importance of this task also becomes manifest in the popular term *multihop routing*, expressing the essential role of network nodes as relay stations.

In wired networks, routing almost always takes place in relatively stable conditions; at least the main neighborhood topology remains identical over weeks, months, or even years. The primary focus of routing in wired networks is on high-performance forwarding of messages; reaction latency in the face of network topology changes, caused by failing hosts or connections, is generally of secondary importance. Considering the stability of wired networks, prompt reaction to topology changes or rapid propagation of according information is often not required, as such events are relatively rare.

D. Wagner and R. Wattenhofer (Eds.): Algorithms for Sensor and Ad Hoc Networks, LNCS 4621, pp. 161–185, 2007.

Wireless ad hoc networks are of a fundamentally different character: To begin with, wireless connections are by nature significantly less stable than wired connections. Effects influencing the propagation of radio signals, such as shielding, reflection, scattering, and interference, inevitably require routing systems in ad hoc networks to be able to cope with comparatively low link communication reliability. More importantly, many scenarios for ad hoc networks assume that nodes are potentially mobile. These two factors, above all in high node mobility, cause ad hoc networks to be inherently more dynamic than wired networks. Traditional routing protocols designed for wired networks therefore generally fail to satisfy the requirements of wireless ad hoc networks.

A considerable number of routing protocols specifically devised for operation in ad hoc networks have consequently been invented. These protocols are usually classified into two groups: *proactive* and *reactive* routing protocols.

Proactive routing protocols resemble protocols for wired networks in that they collect routing information ahead of time. A request for a message to be routed can be serviced without any further preparative actions. As every node keeps a table specifying how to forward a message, information on topology changes is propagated whenever they occur. Similar to routing protocols in wired networks, proactive routing protocols are efficient only if links are stable and node mobility is low compared to the rate of communication traffic. Already if node mobility reaches a reasonable degree, the routing overhead incurred by table update messages can become unacceptably high. Another question is whether lightweight ad hoc network nodes with scarce resources can be expected to maintain routing tables potentially for all possible destinations in the network.

Reactive routing protocols, on the other hand, try to delay any preparatory actions as long as possible. Routing occurs on demand, only. In principle, a node wishing to send a message has to flood the network in order to find the destination. Although there are many tricks to restrict flooding or to cache information overheard by nodes, flooding can consume a considerable portion of the network bandwidth. Attempting to combine the advantages of both concepts, proposals have also been made to incorporate both approaches in hybrid protocols that adapt to current network conditions.

Most of these routing protocols have been described and studied from a system-centric point of view. Simulation appears to be the preferred method of assessment. It appears, however, that a global evaluation of protocols is difficult. Ad hoc networks have many parameters, such as transmission power, signal attenuation, interference, physical obstacles, node density and distribution, degree and type of node mobility, just to mention a few. Therefore simulation cannot cover all the degrees of freedom. For a given set of parameters, certain protocols appear superior to others; for other parameters, the ranking may be reversed. One possible answer to this problem may be found in trying to rigorously analyze the efficiency of proposed protocols and algorithms. However, analyzing the complexity of ad hoc routing algorithms appears to be not only intricate, but virtually impossible. Accordingly, only few attempts have been made to analyze ad hoc routing in a general setting from an algorithmic perspective.

One specific type of ad hoc routing, in contrast, appears to be more easily accessible to algorithmic analysis: geographic routing. Geographic routing, sometimes also called directional, geometric, location-based, or position-based routing, is based on two principal assumptions. First, it is assumed that every node knows its own and its network neighbors' positions. Second, the source of a message is assumed to be informed about the position of the destination. The former assumption is currently becoming more and more realistic with the advent of inexpensive and miniaturized positioning systems. It is also conceivable that position information could be attained by local computation and message exchange with stationary devices. In order to come up to the latter assumption, that is to provide the source of a message with the destination position, several so-called location services have been proposed [1, 169, 253]. For some scenarios it can also be sufficient to reach *any* destination currently located in a given area, sometimes called "geocasting". These are only briefly summarized explanations why the two basic assumptions of geographic routing are reasonable. This issue is discussed in more depth for instance in [408, Chapter 11].

Geographic routing is particularly interesting, as it operates without any routing tables whatsoever. Furthermore, once the position of the destination is known, all operations are strictly local, that is, every node is required to keep track only of its direct neighbors. These two factors—absence of necessity to keep routing tables up to date and independence of remotely occurring topology changes—are among the foremost reasons why geographic routing is exceptionally suitable for operation in ad hoc networks. Furthermore, in a sense, geographic routing can be considered a lean version of source routing appropriate for dynamic networks: While in source routing the complete hop-by-hop route to be followed by the message is specified by the source, in geographic routing the source simply addresses the message with the position of the destination. As the destination can generally be expected to move slowly compared to the frequency of topology changes between the source and the destination, it makes sense to keep track of the position of the destination instead of maintaining network topology information up to date; if the destination does not move too fast, the message is delivered regardless of possible topology changes among intermediate nodes. Finally, from a less technical perspective, it can be hoped that by studying geographic routing it is possible to gain insights into routing in ad hoc networks in general, without availability of position information.

We will start our analysis of geographic routing by describing a simple greedy routing approach in Section 9.4. The main drawback of this approach is that it cannot guarantee to always reach the destination. Geographic routing algorithms that, in contrast, always reach the destination, are based on faces, contiguous regions separated by the edges of planar network subgraphs.

It may however happen that these algorithms take $\Omega(n)$ steps before arriving at the destination, where n is the number of network nodes. In other words, they basically do not perform better than an algorithm visiting every node in the network. In Section 9.5 we will describe the concept of face routing and describe algorithms that not only always find the destination, but are also guaranteed to do so with cost at most $O(c^2)$, where c is the cost of a shortest path connecting the source and the destination. The next section will show that, given an instance of a class of lower bound graphs, no geographic routing algorithm will be able to perform better; in this sense, the presented face routing algorithms are asymptotically optimal in worst-case networks. Despite their asymptotic optimality, these algorithms are relatively inflexible in that they follow the boundaries of faces also in dense average-case networks where greedy routing would reach the destination much faster. Section 9.7 will outline how greedy routing and face routing can be combined, resulting in the GOAFR$^+$ algorithm, which preserves the worst-case guarantees of its face routing components. In addition, simulations, as mentioned in the same section, showed that the GOAFR$^+$ algorithm is—to the best of our knowledge—the currently most efficient geographic routing algorithm also in average-case networks. GOAFR$^+$ particularly outperforms other routing algorithms in a critical node density range, where the network is just about to become connected; networks in this density range form a challenge to any routing algorithm, also non-geographic routing algorithms.

9.2 Related Work

As mentioned earlier, routing protocols for ad hoc networks can be classified as proactive and reactive protocols. Proactive protocols, such as DSDV [318], TBRPF [306], and OLSR [76], distribute routing information ahead of time in order to be able to react immediately whenever a message needs to be forwarded. On the other hand, reactive protocols, such as AODV [319], DSR [203], or TORA [309] do not try to anticipate communication and initiate route discovery as late as possible, as a reaction to a message requested to be routed. As the performance and incurred routing overhead of such protocols highly depend on the type and extent of network mobility, also hybrid protocols, such as [201, 304, 327], have been proposed. Further reviews of routing algorithms in mobile ad hoc networks in general can be found in [51] and [337]. Most of these protocols have been described and studied from a system perspective; performance and efficiency assessment was commonly carried out by means of simulation. To date, only few attempts have been made to analyze routing in ad hoc networks in a general setting from an analytical algorithmic perspective [56, 142, 321].

The early proposals of geographic routing—suggested over a decade ago—were of purely greedy nature: At each intermediate network node the message to be routed is forwarded to the neighbor closest to the destination [128,

186, 355]. This can however fail if the message reaches a local minimum with respect to the distance to the destination, that is a node without any "better" neighbors. Also a "least deviation angle" approach (*Compass Routing* in [225]) cannot guarantee message delivery in all cases.

The first geographic routing algorithm that does guarantee delivery was *Face Routing* introduced in [225] (called *Compass Routing II* there). *Face Routing* walks along faces of planar graphs and proceeds along the line connecting the source and the destination. Besides guaranteeing to reach the destination, it does so with $O(n)$ messages, where n is the number of network nodes. However, this is unsatisfactory, since also a simple flooding algorithm will reach the destination with $O(n)$ messages. Additionally it would be desirable to see the algorithm cost depend on the distance between the source and the destination.

There have been later suggestions for algorithms with guaranteed message delivery [45, 97]; at least in the worst case, however, none of them outperforms original Face Routing. Yet other geographic routing algorithms have been shown to reach the destination on special planar graphs without any runtime guarantees [43]. [44] proposed an algorithm competitive with the shortest path between source and destination on Delaunay triangulations; this is however not applicable to ad hoc networks, as Delaunay triangulations may contain arbitrarily long edges, whereas transmission ranges in ad hoc networks are limited. Accordingly, [148] proposed local approximation of the Delaunay graph, however without improving performance bounds for routing. A more detailed overview of geographic routing can be found in [369].

In [241] *Adaptive Face Routing* AFR was proposed. The execution cost of this algorithm—basically enhancing Face Routing by the employment of an ellipse restricting the searchable area—is bounded by the cost of the optimal route. In particular, the cost of AFR is not greater than the squared cost of the optimal route. It was also shown that this is the worst-case optimal result any geographic routing algorithm can achieve.

Face Routing and also AFR are not applicable for practical purposes due to their strict employment of face traversal. There have been proposals for practical purposes to combine greedy routing with face routing [45, 97, 211], however without competitive worst-case guarantees. In [243] the GOAFR algorithm was introduced; to the best of our knowledge, this was the first algorithm to combine greedy and face routing in a worst-case optimal way; the GOAFR+ algorithm [240] remains asymptotically worst-case optimal while improving GOAFR's average-case efficiency by employing a counter technique for falling back as soon as possible from face to greedy routing.

Lately, first experiences with geographic and in particular face routing in practical networks have been made [215, 216]. More specifically, problems in connection with graph planarization that can occur in practice were observed, documented, and tackled. Another strand of research approaches these issues by allowing the routing algorithm to store certain limited information in the network nodes [251, 252].

The results in this chapter partly rely on the $\Omega(1)$-model, the assumption that the distance between any pair of network nodes is at least a (possibly small) constant. In [240] it was shown that equivalently a clustering technique can be employed for graphs that do not comply with the $\Omega(1)$-model assumption. Clustering for the purpose of ad hoc routing has been proposed by various researchers [71, 226]. A closely related approach is the construction of *dominating sets* (cf. Chapter 3 and [12, 30, 125, 147, 148, 172, 188, 200, 231, 238, 296, 397]), for instance for employment as routing backbones.

In the context of routing in ad hoc and sensor networks the assumption is commonly made that—if the network nodes know their own positions—they are also informed about their neighbors' positions practically for free, that is by local exchange of according messages. [40, 140, 182, 409] studied what can be done without such regularly exchanged beacon messages.

The question what is possible if no position information at all is available to the network nodes was addressed in [235, 290] by computation of virtual node coordinates and in [389] with the focus on geographic routing.

9.3 Models and Preliminaries

At the beginning of every theoretical analysis stands the question of how to model the considered system. An obvious abstraction of a communication network is a graph with nodes representing networking devices and edges standing for network connections. The study of ad hoc networks in this chapter assumes that network nodes are placed in the Euclidean plane. We furthermore model ad hoc networks as unit disk graphs [75]. A unit disk graph (UDG) is defined as follows:

Definition 9.3.1. (Unit Disk Graph) *Let $V \subset \mathbb{R}^2$ be a set of points in the 2-dimensional plane. The graph with edges between all nodes with distance at most 1 is called the unit disk graph of V.*

Accordingly, a unit disk graph models a flat environment with network devices equipped with wireless radio, all having equal transmission ranges. Edges in the UDG correspond to radio devices positioned in direct mutual communication range. Clearly, the unit disk graph model forms a highly idealistic abstraction of ad hoc networks. Nevertheless it admits certain insights based on algorithmic analysis. Discussions of routing in a model that more closely captures the connectivity characteristics of wireless networks can be found in [31, 242].

To measure the quality of a routing algorithm, we attribute to each edge e a cost which is a function of the Euclidean length of e.

Definition 9.3.2. (Cost Function) *A cost function $c\colon\,]0,1] \mapsto \mathbb{R}^+$ is a nondecreasing function which maps any possible edge length d $(0 < d \le 1)$ to a positive real value $c(d)$ such that $d' > d \implies c(d') \ge c(d)$. For the cost of an edge $e \in E$ we also use the shorter form $c(e) := c(d(e))$.*

Note that $]0,1]$ really is the domain of a cost function $c(\cdot)$, that is, $c(\cdot)$ has to be defined for all values in this interval and in particular, $c(1) < \infty$. The cost model thus defined includes all popular cost measures such as the link (or hop) distance metric $(c_\ell(d) :\equiv 1)$, the Euclidean distance metric $(c_d(d) := d)$, energy $(c_E(d) := d^2$, or more generally d^α for $\alpha \geq 2)$, as well as hybrid measures which are positive linear combinations of the above metrics.

For convenience we also define the cost of a path, a sequence of contiguous edges, and of algorithms. The cost $c(p)$ of a path p is defined as the sum of the cost values of its edges. Analogously, the cost $c(\mathcal{A})$ of an algorithm $\mathcal{A}$ is defined as the summed-up cost of all edges which are traversed during the execution of an algorithm on a particular graph. The question whether a node can send a message to several neighbors simultaneously does not affect our results, as the considered algorithms do not send messages in parallel to more than one recipient.

For the sake of simplicity we assume that the distance between any two nodes may not be arbitrarily small:

Definition 9.3.3. ($\Omega(1)$-model) *If the distance between any two nodes is bounded from below by a term of order $\Omega(1)$, i.e. there is a positive constant d_0 such that d_0 is a lower bound on the distance between any two nodes, this is referred to as the $\Omega(1)$-model.*

Graphs with this restriction have also been called *civilized* [104] or λ-*precision* [192] graphs in the literature. As a consequence of the $\Omega(1)$-model, the above-mentioned three metrics are equivalent up to a constant factor with respect to the cost of a path. As shown in the following lemma, this holds for all metrics defined according to Definition 9.3.2.

Lemma 9.3.4. *Let $c_1(\cdot)$ and $c_2(\cdot)$ be cost functions according to Definition 9.3.2 and let G be a unit disk graph in the $\Omega(1)$-model. Further let p be a path in G. We then have*

$$c_1(p) \leq \alpha \cdot c_2(p)$$

for a constant α.

Proof. Assume without loss of generality that p consists of k edges, that is, $c_\ell(p) = k$. As $d_0 \leq c_d(e) \leq 1$ for all edges $e \in E$ and the cost functions being nondecreasing, we have $c_1(p) \leq c_1(1) \cdot k$ and $c_2(d_0) \cdot k \leq c_2(p)$. Since—according to Definition 9.3.2—both $c_1(1)$ and $c_2(d_0)$ are constants greater than 0, the lemma holds with $\alpha = c_1(1)/c_2(d_0)$. $\qquad\square$

Also the distance in a graph of a pair of nodes u and v—defined to be the cost of the shortest path connecting u and v—differs only by a constant factor for the different cost metrics:

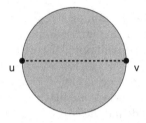

Fig. 9.1. An edge (u, v) in the Gabriel graph exists if and only if the shaded disk (including its boundary) does not contain any third node.

Lemma 9.3.5. *Let G be a unit disk graph with node set V in the $\Omega(1)$-model. Further let $s \in V$ and $t \in V$ be two nodes and let p_1^* and p_2^* be optimal paths from s to t on G with respect to the metrics induced by the cost functions $c_1(\cdot)$ and $c_2(\cdot)$, respectively. It then holds that*

$$c_1(p_2^*) \leq \alpha \cdot c_1(p_1^*) \text{ and } c_1(p_2^*) \geq \beta \cdot c_1(p_1^*)$$

for two constants α and β, that is, the cost values of optimal paths for different metrics only differ by a constant factor.

Proof. By the optimality of p_2^* we have

$$c_2(p_2^*) \leq c_2(p_1^*). \tag{9.1}$$

Applying Lemma 9.3.4 we obtain

$$c_1(p_2^*) \leq \gamma \cdot c_2(p_2^*) \text{ and } c_2(p_1^*) \leq \delta \cdot c_1(p_1^*) \tag{9.2}$$

for two constants γ and δ. Combining Equations (9.1) and (9.2) yields $c_1(p_2^*) \leq \alpha \cdot c_1(p_1^*)$ for $\alpha = \gamma \cdot \delta$. Furthermore, by the optimality of p_1^*, we have $c_1(p_2^*) \geq c_1(p_1^*)$, and therefore the second equation of the lemma holds with $\beta = 1$. $\square$

As this equivalence of cost metrics applies not only to the link, the Euclidean, and the energy metrics, but to all cost functions according to Definition 9.3.2, we sometimes refer to the "cost" of an edge and mean any cost metric belonging to the above class of cost functions. In [240] it was shown that employing clustering techniques a similar result can be achieved without the $\Omega(1)$-model assumption. [240] also describes the existence of two classes of cost functions and discusses their implications on routing. In this chapter we will however adhere to the $\Omega(1)$-model for simplicity.

For our routing algorithms the network graph is required to be *planar*, that is without intersecting edges.[8] A planar graph features *faces*, contiguous regions separated by the edges of the graph. In order to achieve planarity

---

[8] More precisely, the considered planar graphs are planar *embeddings* in the Euclidean plane.

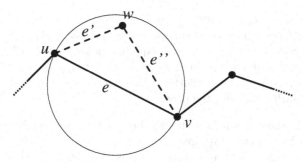

Fig. 9.2. The Gabriel graph contains an energy-optimal path.

on the unit disk graph G, the *Gabriel graph* is employed. A Gabriel graph contains an edge between two nodes u and v if and only if the disk (including its boundary) having $\overline{uv}$ as a diameter does not contain a "witness" node (cf. Figure 9.1). Besides being planar, GG —the Gabriel graph on the unit disk graph G—features two important properties:

- It can be computed locally: A network node can determine all its incident nodes in GG by mere inspection of its neighbors' locations (since G is a unit disk graph).
- The Gabriel graph is a constant-stretch spanner for the energy metric: The construction of the Gabriel graph on G preserves an energy-minimal path between any pair of network nodes. Together with the $\Omega(1)$-model it follows that the distance in GG between any pair of nodes is equal (up to constant factors) to their distance in G for all considered metrics. This is shown in the following lemma.

Lemma 9.3.6. *In the $\Omega(1)$-model the shortest path for any of the metrics according to Definition 9.3.2 on the Gabriel graph intersected with the unit disk graph is only by a constant factor longer than the shortest path on the unit disk graph for the respective metric.*

Proof. We first show that at least one best path with respect to the energy metric on the UDG is also contained in $GG \cap UDG$. Suppose that $e = (u, v)$ is an edge of an energy-optimal path on the UDG. For the sake of contradiction suppose that e is not contained in $GG \cap UDG$. Then there is a node in or on the circle with diameter $\overline{uv}$ (see Figure 9.2). The edges $e' = (u, \ )$ and $e'' = (v, \ )$ are also edges of the UDG and because lies in the described circle, we have $(e')^2 + (e'')^2 \leq (e)^2$. If is inside the circle with diameter $\overline{uv}$, the energy for the path $' := \setminus \{e\} \cup \{e', e''\}$ is smaller than the energy for , and is therefore not an energy-optimal path, contradicting the above assumption. If lies exactly on the above circle, $'$ is an energy-optimal path as well and the argument applies recursively.

According to the optimality of $p^*_{GG\cap UDG}$—defined to be a shortest path on $GG\cap UDG$ with respect to a cost function $c(\cdot)$—we have $c(p^*_{GG\cap UDG}) \leq c(p) \leq \alpha \cdot c_E(p)$ for a constant α, the last inequality holding due to Lemma 9.3.4. Employing Lemma 9.3.5, we furthermore obtain $c_E(p) \leq \beta \cdot c(p^*_{UDG})$ for a constant β, where p^*_{UDG} is a shortest path with respect to $c(\cdot)$ on the unit disk graph, which concludes the proof. $\square$

Unless stated otherwise, it is assumed that every node locally computes its neighbors in the Gabriel graph prior to the start of routing algorithms.

The geographic ad hoc routing algorithms we consider in this chapter can be defined as follows.

Definition 9.3.7. (Geographic Ad Hoc Routing Algorithm) *Let $G = (V, E)$ be a Euclidean graph. The task of a geographic ad hoc routing algorithm $\mathcal{A}$ is to transmit a message from a source $s \in V$ to a destination $t \in V$ by sending packets over the edges of G while complying with the following conditions:*

- *All nodes $v \in V$ know their geographic positions as well as the geographic positions of all their neighbors in G.*
- *The source s is informed about the position of the destination t.*
- *The control information which can be stored in a packet is limited by $O(\log n)$ bits, that is, only information about a constant number of nodes is allowed.*
- *Except for the temporary storage of packets before forwarding, a node is not allowed to maintain any information.*

In the literature, geographic ad hoc routing has been given various other names, such as $O(1)$-memory routing algorithms in [43, 44], local routing algorithms in [225], geometric, position-based, or location-based routing. Due to these storage restrictions, geographic ad hoc routing algorithms are inherently local. In particular, nodes do not store any routing tables, eliminating a possible source of outdated information.

Finally, it is assumed that routing takes place much faster than node movement: A routing algorithm is modeled to run on temporarily stationary nodes. The issues faced when easing or giving up this assumption were discussed in [408, Chapter 12].

9.4 Greedy Routing

The probably most straightforward approach to geographic routing—which has also been studied as the first type of geographic routing algorithms in the related work—is *greedy forwarding*: Every node relays the message to be routed to its neighbor located "best" with respect to the destination. If "best" is interpreted as "closest to the destination", greedy forwarding can be formulated as follows:

Greedy Routing GR

0. Start at s.
1. Proceed to the neighbor closest to t.
2. Repeat step 1 until either reaching t or a local minimum with respect to the distance from t, that is a node v without any neighbor closer to t than v itself.

This formulation clearly reflects the simplicity of such an approach with respect to both concept and implementation. However, as indicated in Step 2 of the algorithm, it shows a big drawback: It is possible that the message runs into a "dead end", a node without any "better" neighbor. If backtracking techniques can overcome local minima in some cases, they fail to serve as a general solution to this problem, especially together with the strict message size limitations imposed on geographic routing (cf. Definition 9.3.7). Also alternative interpretations of "best neighbor" fail to reach the destination; in a "least deviation angle" approach for instance the message can end up in an infinite path loop [225].

If greedy routing however reaches the destination, it generally does so efficiently. Informally, this is due to the fact that—except in degenerate cases— the message stays relatively close to the line connecting the source and the destination. As discussed later in Section 9.7, employment of greedy routing whenever possible is beneficial above all in densely populated average-case networks. But also in worst-case networks the cost expended by greedy routing cannot become arbitrarily high:

Lemma 9.4.1. *If GR reaches t, it does so with cost $O(d^2)$, where $d := |\overline{st}|$ denotes the Euclidean distance between s and t.*

Proof. A detailed proof of this lemma can be found in [148]. For completeness we just give a proof sketch. Let $p := v_1, \ldots, v_k$ be the sequence of nodes visited during greedy routing. According to the definition of greedy routing, no two nodes v_i, v_j with odd indices i, j are neighbors. Further, since the distance to t is decreasing along the path p, all nodes v_i are inside $D(t, d)$, the disk with center t and radius d. $D(t, d)$ contains at most $O(d^2)$ nodes with pairwise distance at least 1. It follows that p consists of $O(d^2)$ nodes. □

In the following sections, greedy routing will be employed as a routing algorithm component for its efficiency in both worst-case and average-case networks.

9.5 Routing with Faces

In the previous section we observed that greedy routing is not guaranteed to always reach the destination. This section introduces a type of geographic routing that, in contrast, always finds the destination if the network contains a connection from the source: routing based on faces.

9.5.1 Face Routing

The first geographic routing algorithm shown to always reach the destination was Face Routing introduced in [225]. Although we will formally describe a variant of Face Routing slightly adapted for our purposes, we will now give a brief overview of the original Face Routing algorithm.

At the heart of Face Routing lies the concept of faces, contiguous regions separated by the edges of a planar graph, that is a graph containing no two intersecting edges. The algorithm proceeds by exploration of face boundaries employing the local *right-hand rule* in analogy to following the right-hand wall in a maze (cf. Figure 9.3). On its way around a face, the algorithm keeps track of the points where it crosses the line $\overline{st}$ connecting the source s and the destination t. Having completely surrounded a face, the algorithm returns to the one of these intersections lying closest to the destination. From here, it proceeds by exploring the next face closer to t. If the source and the destination are connected, Face Routing always finds a path to the destination. It thereby takes at most $O(n)$ steps, where n is the total number of nodes in the network.

9.5.2 AFR

Where the Face Routing algorithm can take up to $O(n)$ steps to reach the destination irrespective of the actual distance between the source and the destination in the given network, the main contribution of the Adaptive Face Routing algorithm AFR—as presented in [241]—consists in limiting the expended cost with respect to the length of the shortest path between s and t. Although the results discussed in the subsequent sections of this chapter go beyond AFR, we will first provide a summary of this algorithm for completeness and continue to give an overview of the employed technique.

As mentioned, the main problem with respect to the performance of Face Routing lies in the necessity of exploring the *complete* boundary of faces. It is thus impossible to bound the cost of this algorithm by the cost of an optimal path between s and t. If, however, we know the length of an optimal path connecting the source and the destination, Face Routing can be extended to *Bounded Face Routing BFR*: The exploration of faces is restricted to a searchable area, in particular an ellipse whose size is chosen such that it contains a complete optimal path. If the algorithm hits the ellipse, it has to "turn back" and continue its exploration of the current face in the opposite direction until hitting the ellipse for the second time, which completes the exploration of the current face. Briefly put—the details will be explained later—, since BFR does not traverse an edge more than a constant number of times, and since the bounding ellipse (together with the $\Omega(1)$-model and graph planarity) does not contain more than $O\big(|\overline{st}|^2\big)$ edges, the cost of BFR is in $O\big(c^2(p^*)\big)$, where p^* is an optimal path connecting s and t.

In most cases, however, a prediction of the length of an optimal path will not be possible. The solution to this problem finally leads to *Adaptive Face*

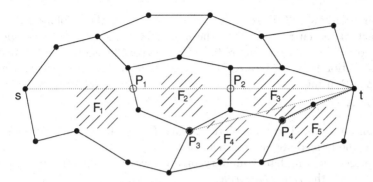

Fig. 9.3. Face Routing starts at s, explores face F_1, finds P_1 on $\overline{st}$, explores F_2, finds P_2, and switches to F_3 before reaching t. OFR, in contrast, finds P_3, the point on F_1's boundary closest to t, continues to explore F_4, where it finds P_4, and finally reaches t via F_5.

Routing AFR: BFR is started with the ellipse size set to an initial estimate of the optimal path length. If BFR fails to reach the destination, which will be reported to the source, BFR will be restarted with a bounding ellipse of doubled size. (It is also possible to double the ellipse size directly without returning to the source.) If s and are connected, AFR will eventually find a path to . This iteration is asymptotically dominated by the cost of the algorithm steps performed in the last ellipse, whose area is at most proportional to the squared cost of an optimal path. Consequently, also the cost of AFR is bounded by $O(\ ^2(\ ^*))$.

Section 9.6 will show that in a lower-bound graph no local geographic routing algorithm can perform better: AFR is asymptotically optimal.

9.5.3 OAFR

As described in Section 9.4, greedy routing promises to find the destination with low cost in all cases where it arrives at the destination. A natural approach to leveraging the potential of greedy routing above all for practical purposes therefore consists in combining greedy routing and face routing. In a first attempt Greedy Routing and AFR can be literally combined: Proceed in a greedy manner and use AFR to escape from potential local minima. It has however been shown in [243] that, employing greedy routing, this algorithm loses AFR's asymptotic optimality. Nevertheless a variant of AFR—named OAFR—was found whose combination with greedy routing does finally yield an algorithm that is both average-case efficient and asymptotically optimal.

Similarly to the above description of AFR, we will explain the OAFR algorithm in three steps: OFR, OBFR, and OAFR.

Other Face Routing OFR differs from Face Routing in the following way: Instead of changing to the next face at the "best" *intersection* of the face boundary with $\overline{st}$, OFR returns—after completing the exploration of the boundary of the current face—to the boundary point (or one of the points) closest to the destination (Figure 9.3). Conserving the headway made towards the destination on each face, OFR in a sense uses a more natural approach than Face Routing.

Other Face Routing OFR

0. Begin at s and start to explore the face F containing the connecting line $\overline{st}$ in the immediate environment of s.
1. Explore the complete boundary of the face F based on local decisions employing the *right-hand rule*.
2. Having accomplished F's exploration, advance to the point p closest to t on F's boundary. Switch to the face containing $\overline{pt}$ in p's environment and continue with step 1. Repeat these two steps until reaching t.

The number of steps taken by OFR is bounded as shown in the following lemma:

Lemma 9.5.1. *OFR always terminates in* $O(n)$ *steps, where n is the number of nodes. If s and t are connected, OFR reaches t; otherwise, disconnection will be detected.*

Proof. Let $F_1, F_2, \ldots, F_k$ be the sequence of the faces visited during the execution of OFR. We will first assume s and t to be connected. Since the switch between two faces always happens at the point on the face boundary closest to t and because the next face is chosen such that it always contains points which are nearer to t, no face is visited twice. Let further $p_0, p_1, p_2, \ldots, p_t$ be the *trace* of OFR's execution, where $p_i, i \geq 1$ is the point with minimum distance from t on the boundary of F_i. Because no face is visited more than once, we have that $\forall i > j : |\overline{p_i t}| < |\overline{p_j t}|$. Hence, if s and t are connected, we eventually arrive at a face with t on its boundary. (Otherwise, there is an i for which $p_i = p_{i+1}$, which means that the graph is disconnected.)

Since each face is explored at most once, each edge is visited at most four times. As every planar graph corresponds to the projection of a polyhedron on the plane, Euler's polyhedron formula can be employed: $n - m + f = 2$, where n, m, and f stand for the number of nodes, edges, and faces in the graph, respectively. Furthermore, the observations that (for $n > 3$) every face is delimited by at least three edges and that each edge is adjacent to at most two faces yield $3f \leq 2m$. Using Euler's formula we have $3m - 3n + 6 = 3f \leq 2m$ and therefore $m \leq 3n - 6$. Thus, OFR terminates after $O(n)$ steps.   □

If the algorithm detects graph disconnection (finding $p_i = p_{i+1}$ for some $i \geq 0$), this can be reported to the source by again using OFR in the reverse direction.

Remark (Gabriel graph). When applying OFR on a Gabriel graph—as we will do for the routing on unit disk graphs—OFR can be simplified in the following way: Instead of changing faces at the *point* on the face boundary

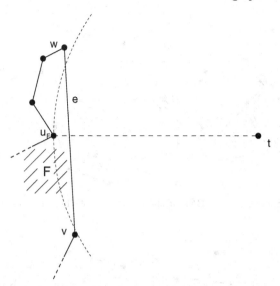

Fig. 9.4. Also OFR modified to switch to the next face at the *node* closest to t (instead of the *point* closest to t) progresses with each face switch if it runs on the Gabriel graph. In particular it cannot happen that the algorithm is caught in an infinite loop: Having arrived at u_F, the node of face F located closest to t, the algorithm always switches to a face other than F. The only possible way of constructing a counterexample fails: A constellation forcing the modified OFR algorithm to again select F as the next face at u_F has at least one edge $e = (v, w)$ on F's boundary intersecting the line segment $\overline{u_F t}$; otherwise t would lie inside F, implying that s and t would be disconnected. The fact that both v and w are not closer to t than u_F—u_F is the node on F's boundary closest to t—implies that at least one of u_F and t are located within the disk with diameter $\overline{vw}$, which contradicts the existence of the edge e in the Gabriel graph.

which is closest to , it is possible to take the *node* which is closest to . This modification leaves the property described in Lemma 9.5.1 unchanged, as also the modified OFR algorithm always switches to a new face in Step 2 if it is run on the Gabriel graph. This is illustrated in Figure 9.4. Since definitions and explanations become clearer, we will use this modified form of the OFR algorithm for the description of the subsequent algorithms. Equivalent results can be achieved with the original version of the algorithm.

When trying to formulate a statement on OFR's cost, the main problem arising is its traversal of complete boundaries of faces: Informally put, OFR can meet an incredibly big face whose total exploration is prohibitively expensive compared to an optimal path from s to . In order to solve this, AFR's trick to bound the searchable area by an ellipse containing an optimal path can be borrowed. Consequently we obtain *Other Bounded Face Routing OBFR*.

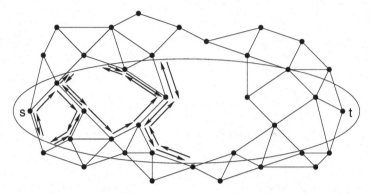

Fig. 9.5. Execution of OBFR if the ellipse is not chosen sufficiently large.

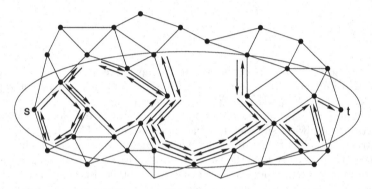

Fig. 9.6. Execution of OBFR if the ellipse is chosen large enough to contain a path from s to t.

For the sake of simplicity we assume for the following description of OBFR that s and are connected. If ˜ is an estimate of the Euclidean length of a shortest path between s and , let be the ellipse with foci s and and with the length of the major axis being ˜ (in other words, contains all paths from s to of Euclidean length at most ˜).

Other Bounded Face Routing OBFR

0. Step 0 of OFR.
1. Step 1 of OFR, but do not leave : When hitting , continue the exploration of the current face in the opposite direction. 's exploration will afterwards be complete when hitting for the second time.
2. Step 2 of OFR with one modification: If the node closest to on 's boundary is the same one as in the previous iteration, that is, no progress has been made in Step 1, report failure back to s by means of OBFR.

Figures 9.5 and 9.6 illustrate the execution of OBFR if the ellipse is chosen too small and if the ellipse contains a path from s to t, respectively. The cost expended by OBFR can be bounded as follows:

Lemma 9.5.2. *If the length $\tilde{c}$ of the major axis of $\mathcal{E}$ is at least the length of a—with respect to the Euclidean metric—shortest path between s and t, OBFR reaches the destination. Otherwise OBFR reports failure to the source. In any case, OBFR expends cost at most $O(\tilde{c}^2)$.*

Proof. We first assume that $\tilde{c}$ is at least the length of a shortest (Euclidean) path p^*, that is, p^* is completely contained in $\mathcal{E}$. Since OBFR stays within $\mathcal{E}$ while routing a message, we only look at the part of the graph which lies inside $\mathcal{E}$. We define the faces to be those contiguous regions which are separated by the edges of the graph and by the boundary of $\mathcal{E}$. (Hence, if a face is cut into several pieces by the boundary of $\mathcal{E}$, now each such piece is denoted a face.) Assume for the sake of contradiction that OBFR reports failure, that is, the algorithm does not make progress in Step 2. This is only possible if the currently traversed face boundary cuts the area enclosed by the ellipse into a region containing s and a second region containing t (cf. Figure 9.5). In this case however, $\mathcal{E}$ does not contain any path connecting s and t, which contradicts our assumption and therefore proves the first claim of the lemma.

If no path connects s and t within $\mathcal{E}$, a face boundary separating s from t as described in the previous paragraph exists. This is detected by OBFR, making no progress beyond a node v. As OBFR reached v starting from s, $\mathcal{E}$ contains a path from v to s and OBFR can be restarted in the opposite direction with the same ellipse, eventually reaching s and reporting failure.

Finally, it remains to be shown that the cost expended does not exceed $O(\tilde{c}^2)$. If $\mathcal{E}$ contains a path connecting s and t, every face is—for the same reasons as for OFR—visited at most once. Otherwise, every face is visited at most twice (the face where failure is detected can be visited an additional time). Furthermore, during the traversal of a face boundary, each edge can be visited at most four times. Consequently, any edge is traversed at most a constant number of times during the complete execution of OBFR.

Due to the planarity of the considered graph, the number of edges is linear in the number of nodes (cf. proof of Lemma 9.5.1). Furthermore—according to the employed $\Omega(1)$-model—the circles of radius $d_0/2$ around all nodes do not intersect each other. Since the length of the semimajor axis a of the ellipse $\mathcal{E}$ is $\tilde{c}/2$, and since the area of $\mathcal{E}$ is smaller than πa^2, the number of nodes n' inside $\mathcal{E}$ is bounded by

$$n' \leq \frac{\pi a^2}{\pi \left(\frac{d_0}{2}\right)^2} = \frac{\tilde{c}^2}{d_0^2} \in O(\tilde{c}^2).$$

Having thus found an upper bound for the number of messages sent, the last statement of the lemma follows with Lemma 9.3.6 for all cost metrics defined according to Definition 9.3.2. □

Note that the above specification of OBFR omits an important point for clarity of representation: The algorithm can distinguish between the case where the chosen ellipse is not sufficiently large to contain a path connecting s with t and the case where s and t are disconnected. As described above, the first case is detected if no progress is made after hitting the ellipse. Also in the second case there exists a node beyond which no progress is made; however, this node is detected as such without hitting the ellipse, that is, after traversing the complete boundary of the network component containing the source.

Since there is usually no a priori information on the optimal path length, initially—in analogy to AFR—a small estimate for the ellipse size is used and iteratively enlarged until reaching the destination.

Other Adaptive Face Routing OAFR

0. Initialize $\mathcal{E}$ to be the ellipse with foci s and t the length of whose major axis is $2 \cdot |\overline{st}|$.
1. Start OBFR with $\mathcal{E}$.
2. If the destination has not been reached, double the length of $\mathcal{E}$'s major axis and go to step 1.

Exploiting that OBFR is able to distinguish between insufficient ellipse size and graph disconnection between s and t, also OAFR detects graph disconnection. Furthermore OAFR's cost is bounded as follows:

Theorem 9.5.3. *If s and t are connected, OAFR reaches the destination with cost $O(c^2(p^*))$, where p^* is an optimal path. If s and t are disconnected, OAFR detects so and reports to s.*

Proof. We denote the first estimate $\tilde{c}$ on the optimal path length by $\tilde{c}_0$ and the consecutive estimates by $\tilde{c}_i := 2^i \tilde{c}_0$. Furthermore, we define k such that $\tilde{c}_{k-1} < c_d(p^*) \leq \tilde{c}_k$. For $c(\text{OBFR}[\tilde{c}])$, the cost of OBFR with the length of $\mathcal{E}$'s major axis set to $\tilde{c}$, we have $c(\text{OBFR}[\tilde{c}]) \in O(\tilde{c}^2)$ and therefore

$$c_\ell(\text{OBFR}[\tilde{c}]) \leq \lambda \cdot \tilde{c}^2$$

for a constant λ (and sufficiently large $\tilde{c}$). The total cost of OAFR can therefore be bounded by

$$c_\ell(\text{OAFR}) \leq \sum_{i=0}^{k} c_\ell(\text{OBFR}[\tilde{c}_i]) \leq \sum_{i=0}^{k} \lambda \left(2^i \tilde{c}_0\right)^2$$

$$= \lambda \tilde{c}_0^2 \frac{4^{k+1} - 1}{3} < \frac{16}{3} \lambda \left(2^{k-1} \tilde{c}_0\right)^2$$

$$< \frac{16}{3} \lambda \cdot c_d^2(p^*) \in O(c_d^2(p^*)).$$

$\square$

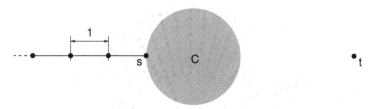

Fig. 9.7. If $n'/2$ nodes are located in the node cluster C (represented as a gray disk) and $n'/2$ nodes form the spike on the left, OAFR executes its component OBFR $\Theta(\log n')$ times, each OBFR execution having cost in $\Theta(n')$ if the nodes in C are placed in a maze-like structure, before detecting that s and t are disconnected.

Remark. It can be shown that for OAFR (and also AFR) the cost of detecting disconnection between s and   is bounded by $O(n' \log n')$, where n' is the number of nodes in the network component containing s: The number of OBFR executions is at most in $O(\log n')$, while the cost expended by OBFR in each of these executions is at most linear in n'. As illustrated in Figure 9.7, there exist graphs for which OAFR expends cost $\Theta(n' \log n')$.

9.6 A Lower Bound

As presented in the previous section, the OAFR algorithm reaches the destination with cost $O(\;^2)$, where   is the cost of the shortest path between the source and the destination. A natural question arising is whether this guarantee is good or if there are algorithms that can perform better. The following will give an answer to this question by showing that no geographic routing algorithm according to Definition 9.3.7 can find the destination with lower cost. In particular a constructive lower bound is given:

Theorem 9.6.1. *Let the cost of a best route for a given source-destination pair be . There exist graphs where any deterministic (randomized) geographic ad hoc routing algorithm has (expected) cost $\Omega(\;^2)$ for any cost metric according to Definition 9.3.2.*

Proof. A family of networks is constructed as follows. We are given a positive integer k and define a Euclidean graph G (see Figure 9.8): On a circle we evenly distribute $2k$ nodes such that the distance between two neighboring points is exactly 1; thus, the circle has radius r k/π. For every second node of the circle we construct a chain of $\lceil r/2 \rceil - 1$ nodes. The nodes of such a chain are arranged on a line pointing towards the center of the circle; the distance between two neighboring nodes of a chain is exactly 1. Node   is one arbitrary circle node with a chain: The chain of   consists of $\lceil r \rceil$ nodes with

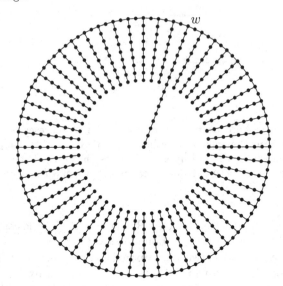

Fig. 9.8. Lower-bound graph.

distance 1. The last node of the chain of   is the center node; the edge to the center node does not need to have length 1.

The unit disk graph consists of the edges on the circle and the edges on the chains only. In particular, there is no edge between two chains because all chains except the   chain end strictly outside radius $r/2$. The graph has k chains with $\Theta(k)$ nodes each.

We route from an arbitrary node on the circle (the source s) to the center of the circle (the destination   ). An optimal route between s and   follows the shortest path on the circle until it hits node   , and then directly follows 's chain to   with link cost $\leq k + r + 1 \in O(k)$. A routing algorithm with routing tables at each node will find this best route.

A geographic routing algorithm, in contrast, needs to find the "correct" chain   . Since there is no routing information stored at the nodes, this can only be done by exploring the chains. Any deterministic algorithm needs to explore the chains in a deterministic order until it finds the chain   . Thus, an adversary can always place   such that 's chain will be explored as the last one. The algorithm will therefore explore $\Theta(k^2)$ (instead of only $O(k)$) nodes.

The argument is similar for randomized algorithms. By placing   accordingly (randomly!), an adversary forces the randomized algorithm to explore $\Omega(k)$ chains before chain   with constant probability. Then the expected link cost of the algorithm is $\Omega(k^2)$.

As all edges (but one) in our construction have length 1, the cost values in the Euclidean distance, the link distance, and the energy metrics are equal.

As for any fixed cost metric $c'(\cdot)$ according to Definition 9.3.2 $c'(1)$ is also a constant, the $\Omega(c^2)$ lower bound holds for all according metrics. □

Note that this lower bound holds generally, not only for $\Omega(1)$-graphs. (As shown in [240], however, a similar lower bound proves that in general graphs the cost metrics according to Definition 9.3.2 fall into two classes.)

Given this lower bound, we can now state that OAFR is asymptotically optimal for unit disk graphs in the $\Omega(1)$-model.

Theorem 9.6.2. *Let c be the cost of an optimal path for a given source-destination pair on a unit disk graph in the $\Omega(1)$-model. In the worst case, the cost for applying OAFR to find a route from the source to the destination is $\Theta(c^2)$. This is asymptotically optimal.*

Proof. This theorem is an immediate consequence of Theorems 9.5.3 and 9.6.1. □

9.7 Combining Greedy and Face Routing

A greedy routing approach as presented in Section 9.4 is not only worth being considered due to its simplicity in both concept and implementation. Above all in dense networks such an algorithm can also be expected to find paths of good quality; here, the straightforwardness of a greedy strategy contrasts highly the inflexible exploration of faces inherent to face routing. For practical purposes it is inevitable to improve the performance of a face routing variant, for instance by leveraging the potential of a greedy approach.

In this section we will briefly outline the GOAFR^{+}[9] algorithm combining greedy and face routing [240]. GOAFR^{+} is a combination of greedy routing and face routing in the following sense: Whenever possible, the algorithm tries to route in a greedy manner; in order to overcome local minima with respect to the distance from the destination, face routing is employed. In face routing mode, GOAFR^{+} restricts the searchable area in a similar way as OAFR. Simulations showed that choosing a circle centered at the destination t instead of an ellipse and above all gradually reducing its radius while the message approaches t improves the average-case performance.

More importantly, for average-case considerations, the algorithm should fall back to greedy routing as soon as possible after escaping the local minimum. This is suggested by the observation that greedy forwarding is—especially in dense networks—more efficient than face routing in the average case. Accordingly, GOAFR^{+} tries to return to greedy routing as early as

---

[9] Expressing the combination of the greedy and face routing approaches, the acronym GOAFR stands for Greedy Other Adaptive Face Routing. The "+" sign indicates that GOAFR^{+} (pronounced as "gopher-plus") is an improvement over a previously defined similar algorithm with the name GOAFR.

possible. However, it was shown [240, 243] that this must not be done too simplistically—such as whenever the algorithm in face routing mode is closer to the destination than the escaped from local minimum—, as this would happen at the expense of the algorithm's asymptotic optimality. In order to preserve this property, the GOAFR$^+$ algorithm employs two counters p and q to keep track of how many of the nodes visited during the current face routing phase are located closer to the destination (counted with p) and how many are at least as far from the destination (counted with q) than the starting point of the current face routing phase; as soon as a certain fallback condition holds, GOAFR$^+$ continues in greedy mode.

It can be proved that the GOAFR$^+$ algorithm retains OAFR's asymptotic optimality, in particular, that it is guaranteed to reach the destination with cost at most $O(c^2(p^*))$, where p^* is an optimal path between the source and the destination [240].

On the other hand it was shown by simulation in networks generated by randomly placing nodes in a given field that GOAFR$^+$'s combination of face routing with greedy forwarding is beneficial for routing also in average-case networks [240, 243, 408]. Particularly, in order to judge the practicability of a geographic routing algorithm, the *normalized cost* of an algorithm A in a network N given a source s and a destination t, defined as

$$cost_A(N, s, t) := \frac{s_A(N, s, t)}{c_\ell(p_\ell^*(N, s, t))},$$

was measured, where $s_A(N, s, t)$ is the number of steps taken by algorithm A in network N finding a route from s to t; $c_\ell(p_\ell^*(N, s, t))$ is the (hop) length of the shortest path (with respect to the hop metric) between the source s and the destination t in N.

The GOAFR$^+$ algorithm was compared with the FR algorithm (Face Routing, see Section 9.5.1), AFR, OAFR, and GFG/GPSR (a combination of greedy and face routing introduced in [45] and [211]). Figure 9.9 shows for each simulated algorithm A its mean cost value, that is

$$\overline{cost_A} := \frac{1}{k} \sum_{i=1}^{k} cost_A(N_i, s_i, t_i),$$

over $k = 2000$ generated triples (N_i, s_i, t_i) for network densities ranging from 0.3 to 20 nodes per unit disk.

In the context of routing in networks formed by randomly placed nodes, node density plays an important role. For very low network densities, the network will almost always be disconnected; for high densities, on the other hand the network can be expected to consist of one connected component with high probability. The transition between these two extremes is astonishingly narrow. This can be observed in Figure 9.9, where, in addition to the algorithm cost values, the network connectivity and greedy success rates are plotted

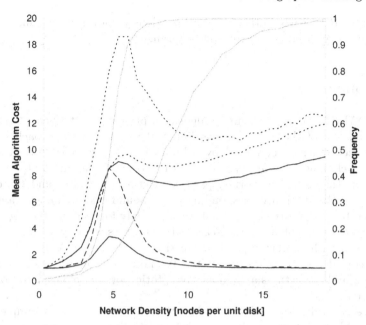

Fig. 9.9. Algorithm performance in network densities including the critical density range around 4.5 nodes per unit disk. Mean cost values of FR (upper dotted line), AFR (lower dotted), OAFR (upper solid), GFG/GPSR (dashed), and GOAFR$^+$ (lower solid). The network connectivity and greedy success rates are plotted in gray against the right y-axis.

against the right y-axis. The network connectivity rate is defined as the portion of triples (N_i, s_i, t_i) in which s_i and t_i are connected in network N_i of a given node density; the greedy success rate represents the portion of randomly generated triples (N_i, s_i, t_i) in which—starting from s_i—the destination t_i can be reached using greedy forwarding alone.

In Figure 9.9, the cost value series for both combinations of greedy and face routing display favorable values for low and high values: With very low densities, if the source and the destination are in the same connected component, they are very likely close together; with high densities, greedy routing alone will efficiently reach the destination with high probability. In between, all simulated algorithms are more or less bell-shaped around the critical density range of about 4.5 nodes per unit disk. It was shown that this behavior is due to the fact that in networks of critical density the shortest path connecting the source and the destination is significantly longer than their Euclidean distance. As reflected by the cost values, this critical density range appears to be a challenge for any routing algorithm. Not only this effect is best tolerated by the GOAFR$^+$ algorithm. More generally, the simulation results show that GOAFR$^+$ clearly outperforms all previously known geographic routing algorithms over a broad network density range.

9.8 Conclusion

The GOAFR$^+$ geographic routing algorithm presented in this chapter forms a combination of the greedy forwarding and face routing approaches. Employing face routing—enhanced by a limitation to a searchable area and a counter technique—the algorithm is proved to require at most $O(c^2)$ steps where c is the cost of a shortest path connecting the source and the destination; together with a corresponding lower bound graph, this guarantee is shown to be asymptotically optimal. Using greedy forwarding, the algorithm also becomes efficient in average-case networks, as shown by simulation and comparison with similar algorithms. In this sense, the GOAFR$^+$ algorithm can be considered a synthesis of simplicity and average-case efficiency on the one hand and correctness as well as asymptotic worst-case optimality on the other hand.

More generally, the GOAFR$^+$ algorithm may stand as an example that it is possible to design algorithms with theoretically proved worst-case guarantees and more practically relevant average-case efficiency. Maybe, even the general conclusion can be drawn that accounting for worst-case behavior before studying the average case appears to be easier than conversely, which may serve as a design principle beyond the scope of ad hoc and sensor networks.

9.9 Chapter Notes

The early proposals of geographic routing—suggested over a decade ago—were of purely greedy nature [128, 186, 355]. All these proposals (including a contribution made later in [225]) have however in common that they can fail to reach the destination.

The first geographic routing algorithm that does guarantee delivery was *Face Routing* based on graph planarization and introduced in [225]. There have been later suggestions for algorithms with guaranteed message delivery [43, 44, 45, 97, 148]; however, each of them has its issues or—at least in the worst case—does not outperform original Face Routing. A more specific overview of related geographic routing algorithms can be found in [369].

In [241] *Adaptive Face Routing* AFR was proposed, whose execution cost is bounded with respect to the cost of the optimal route. There have been proposals for practical purposes to combine greedy routing with face routing [45, 97, 211], without competitive worst-case guarantees, though. In [243] the GOAFR algorithm was introduced; to the best of our knowledge, this was the first algorithm to combine greedy and face routing in a worst-case optimal way; the GOAFR$^+$ algorithm [240] remains asymptotically worst-case optimal while improving GOAFR's average-case efficiency.

Lately, first experiences with geographic and in particular face routing in practical networks have been made [215, 216]. Other approaches study geographic routing with eased [251, 252] or strengthened [235, 290, 389] assumptions made for geographic routing. A more detailed description of related work can be found in Section 9.2.

Compact Routing

Michael Dom

10.1 Introduction

In a distributed network, the delivery of messages between its nodes is a basic task, and a mechanism is required that is able to deliver packages of data from any node of the network to any other node. To this end, usually a distributed algorithm runs on every node of the network, which properly forwards incoming packages, employing some information that is stored in the local memory of the node.

In sensor and ad hoc networks, the resources of the nodes are typically strongly limited—in particular, small memory sizes are common. However, the number of nodes in these networks can be very high, such that it is normally too expensive to use $\Omega(n)$ bits of local memory in each node: Storing at each node a routing table that contains for every destination node one entry of routing information is not possible. Therefore, distributed algorithms and data representations are required that allow to use only $o(n)$ bits of local memory, typically at the cost of increasing the lengths of the routing paths. *Compact routing* is the field of research that deals with the theoretical basics of this problem. It analyzes how "good" routing algorithms can be when the available resources, such as memory and computational power, are limited, and provides the fundamentals for memory-efficient routing protocols.

The model that is used assumes that a distributed algorithm, the so called *routing scheme*, runs at each node of the network and decides for any incoming data package to which immediate neighbor of the node the package shall be forwarded in order to make it reach its destination.

The network is usually considered as an undirected and edge-weighted graph with labeled nodes in which every edge has a number—the *port number*—at each of its endpoints. Each data package has a *header* which contains information about the destination of the package—for example the label of the destination node—and possibly additional information that is used during the routing. A node that receives a data package has to check whether the node itself is the destination of the package. If this is not the case, it has to

D. Wagner and R. Wattenhofer (Eds.): Algorithms for Sensor and Ad Hoc Networks, LNCS 4621, pp. 187–202, 2007.

send it immediately to another node in such a way that the package eventually reaches its destination.

The decision process which determines in which direction a data package has to be forwarded can be regarded as a *routing function*, which takes as input the current node, the port number of the edge through which the data package has reached the node, and the header of the data package, and which computes two outputs: the port number of the edge through which the package has to be sent and, if the model of the network allows to change headers of data packages during the routing, a new header for the package. A routing scheme is the implementation of a routing function.

Most routing schemes need some information, which is stored in local *routing tables*, and these tables have to be initialized before any data packages can be routed. Hence, in order to run a distributed network, a *routing strategy* is needed, which takes care of all tasks associated with routing and which consists of

- a global preprocessing algorithm, which initializes the local data structures of all nodes and which—if this is allowed by the model of the network— assigns labels to the nodes and port numbers to the edges (otherwise, the node labels and port numbers, respectively, are part of the input graph and cannot be changed), and
- a distributed algorithm, called the routing scheme, which implements an adequate routing function.

Consider now the following straightforward routing strategy: In the preprocessing phase, the shortest paths between all pairs of nodes are computed, and at each node v of the network a routing table is stored which contains for each node w of the network the port number of the edge leading from v to the next node on the shortest path from v to w. If a data package whose header contains the label of a node w as destination address arrives at a node v, the routing algorithm at v searches in the local routing table of v for the entry belonging to w and sends the package through the edge determined by the port number found in the table. This routing strategy needs no rewritable package headers and it routes every data package on the shortest path to its destination.

However, each such routing table needs $n \log(\deg(v))$ bits of local memory space in every node v, where n is the number of nodes in the network and $\deg(v)$ is the number of edges incident to v. In a large network, such a demand for memory can be too expensive, and more intricate routing strategies are needed. A routing strategy (and also its routing scheme) is called *compact* if, for a network consisting of n nodes, it needs only $o(n)$ bits of local information at each node (i.e., it needs *less* than $c \cdot n$ bits of local memory at each node for every constant c if n is big enough).

In order to save memory space, one often accepts that packages are routed towards their destination not on the shortest possible path. Therefore, besides the memory consumption, one of the most important performance measurements of a routing strategy is its *stretch*, which is defined as the maximum ratio, taken over all node pairs (v, w), between the length of the path a data package between v and w is routed and the length of the shortest path between v and w in the network. The trade-off between the memory space that is needed and the maximum stretch guaranteed by a routing strategy is a very extensively studied topic in the area of routing in networks.

Of course, there are also other properties of routing strategies that have to be optimized and that are considered in literature. The most common performance measurements for routing strategies are

- the memory space needed in each node (called *local memory*),
- the total memory space used by all nodes (called *total memory*),
- the stretch,
- the sizes of the addresses and package headers,
- the time needed to compute the routing function (called *routing time* or *latency*), and
- the time needed for the preprocessing.

Note that the memory consumption of a routing strategy cannot be considered without also considering the sizes of the addresses and package headers, because if routing strategies are allowed to assign arbitrary addresses to the nodes, the addresses can be used for storing routing information.

In this chapter, we will first give an overview over some important results and then describe three exemplary routing strategies.

10.2 Definitions

A network is modeled as an undirected, connected graph $G = (V, E)$ with $n := |V|$, $m := |E|$ and a weight function $r : E \rightarrow \mathbb{R}^+$ representing distances between the nodes that are connected by edges. (Sometimes networks are also modeled as unweighted graphs, i.e., $\forall e \in E : r(e) = 1$.) The nodes of the graph are labeled with numbers or, more generally, with arbitrary data types that serve as *addresses*. We denote such an address of a node v with $a(v)$. Moreover, there exists a numbering $p : \{(v, w) \mid \{v, w\} \in E\} \rightarrow \mathbb{N}$ representing the *port numbers*: The edge $\{v, w\}$ is labeled with $p(v, w)$ at its endpoint v and with $p(w, v)$ at its endpoint w. The node addresses and port numbers are either fixed (and, therefore, part of the input for the preprocessing algorithm of the routing strategy) or can be assigned arbitrarily by the routing strategy, which can possibly lead to more efficient routing schemes. In order to prevent port numbers from being abused for storing hidden routing information, the port numbers of the edges incident to a node v must lie between 1 and $\deg(v)$, where $\deg(v)$ stands for the number of neighbors of v, i.e., $\forall \{v, w\} \in E : 1 \leq p(v, w) \leq \deg(v)$.

To every data package a *header* is attached containing some information (e.g., the address of the destination node), which is used by the routing scheme.

Some routing schemes need rewritable headers, that is, the headers can be modified by the nodes that route the data packages.

A *shortest path* between two nodes v and w is a path where the sum of the edge weights is minimum; this sum is denoted as the *distance* $d(v, w)$ between v and w. The *stretch* of a routing scheme is defined as

$$\max_{v,w \in V} \frac{\sum_{e \in P(v,w)} r(e)}{d(v, w)},$$

where $P(v, w)$ is the edge set traversed by a package that is routed by the routing scheme from v to w.

While some routing strategies can only provide routing schemes for special classes of graphs, a *universal* routing strategy works on every arbitrary graph.

A *direct* routing function depends only on the address of the destination node (and, of course, on the current node)—this makes direct routing schemes (i.e., routing schemes implementing direct routing functions) quite simple. A direct routing scheme does not change the package header (which only contains the address of the destination node), which implies that it has to route any package on a loop-free path (i.e., a path that passes each of its nodes exactly once). A routing scheme with stretch 1 is called a *shortest path routing scheme*.

Depending on whether a routing strategy is allowed to assign addresses to the nodes and port numbers to the edges, the following concepts are distinguished: In the case of *labeled routing*, the node addresses can be chosen arbitrarily by the routing strategy, such that routing information can be stored in the node addresses, whereas *name-independent routing* handles with node addresses that are fixed—mostly numbers between 1 and n. In the *fixed port model*, the port numbers are fixed before the labels of the nodes are given. In the *designer port model* port numbers from 1 to $\deg(v)$ can be assigned by the routing strategy arbitrarily to the edges incident to every node v.

A *node coloring* for a graph is a mapping $c : V \to \{1, \ldots, b\}$ where b stands for the number of the colors that are used.

With log we denote the logarithm to the base 2; the notation $\tilde{O}()$, similar to $O()$, omits constant and poly-logarithmic factors. We generally omit rounding notations $\lceil \ldots \rceil$ and $\lfloor \ldots \rfloor$, e.g., we write $\log n$ instead of $\lceil \log n \rceil$ or $\lfloor \log n \rfloor$.

10.3 Overview

In this section we will give an overview over a selection of results concerning routing with low memory consumption; this overview will contain fundamental past works as well as up to date results. While in Sect. 10.3.1 we will consider universal routing strategies, in Sect. 10.3.2 we will turn our attention to routing strategies that work only on special graph classes, for example on trees. Some of the presented results are summarized in Tables 10.3 and 10.4.

Universal routing strategies			
Labeled routing			
Port model	**Stretch**	**Addr. size**	**Local memory**
designer port	5	$\log n$	$O(\sqrt{n}(\log n)^{3/2})$ [108]
fixed port	3	$3\log n$	$O(n^{2/3}(\log n)^{4/3})$ [86]
designer port	3	$(1+o(1))\log n$	$O(\sqrt{n\log n})$ [363, 362]
designer port	$4k-5$	$o(k(\log n)^2)$	$\tilde{O}(n^{1/k})$ [363, 362]
Name-independent routing			
Port model	**Stretch**	**Local memory**	
fixed-port	5	$\tilde{O}(\sqrt{n})$ [17]	
fixed-port	3	$O(\sqrt{n}(\log n)^3/\log\log n)$ [7]	
Labeled routing in trees			
Port model	**Stretch**	**Addr. size+local memory**	
designer-port	1	$(1+o(1))\log n$ [362]	
fixed-port	1	$O((\log n)^2/\log\log n)$ [133, 362]	

Table 10.3. Performance data of several routing strategies.

10.3.1 Universal Routing Strategies

Universal routing strategies are able to route in any network, without any restriction on the topology. For routing on shortest paths it is optimal with respect to memory consumption to simply store a routing table in each node with one entry for each destination node [157]. However, by routing on almost-shortest paths the memory consumption can be reduced; this issue was first raised by Kleinrock and Kamoun [219]. For overviews see [152, 153, 155, 313, 374].

Labeled Routing. Labeled routing strategies are allowed to arbitrarily assign addresses to the nodes. Using thereby addresses that contain routing information often results in a smaller memory consumption compared to name-independent routing, where the node names are given and cannot be changed.

Interval Routing. An often used way to save memory space is to assign the node labels in such a way that, at each node, packages for several consecutive destination addresses can be routed through the same edge—this technique is called *interval routing*. In such interval routing schemes typically each edge of a node serves as output port for a set of consecutive destination addresses, i.e., for an interval of addresses. A typical routing table then consists of a mapping from address intervals to port numbers. This idea was introduced by Santoro and Khatib [340]; the term interval routing was introduced by van Leeuwen and Tan [372] who showed that for every network there is a (not necessarily shortest path) interval routing scheme that uses every edge of the network. Characterizations of networks that have shortest path interval routing schemes can be found in [137, 152, 340, 373]. There are also interval routing schemes using more than one address interval per port [373]; the *compactness* of a network is then defined as the maximum number of intervals, taken over all nodes, that have to be mapped to the same output port [152]. For an overview about interval routing schemes see [152, 155].

Upper Bounds for Labeled Routing. A labeled routing strategy for the fixed port model using addresses of size $O((\log n)^2)$ and headers of size $O(\log n)$ was presented by Peleg and Upfal [315]; their strategy needs a total memory space of $O(k^3 n^{1+1/k} \log n)$ and guarantees a stretch of $s = 12k + 3$ for $k \geq 1$. However, it neither gives a guarantee for the local memory consumption nor is it able to handle weighted edges. The strategy is based on a technique called *hierarchical routing*—the idea here is to cover the network with several levels, that is, with subgraphs of increasing radii. A package is first sent on the lowest level; if it does not reach its destination because it is out of the radius the package will bounce back to the sender, which tries to send it on a higher level.

A much more simple scheme of Cowen [86] uses some selected nodes as "landmarks" which results in a direct routing scheme with a stretch of three in the fixed port model; it has addresses and headers of size $3 \log n$ bit and needs a local memory of size $O(n^{2/3}(\log n)^{4/3})$.

By using the technique of interval routing, the memory demand can be further reduced to $O(\sqrt{n}(\log n)^{3/2})$ while attaining a maximum stretch of five and an average stretch of three [108]; this scheme for the designer port model uses numbers from 1 to n as addresses and takes $O(\log n)$ routing time.

An improved stretch-three strategy in the designer port model was given by Thorup and Zwick [362, 363]; their scheme uses size-$(1 + o(1)) \log n$ addresses and headers and needs $O(\sqrt{n \log n})$ bits of local memory; the routing time is constant. A variation of the algorithm has $(\log n)$-bit headers and takes $O(\log \log n)$ time. Thorup and Zwick also generalize their result: Stretch $4k - 5$ is possible with $o(k(\log n)^2)$-bit addresses, $o((\log n)^2)$-bit headers and $\tilde{O}(n^{1/k})$ bits of local memory. If handshaking—that is, communication between nodes in order to get routing information—is allowed, the stretch can be reduced to $2k - 1$.

Lower Bounds for Labeled Routing. Gavoille and Pérennès [157] showed that every shortest path routing scheme needs $\Omega(n \log(\deg(v)))$ bits of local memory per node v if addresses of size $O(\log n)$ are used, independent of the header size of the packages and even in the designer port model. This implies that there is no better universal strategy for routing on shortest paths than simply storing a routing table in each node with one entry for every destination node.

For routing on non-shortest paths, Peleg and Upfal [315] showed that every universal strategy of a stretch factor $\Theta(s)$ requires a total memory space of $\Omega(n^{1+1/\Theta(s)})$ bits (a more detailed look at the hidden constants shows a lower bound of $\Omega(n^{1+1/(2s+4)})$ bits total memory for a stretch of $s \geq 1$ [153]).

Eilam et al. [108] proved that even in the designer port model there is no loop-free universal strategy (in particular, no strategy with a direct routing scheme) with addresses chosen from $\{1, \ldots, n\}$ that uses a local memory space smaller than $c\sqrt{n}$ bits on every network for a constant $c = \pi\sqrt{2/3}/\ln 2$. This

Universal routing strategies			
Labeled routing with node addresses from $\{1,\ldots,n\}$			
Port model	**Stretch**	**Local memory**	**Remarks**
designer port	< 3	$\Omega(n)$ [154]	
designer port	1	$\Omega(n \log n)$ [157]	
designer port	every	$> \pi\sqrt{2/3}/\ln 2 \cdot \sqrt{n}$ [108]	loop-free strategies
Name-independent routing			
Port model	**Stretch**	**Local memory**	**Remarks**
designer port	< 5	$\Omega(\sqrt{n})$ [362]	
fixed port	$< 2k+1$	$\Omega((n \log n)^{1/k})$ [5]	for any $k \geq 1$
Labeled routing in trees			
Port model	**Stretch**	**Addr. size+local memory**	**Remarks**
fixed port		1	$\Omega((\log n)^2/\log \log n)$ [134]	

Table 10.4. Some lower bounds for the demand of local memory.

holds for almost every family of graphs, even for trees, and, therefore, for every stretch (note that every non-shortest path between two nodes in a tree has to go through at least one node twice). This explains why many of the strategies mentioned in this chapter (for example [7, 28]) have to rewrite the headers of the packages at least once during the routing.

Another lower bound for universal routing strategies using integers from 1 to n as addresses involves a total memory space of $\Omega(n^2)$ bits (i.e., $\Omega(n)$ bits local memory in at least one node) for a stretch smaller than three [154] even in the designer port model; this result implies that there is no compact universal routing strategy with a stretch smaller than three.

Name-Independent Routing. In the case of name-independent routing the node addresses are part of the input network and cannot be changed in the preprocessing phase of the routing strategy.

Upper Bounds for Name-Independent Routing. The first to distinguish between labeled and name-independent routing were Awerbuch et al. [26] who presented a name-independent routing strategy that uses the technique of hierarchical routing and needs $O(kn^{2/k} \log n)$ bits of local memory per node and attains a stretch factor of $O(k^2 3^k)$ for any integer $k \geq 1$. A series of improvements [3, 27, 28] resulted in a stretch-five routing strategy using $\tilde{O}(\sqrt{n})$ bits of memory per node by Arias et al. [17] and a stretch-three routing strategy using rewritable headers of size $O((\log n)^2/\log \log n)$ and $O(\sqrt{n}(\log n)^3/\log \log n)$ bits of memory per node by Abraham et al. [7]. We will describe the algorithm of Abraham et al. in Section 10.4.

Lower Bounds for Name-Independent Routing. Obviously, all lower bounds given for the case of labeled routing also hold for the case of name-independent routing, but there are also stronger bounds.

One of them can be seen by considering the complete bipartite graph $K_{n/2,n/2}$: Every universal name-independent routing strategy in the fixed port model with a stretch smaller than three has to use a local memory of $\Omega(n \log n)$ bits [7]. This result was generalized by Abraham et al. [5], who proved a lower bound of $\Omega((n \log n)^{1/k})$ bits of local memory for any stretch stretch smaller than $2k + 1$ for any integer $k \geq 1$. Moreover, it is known [362] that there is no universal name-independent routing strategy with stretch smaller than five using $o(\sqrt{n})$ bits of local memory.

By considering the center node of a star, one can see that there is no universal loop-free name-independent routing strategy that uses $o(n)$ bits of memory in every node [7].

10.3.2 Special Graph Classes

There is a large portion of work on special families of graphs, especially for trees. Routing strategies for trees are also used as subroutines in some universal routing strategies (see Sect 10.4.3).

Trees. The first interval routing scheme for trees using $O(\deg(v) \log n)$ bits of memory in each node v was presented by Santoro and Khatib [340]. Gavoille [152] showed that labeled routing with $3.71\sqrt{n}$ bits local memory per node is possible in the designer port model, although with exponential routing time. If addresses of size $5 \log n + o(\log n)$ are allowed, there is a direct routing scheme in the designer port model that needs $3 \log n + O(\log \log n)$ bits per node and has constant routing time and a preprocessing time of $O(n \log n)$ [133]. It is even possible to reduce the address size to $2.8 \log n$ bit while increasing the routing time to $n^{O(1)}$ and the time for preprocessing to $n^{O(1)}$ [133]. A similar result was achieved by Thorup and Zwick [362] whose routing scheme uses addresses and local memory of size $(1 + o(1)) \log n$ bit and has a constant routing time.

For the fixed port model, Peleg [312] presented a direct labeled routing scheme for trees that needs addresses and local memory of size $O((\log n)^2)$ and whose routing time is $O(\log n)$. This scheme is based on a technique called *distance labeling*: This labeling of the nodes allows, given the addresses of two nodes, to compute their distance (see also [48, 49, 156]). Fraigniaud and Gavoille [133] as well as Thorup and Zwick [362] improved this result by giving direct routing schemes with addresses, headers and local memory of size $O((\log n)^2 / \log \log n)$; their routing time is constant and the preprocessing time is $O(n \log n)$. Complementing this result, Fraigniaud and Gavoille [134] showed that for every shortest path routing scheme for trees in the fixed port model the sum of the address size and the local memory size is $\Omega((\log n)^2 / \log \log n)$ for some node in some tree. Moreover, it is known [133] that any shortest path routing scheme with addresses from $\{1, \ldots, n + o(\sqrt{n} / \log n)\}$ requires a local memory of size $n - o(n)$ bit in the fixed port model and a local memory of size $c\sqrt{n} - o(n)$ bit for some constant c in the designer port model, which points out how even a small a change of the header size has a big impact on the size of the local memory (note that by attaching the routing information of a node into its address, it is even possible to create routing strategies that need no local memory but that require bigger addresses).

Further Graph Classes. Other graph classes are considered as well; a large list of literature is given, e.g., by Abraham et al. [4]. For planar graphs, the best known (labeled) shortest path routing scheme is from Lu [265], who uses addresses from 1 to n and $7.181n + o(n)$ bits of local memory in the designer port model. No lower bound for the memory size of shortest path routing schemes on planar graphs is known. For a stretch greater than one, Thorup [361] presented a stretch-$(1 + \epsilon)$ labeled routing scheme that uses addresses and routing tables of size $O((1/\epsilon)(\log n)^2)$. Abraham et al. [4] gave a name-independent routing strategy for unweighted graphs excluding any fixed minor (which is, therefore, applicable on planar graphs) in the fixed port model. They use headers of size $O((\log n)^2/\log\log n)$ and need a local memory of size $\tilde{O}(1)$.

Other results regard, e.g., routing on euclidean metrices [8], routing in growth bounded networks [9], routing in graphs with low doubling dimension [2, 221], routing in power law graphs [47], or routing in interval graphs and circular arc graphs [49].

10.4 Algorithms

In this section we describe three routing strategies. First, in Sect. 10.4.1, we present a straightforward labeled routing strategy for trees as an easy to understand example for interval routing. This strategy needs $O(\deg(v)\log n)$ bits of memory in each node v. The memory requirement is improved to $6\log n$ bits of local memory in Sect. 10.4.2, where the idea of interval routing is combined with that of moving part of the routing information from the local memories into the node addresses. Finally, we sketch a state of the art name-independent routing strategy for arbitrary graphs in Sect. 10.4.3 which needs $O(\sqrt{n}(\log n)^3/\log\log n)$ bits of local memory per node and guarantees a stretch of three. Although this routing strategy is name-independent, it uses as a subroutine a compact labeled routing strategy for trees, for example the one of Sect. 10.4.2.

10.4.1 An Interval Routing Scheme for Trees

The direct labeled routing strategy of Santoro and Khatib [340] for trees in the fixed port model uses the technique of interval routing and needs $O(\deg(v)\log n)$ bits of local memory in each node v. The strategy is based on a depth first search (DFS) numbering of the nodes, and because of its simplicity we will use it as an example for interval routing on trees.

In the preprocessing phase of the strategy, the tree is rooted at an arbitrary node, and the nodes are labeled via a depth first search (DFS). Thus, for each

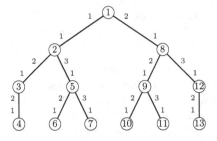

Local memory of node 2	
Own address	2
f_2 (= highest addr. in subtree)	7
Port to parent	1
Port to subtree	(4, 2)
Port to subtree	(7, 3)

Fig. 10.1. Example for the interval routing strategy for trees of Sect 10.4.1. The picture on the left shows a tree whose edges are labeled with port numbers. The routing strategy performs a DFS to determine the node addresses. The table on the right shows the local memory of node 2.

node v with address (v), the addresses of the nodes in the subtree rooted at v form a complete interval, that is, if is the number of the nodes in this subtree, then the addresses of the nodes in the subtree are exactly the numbers from (v) to $(v) + - 1$. Now the following information is stored at each node v (see Fig. 10.1 for an example):

- its address (v),
- the highest address occurring in the subtree rooted at v, denoted with ,
- the port number of the edge leading to the parent of v, and
- a table with one entry ($_i$, $_i$) for each child v_i of v, where $_i$ is the highest node address occurring in the subtree rooted at v_i and $_i$ is the port number (v, v_i) of the edge leading from v to v_i.

The header of a data package contains nothing but the address of its destination node. A node v that has to route a package with destination address ( ) performs the following steps:

1. If ( ) $=$ (v): The package has reached its destination. Stop.
2. If ( ) (v) or ( ) $>$: The destination is not a descendant of v. Send the package to the parent of v and stop.
3. Otherwise, the destination node lies in a subtree rooted at a child of v. Search in the local memory the entry ($_i$, $_i$) with the smallest $_i \geq$ ( ) and send the package through the port numbered with $_i$.

Clearly this strategy uses $O(\deg(v) \log n)$ bits local memory at each node v: three entries of size at most $\log n$ bit and $\deg(v) - 1$ entries of size $\log(n) + \log(\deg(v))$ bit. Although the total memory over all nodes is only $O(n \log n)$ bits, such a local memory requirement can be impractical when the node degrees are not distributed equally over the network and there are nodes with very high degree. The routing time also depends on the degree of the current node, but can be bounded from above by $O(\log \log n)$ [370].

10.4.2 An Improved Labeled Routing Scheme for Trees

Thorup and Zwick [362] introduced a direct labeled routing scheme for trees in the fixed port model that uses only $6 \log n$ bits of local memory in each node. The main idea is to enlarge the addresses of the nodes by storing some routing information in the addresses. In contrast to the algorithm of Sect. 10.4.1, the local memory of a node u here does not contain the port numbers of all edges leading to child nodes of u, but only the port number of at most one edge leading to a child node that has a "big" number of successors. The intuition is that then the depths of the subtrees rooted at the other children v of u cannot exceed a certain value. More specifically, the depths of these subtrees are bounded from above by $\log n$ (we will prove this bound later), and, therefore, one can afford for every successor w of such a child v to put the port numbers used on the path from u to w into the address of w. Altogether, the scheme uses addresses (and headers) of size $(\log n)^2 + \log n$ bit (compared to $(\log n)$-bit addresses in the scheme introduced in Sect. 10.4.1). The routing takes only constant time in each node.

We will now have a closer look at the details. Like in the strategy of Sect. 10.4.1, the strategy here also uses a DFS labeling for the nodes, but unlike in Sect. 10.4.1, the numbers obtained by this traversal are not directly used as addresses. We call such a number an *index*, and we denote the index of a node v with $\mathrm{dfs}(v)$. Before performing this DFS, the nodes of the tree are partitioned (also by a DFS) into *heavy* and *light* nodes: A node v is called heavy if the subtree rooted at v contains at least half of the nodes of the subtree rooted at the parent of v. Nodes that are not heavy are called light; the root is defined to be heavy. The DFS that determines the indices of the nodes is performed in such a way that at each node the light children are visited before the heavy child (clearly every node has at most one heavy child). The following information is stored at each node v:

- the index dfs_v of v,
- the highest address occurring in the subtree rooted at v, denoted with f_v,
- if v has a heavy child, the index of the heavy child, denoted with h_v; otherwise $h_v = f_v + 1$,
- the number of light nodes (including v itself if v is light) lying on the path from the root to v, called the *light level* of v and denoted with ℓ_v,
- the port number of the edge to the parent of v, denoted with $P_v[0]$, and
- the port number of the edge leading to the heavy child of v, denoted with $P_v[1]$ (if v has no heavy child, $P_v[1]$ contains an arbitrary entry).

As mentioned before, some routing information is stored in the addresses of the nodes: The address $a(v)$ of a node v consists of an array $(\mathrm{dfs}_v, L_{v,1}, L_{v,2}, \ldots, L_{v,l_v})$, where $L_{v,i}$ denotes the port number $p(x, y)$ with y being the i-th light node on the path from the root to v and x being the parent of y. Fig. 10.2 shows an example for the addresses and the local memory used by the scheme.

The routing scheme at a node v performs the following steps to route a package with destination address $(\mathrm{dfs}_w, L_{w,1}, L_{w,2}, \ldots, L_{w,\ell_w})$:

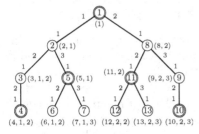

Local memory of node 2	
dfs_2 (= own index)	2
f_2 (= highest index in subtree)	7
h_2 (= index of heavy child)	5
ℓ_2 (= #light nodes on path from root)	1
$P_2[0]$ (= port to parent)	1
$P_2[1]$ (= port to heavy child)	3

Fig. 10.2. Example for the improved routing strategy for trees of Sect 10.4.2. The picture on the left shows a tree whose edges are labeled with port numbers. The routing strategy performs a DFS—visiting always the light children before the heavy one—to determine the indices of the nodes. The addresses of the nodes are written in brackets; heavy nodes are drawn with two bordering circles. The table on the right shows the local memory of node 2.

1. If dfs $=$ dfs : The package has reached its destination. Stop.
2. If dfs $<$ dfs or dfs $>$: The destination is not a descendant of v. Send the package through the edge labeled with $P[0]$ to the parent of v and stop.
3. If dfs $\geq$: The destination is a node in the subtree rooted at the heavy child of v (because the heavy child is the last child visited by the DFS). Send the package through the edge labeled with $P[1]$ to the heavy child of v and stop.
4. Otherwise, the destination must be a node in a subtree rooted at a light child of v. Send the package through the edge labeled with ℓ_v+1 to the light child of v who lies on the path from v to the destination node (the port number ℓ_v+1 can be extracted from the destination address).

Clearly every package that is routed by this algorithm reaches its destination. Let us discuss the performance—in particular, the memory consumption and address size—of this strategy. The information stored in each node consists of at most $\log n$ bits for each of the entries dfs , , , , $P[0]$, and $P[1]$, which is altogether $6 \log n$ bits of local memory per node.

To determine the size of the addresses and headers, it suffices to give a bound for , because clearly the sizes of the addresses are bounded from above by $(1 +) \log n$ bit. Recall the definition of a light node: The subtree rooted at a light node v contains less than half of the nodes of the subtree rooted at the parent of v. Hence, if a node v is a descendant of a node u and $= + 1$, the subtree rooted at v contains less than half of the nodes of the subtree rooted at u. Therefore, the light level of each node v can be at most $\log n$. Altogether, the addresses have a size of $(\log n)^2 + \log n$ bit as claimed. Note that the routing can be performed in constant time and that it is independent from the numbering of the ports (i.e., it accords to the fixed port model).

By a slight modification the size of the addresses can be further reduced [362]. To this end, the definition of heavy (and light) nodes is modified: A node v is called *heavy* if the subtree rooted at v contains at least $1/b$ (instead of one half) of the nodes of the subtree rooted at the parent of v for a fixed integer $b \geq 3$. The information stored locally now contains not only one entry $P_v[1]$ with the port number of the edge to one heavy child, but one entry for each of the at most $b - 1$ heavy children. Moreover, there is a new field $H_v[0]$ containing the number of heavy children, and the field h_v is replaced by new fields $H_v[i], 1 \leq i < b$ containing the indices of all heavy children of v. This information allows to route every package again in constant time. The modified strategy needs $(2b + 3) \log n$ bits of local memory, but has addresses consisting of only $(1 + \log_b n) \log n$ bits.

With much more effort, Thorup and Zwick [362] and Fraigniaud and Gavoille [133] could also give labeled routing strategies for trees in the fixed port model that need only $O((\log n)^2 / \log\log n)$-bit addresses.

10.4.3 A Universal Compact Name-Independent Routing Scheme

The algorithms presented in Sect. 10.4.1 and Sect. 10.4.2 have been labeled shortest path routing strategies for trees. Now we will use such a strategy as a subroutine for a universal name-independent routing strategy. The algorithm of Abraham et al. [7] that we are going to describe provides a stretch of three for arbitrary networks in the fixed port model and needs only $O(\sqrt{n}(\log n)^3 / \log\log n)$ bits of local memory and $O((\log n)^2 / \log\log n)$-bit rewritable headers. The preprocessing can be done in polynomial time, and the routing time is constant.

The main idea behind the routing strategy is to cover the network with several single source shortest paths spanning trees and to store in every node of the network for each of these trees an $O((\log n)^2 / \log\log n)$-bit routing table according to one of the labeled routing strategies mentioned in Sect. 10.4.2 [133, 362]. A data package will then be routed along one of these spanning trees. In order to guarantee a stretch of three, only $\sqrt{n} \log n$ spanning trees are necessary, and, consequently, the local memory of each node seemingly needs to store only $\sqrt{n} \log n$ routing tables, each of size $O((\log n)^2 / \log\log n)$ bit. Unfortunately, the compact shortest path routing scheme used for the routing in the spanning trees is a labeled routing scheme. This means that the routing tables stored in the local memories of the nodes are useless if one does not additionally know the node addresses that are used by the labeled routing strategy and that differ from the node addresses of the input network for the name-independent routing scheme.

In order to overcome this difficulty, the routing strategy combines several ingredients. First of all, we define the vicinity $B(v)$ of a node v as the set of the $b\sqrt{n} \log n$ nodes closest to v (ties are broken by choosing nodes with small addresses first) with b being a fixed constant large enough. The second ingredient is a node coloring $c : \{1, \ldots, n\} \rightarrow \{1, \ldots, \sqrt{n}\}$ with the following two properties: At most $2\sqrt{n}$ nodes are colored with the same color, and for each node v there is at least one node from each color in its vicinity $B(v)$. (Note that, in contrast to the colorings occurring in many common graph problems, it is not required that two adjacent nodes have different colors.) A random coloring would satisfy these two properties with high probability, but the coloring can also be performed deterministically in polynomial time [7].

The third ingredient is a hash function $h : \{1, \ldots, n\} \to \{1, \ldots, \sqrt{n}\}$ that maps node addresses to colors such that at most $O(\sqrt{n} \log n)$ addresses are mapped onto the same color. Such a hashing can be computed in constant time [7], for example by extracting $(1/2) \log n$ bits from the node addresses.

For the sake of a clear presentation of the main ideas, we will present a slightly simplified version of the algorithm using only the first two ingredients; the purpose of the hash function will be explained at the end of the section. Moreover, we will denote the colors of the nodes with color names instead of numbers.

After the network is colored such that each node v has a color $c(v)$, w.l.o.g. the color *red* is designated as a special color, and the following steps are performed: For each node w whose color $c(w)$ is red, a spanning tree T_w is constructed that connects the node w to all other nodes of the network on shortest paths. On this tree T_w, we run the preprocessing algorithm of the labeled routing strategy mentioned in Sect. 10.4.2, which computes for each node v of the network an address and a routing table for routing in T_w. We denote this address with $\lambda(T_w, v)$ and the routing table with $\mu(T_w, v)$.

Now the following informations are stored in each node v:

- its address $a(v)$,
- for each red node $w \in V$: the routing table $\mu(T_w, v)$,
- for each node $w \in B(v)$: the port number of the edge leading on the shortest path from v to w, and
- for each node $w \in V$ that has the same color as v (i.e., $c(w) = c(v)$): the number of a red node $u \in B(w)$ and the address $\lambda(T_u, w)$ of w in the spanning tree T_u of u.

Fig. 10.3 shows an example for a network and the resulting local memory of one of its nodes.

To route a data package with destination address $a(w)$, the routing scheme running at a node v performs the following steps:

1. If $a(w) = a(v)$: The package has reached its destination. Stop.
2. If $w \in B(v)$: Send the package to the next node on the shortest path to w, using the port number stored in the local memory. Stop.
3. If $c(w) = c(v)$: Overwrite the header of the data package with the number of a red node $u \in B(w)$ and the address $\lambda(T_u, w)$ (both values are stored in the local memory of v) and send the package according to $\mu(T_u, w)$. Stop. (Each subsequent node x can now use its routing table $\mu(T_u, x)$ and route the package on the tree T_u.)

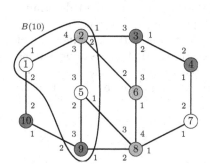

Local memory of node 10	
Own address	10

$(T_2\ 10)$ $(=$ local mem. of node 10 in $T_2)$	
...	...
...	...

$(T_6\ 10)$ $(=$ local mem. of node 10 in $T_6)$	
...	...
...	...

$(T_8\ 10)$ $(=$ local mem. of node 10 in $T_8)$	
...	...
...	...

Port to node $1 \in B(10)$	2
Port to node $2 \in B(10)$	2
Port to node $5 \in B(10)$	1
Port to node $9 \in B(10)$	1
Red node in $B(3)$	2
$\lambda(T_2, 3)$ $(=$ addr. of node 3 in $T_2)$	...
Red node in $B(4)$	6
$\lambda(T_6, 4)$ $(=$ addr. of node 4 in $T_6)$	...

Fig. 10.3. Example for the universal name-independent routing strategy of Sect 10.4.3. The picture on the left shows a network with node labels and port numbers. We assume that the vicinity $B(v)$ of each node v contains four nodes and that the network is colored with three colors in the preprocessing phase. The red color is displayed as light grey (nodes 2, 6, and 8 are red); the four nodes 1, 2, 5, and 9 are the vicinity of node 10. The table on the right shows the local memory of node 10.

4. Otherwise: Send the package to a node $u \in B(v)$ with $(u) = (\ )$. Then u can look up the adequate red node and start routing the package on the corresponding spanning tree.

Let us shortly discuss the stretch of this scheme. In the worst case, the node v sends a package with destination   to a node $u_1 \in B(v)$ with $(u_1) = (\ )$, and from there the package is routed on a spanning tree T_2 with $u_2 \in B(\ )$ to .

Assume the case that on every shortest path from v to   there is a node $\notin B(v) \cup B(\ )$. If we denote with $b(v)$ the "radius" of $B(v)$, that is, $b(v) = \max_{\in (\ )} d(v,\ )$, then it clearly holds that $b(v) + b(\ ) \le d(v,\ )$. The length of the path that is tracked by the package is bounded from above by

$$d(v, u_1) + d(u_1, u_2) + d(u_2,\ )$$
$$\le b(v) + d(u_1, u_2) + b(\ )$$
$$\le b(v) + (d(u_1, v) + d(v,\ ) + d(\ , u_2)) + b(\ )$$
$$\le b(v) + (b(v) + d(v,\ ) + b(\ )) + b(\ )$$
$$\le 3d(v,\ )$$

For the other case (i.e., there is a shortest path from v to w that contains only nodes from $B(v) \cup B(w)$) we cannot prove a stretch of three due to our simplification. In the original algorithm, not only the spanning trees consisting of shortest paths to the red nodes are used, but also spanning trees consisting of shortest paths to all nodes $w \in B(v)$ for every node v. A data package from v to w can then be routed along a path that is composed of two such spanning trees.

Our second simplification concerns the following question: How can a node v that has to route a package with destination $w \notin B(v)$ determine the color of the node w? To overcome this problem, the original algorithm uses the mentioned hash function h. Instead of storing in the local memory of a node v informations for each node w with $c(w) = c(v)$, it stores informations for each node w with $h(w) = c(v)$. During the routing, a node v then does not have to compute the color $c(w)$ of a destination node w, but only its hash value $h(w)$.

10.5 Chapter Notes

In Section 10.3 we have given an overview over some of the most important results and their sometimes subtle differences. For other overviews see [152, 153, 155, 313, 374]. The algorithms considered in Section 10.4 have been originally presented by Santoro and Khatib [340], Thorup and Zwick [362], and Abraham et al. [7], respectively. Recent work includes, e.g., results of Abraham et al. [2, 4, 5, 6], Arias et al. [17], Brady and Cowen [48, 49, 47], and Konjevod et al. [221].

11

Pseudo Geometric Routing for Sensor Networks

Olaf Landsiedel

11.1 Introduction

The first sensor network deployments mainly focused on data collection [274, 308]. Corresponding implementations [184, 161] commonly offer tree based routing, such as many-to-one and one-to-many communication patterns. Routing protocols to mention for this purpose are Directed Diffusion [194] and TAG [271]. These protocols build trees to broadcast commands from a base station to the sensor network and to collect and aggregate data along the path to the base station.

However, recently proposed applications for sensor networks require more advanced routing schemes, mostly efficient any-to-any routing. Examples for such applications are data-centric storage [346], PEG (persuader-evader games[10], [100]), and various data query methods, such as multi-dimensional queries [145], multi-dimensional range queries [257] and spatial range queries [168].

Above listed sensor network applications rely heavily on any-to-any routing algorithms. Thus, scalable any-to-any routing is necessary to evaluate and deploy these applications and the ones to come in the future. However, as we discuss in the next section, today's available any-to-any routing algorithms have limitations in their applicability to sensor networks. The complexity of these proposed solutions is in strong contrast to the requirements of sensor network implementations. On the one hand such implementations have to meet demanding energy efficiency, scalability, robustness and flexibility requirements, but on the other hand they have to deal with the limited resources (memory and computational power) of the sensor nodes.

In this chapter we discuss and evaluate a routing algorithm which tries to meet the above described requirements: "Beacon Vector Routing" (BVR) [132]. This algorithm makes very limited assumptions on resources, radio quality

---

[10] In this kind of game a large sensor network monitors the movements of evading robots.

D. Wagner and R. Wattenhofer (Eds.): Algorithms for Sensor and Ad Hoc Networks, LNCS 4621, pp. 203–213, 2007.
© Springer-Verlag Berlin Heidelberg 2007

and the availability of GPS or other positioning systems. Therefore, it is an interesting algorithm for sensor network routing. As the presented algorithm bases on virtual coordinates, we conclude the chapter with a discussion of virtual coordinate based routing from an algorithmic point of view.

The remainder of this chapter is structured as follows: We discuss the shortcomings and limitations of today's routing algorithms in Section 11.2. Section 11.3 introduces virtual coordinate based routing. Next, section 11.4 discusses the proposed algorithm "Beacon Vector Routing" and presents simulation results. Section 11.5 discusses bounds for the proposed technique to provide an algorithmic background. Section 11.6 addresses related work and section 11.7 concludes.

11.2 Routing Algorithms for Sensor Networks

In this section we discuss various algorithms for any-to-any routing in wireless networks. We identify three different classes of routing techniques and discuss the advantages and disadvantages for their application to sensor networks.

11.2.1 Shortest Path Routing

The Dijkstra algorithm – and its distributed versions – are the classic approach to shortest path routing. This algorithm is used to determine the shortest path between two nodes. Thus, protocols like Distance-Vector, DSDV [318], and Link-State compute the shortest path between any two nodes and each node stores its successor in the path. The algorithm results in $O(n)$ routing states and $O(n^2)$ exchanged messages for a network with n nodes. As result, the overhead in terms of node state and message exchange limits the scalability of this approach.

To address this overhead, on-demand routing has been proposed, such as AODV [319]. The simplest form of on-demand routing is route discovery via flooding: whenever a new route needs to be established, the sender broadcasts a route request to all nodes in the network [202]. On-demand routing works well for systems of small and medium size, as well as for large systems with limited communication patterns and stable routes. However, research shows that links in sensor networks are highly dynamic [65], resulting in frequent route changes. Furthermore, for sensor networks, which commonly have bursty and dynamic communication patterns, like any-to-any communication, the overhead and latency of repeated route discovery can be huge [210].

11.2.2 Hierarchical Addressing

Internet routing uses hierarchical addressing, e.g. address aggregation techniques like longest prefix matching. A route advertisement contains route information for many destinations at the same time. Such a technique ensures high scalability, but is not useful in wireless sensor networks. As in sensor networks nodes are commonly deployed randomly, nodes geographically or topologically close to each other have no similarity in their identifiers.

Landmark Routing [366] addresses this problem by the use of a hierarchical addressing system. A set of hierarchical landmark nodes send scoped flooding messages for route discovery. The concatenation of the addresses of the closest landmark node at each hierarchy level forms a node's address. On the one hand a node only needs to maintain a table of its physical neighbors and the direction, e.g. the next hop, to the next landmark. Furthermore, route setup overhead is reduced to $O(n \log n)$. On the other hand, the proposed routing algorithm needs a complex protocol to set up and maintain the landmark hierarchy and additionally to tune the scoped flooding for landmark and route discovery. This has not been discussed in the original publication, and the algorithms proposed to address these challenges seem quite complex [244, 69].

11.2.3 Geographic Routing

The third class of routing algorithms are geographic routing algorithms. Thus, geographic node locations are used as addresses and packets are forwarded greedily towards the destination [45, 211, 240], for more details see the previous chapter. One challenge in geographic routing are voids in the network: greedy forwarding fails in dead ends, when a node has no neighbor closer to the destination than itself. Various techniques, such as perimeter routing [45, 211], have been proposed to address this. Geographic routing is highly scalable and has very low overhead. For route discovery, $O(1)$ messages need to be send and $O(1)$ entries, e.g. the physical one-hop neighbors, need to be stored in a routing table. Furthermore, algorithms operate locally and path length are close to the shortest path. Therefore, it is an attractive approach for sensor networks.

Nonetheless, geographic routing has three major limitations. First, GPS localization requires a line of sight to at least three of the orbiting satellites. Thus, for outdoor deployments with limited line of sight and indoor deployments in general GPS based routing is not an option. The correctness of the perimeter routing relies on the assumption, that the radio propagation is unit disk like. Meaning, a node hears all transmissions in a fixed range and none from outside this range. However, measurements [395, 404, 85] have shown, that this assumption does not hold in practice. Finally, a GPS receiver is an additional unit on the node, increasing production cost and energy consumption.

11.2.4 Concluding the Overview

Concluding, it can be said, that all three classes of routing algorithms have their limitations for the use in sensor networks. In the following chapter we introduce and discuss a fourth class, namely Virtual Coordinate Based Routing, as a solution to the above described challenges in ad hoc and sensor network routing.

11.3 Virtual Coordinate Based Routing

Geographic routing schemes have ideal scaling properties, but their need to know the physical node locations and its assumptions on radio propagation make geographic routing not usable in sensor network routing. However, recent proposals try to provide similar scaling properties without the need to know the physical locations. The basic technique of such virtual coordinate routing schemes is to provide each node with a virtual coordinate system and its corresponding position in it. To provide this topology awareness, typically a number of nodes flood the network and each node determines its position based on hop-counts relative to the flooding nodes. Note, that this position only represents the node's position in the network topology and not the physical position. However, unlike other approaches, the algorithms discussed in this chapter do not focus on the physical node positions. The consistent virtual coordinate system provided by the described algorithms is sufficient for efficient routing.

11.4 Beacon Vector Routing

Beacon Vector Routing (BVR) [132] is a scalable point-to-point routing algorithm. The algorithm assigns coordinates to a sensor node based on the vector of hop counts to a set of beacon nodes. BVR uses the resulting virtual coordinate system and a distance metric function for greedy forwarding. A small set of beacon nodes is chosen randomly from the node set. Via a standard reverse path construction algorithm, all nodes learn their hop-distance to the beacon node set. Based on the hop-count, a node's position in the virtual coordinate system is the distance to all beacon nodes. BVR uses the virtual coordinates for greedy forwarding, just like GPSR, to route to the destination. Additionally, BVR provides a correction algorithm in case greedy routing fails.

Let q_i denote the hop distance from a node q to a beacon node i. Furthermore, let r denote the total number of beacon nodes in the network. BVR defines the position of a node, denoted q, as $P(q) = \langle q_1, q_2, ..., q_r \rangle$. Additionally a unique node identifier is used to distinguish between nodes which have the same position. To make a routing decision, a node needs to know the position of its neighbors. Thus, nodes periodically announce their coordinates by local broadcast messages.

To decide which neighbor to select as next hop in a routing process, BVR introduces a distance function $\delta(p, d)$. The goal of such a distance function is straightforward: It should select the next hop greedily in a way that it succeeds in packet delivery with a high probability. The function should favor nodes as next hop which are closer to the destination than the forwarding node. For each beacon node, the function metric has to decide whether to move towards or away from this beacon node. Although there is only one direction a packet can be routed towards a beacon node, there are – for a two dimensional topology – three directions a packet can be routed away from a beacon node. Implying, the information to route towards a beacon node is more valuable than the information to route away from a beacon node. BVR values this by the use of two metric functions, one for routing towards the beacon node and the other one to route away from it. Their results are combined by a weighted function δ_k:

$$\delta_k^+ = \sum_{i \in C_k(d)} max(p_i - d_i, 0) \tag{11.1}$$

$$\delta_k^- = \sum_{i \in C_k(d)} max(d_i - p_i, 0) \tag{11.2}$$

$$\delta_k = A\delta_k^+ + \delta_k^- \tag{11.3}$$

For this, let $C_k(d)$ denote the set of the k beacon nodes closest to the destination d. δ_k^+ represents the sum of the differences for the beacons that are closer to the destination d than the node p that has to make the routing decision, while δ_k^- represents the sum of the distances to the beacon nodes further away. Finally, δ_k equals to the weighted sum of both. Thus, the routing decision is the minimum value of δ_k of all neighboring nodes with a sufficiently large A. In practise, the BVR implementation uses $A = 10$. To ensure a lightweight implementation, BVR only considers the k closest beacon nodes for the routing decisions.

To ensure that every packet reaches its destination, a BVR packet header contains three header fields:

- The destination's unique identifier.
- The destination's current position $P(dst)$, defined over the beacon set $C_k(dst)$.
- δ_{min} a vector of k-positions with δ_i^{min} as the minimum δ that the packet as seen so far using $C_i(dst)$, the i closest beacons to dst. Note, that δ_i^{min} can guarantee that the route never loops.

Algorithm 28 describes the pseudo code for BVR packet forwarding. Parameters are r, the total number of beacon nodes, and $k \leq r$, the number of beacons, that define the destination's position. Packet forwarding consists of three modes:

1. First, BVR tries greedy forwarding. The current node (denoted $curr$) chooses among its neighbors the node $next$ that minimizes the distances to destination. BVR uses the k closest nodes, and drops successively beacons from the list when no improvement can be seen.
2. When greedy routing fails, meaning that no neighbor will improve the minimum distance δ_i^{min} for any given i, BVR uses a fallback mechanism. In this mode, the node forwards the packet to a beacon node close to the destination. From this beacon node the packet can be send to the destination, as the destination is a child in the tree of this beacon node.

Algorithm 28: The BVR forwarding algorithm

1: // update packet header
2: **for** $i = 1$ to k **do**
3:      $P.\delta_i^{min} = min(P.\delta_i^{min}, \delta(curr, P.dst))$
4: **end for**
5:
6: // first try greedy forwarding
7: **for** $i = k$ to 1 **do**
8:      $next \Leftarrow argmin_{x \in NBR(curr)} \delta_i(x, P.dst)$
9:      **if** $\delta_i(next, P.dst) < P.\delta_i^{min})$ **then**
10:         unicast P to $next$
11:     **end if**
12: **end for**
13:
14: // greedy routing failed, use fallback
15: $fallback\_bcn \Leftarrow$ closest node to $P.dst$
16: **if** $fallback\_bcn \neq curr$ **then**
17:     unicast P to $PARENT(fallback\_bcn)$
18: **end if**
19:
20: // fallback failed, use scoped flooding
21: broadcast P with scope $P.P(dst)[fallback\_bcn]$

3. If even this beacon node cannot route to the destination, it will initiate a scoped flood. The flood range can be determined precisely from the distance between destination and beacon position.

11.4.1 Beacon Maintenance

Sensor networks are prone to failure and so BVR needs to deal with nodes going down unexpectedly. Especially, beacon node failure needs to be addressed. The described BVR forwarding algorithm can work with less than r – a configurable parameter – beacon nodes. Thus, when there are inconsistencies in the sets of beacon nodes that nodes are aware of, BVR routing bases on the beacon nodes that the nodes have in common. To detect beacon node failure, each beacon node sends a "still-alive" message with an increasing sequence number. Furthermore, based on algorithms described in [130], a node elects itself as beacon node, when it sees less than r beacon nodes. And it stops to act as beacon node, when it detects more than r beacon nodes.

11.4.2 Location Directory

To route to a node, the sender needs to know its position in the network. Thus, BVR provides a location directory, which maps the unique node IDs on network positions. It uses the beacon nodes to provide this service. A

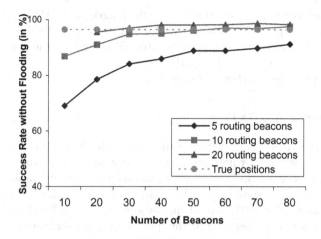

Fig. 11.1. Success rate of BVR routing without flooding in a 3200 node network.

consistent hash function [207], H, maps a node ID on a beacon ID, this beacon then stores the corresponding node ID and its coordinates in the network. Standard replication techniques may be used to ensure redundancy in case of beacon node failure.

11.4.3 BVR Evaluation

For the evaluation of the routing protocol described in this chapter, we focus on two metrics. For more details, please refer to the corresponding papers.

- (Greedy) success rate: This metric is commonly used for protocol evaluation, as it allows to compare the routing success without the impact of fallback techniques like flooding. Flooding ensures that all packets reach their destinations and so will always succeed.
- Transmission stretch: This metric compares the number of hops for BVR routing with the optimal solution of Dijkstra's algorithm, e.g. the shortest path.

Figure 11.1 shows the success rate of greedy routing depending on the number of beacons r for three different values of routing beacons k ($k = 5$, 10 and 20). Obviously, the success rate increases with the number of total beacons r as well as with the number of routing beacons k. The simulations show, that with ten routing beacons ($k = 10$) BVR achieves a routing performance similar to using the true positions, while $k = 20$ increases the routing quality only marginally. Thus, the authors propose ten routing beacons as a good compromise between routing performance and the per-packet routing overhead. For $k = 10$ the simulations show, that values of 20 to 30 beacons r result in a routing performance comparable to using true positions. Only 1% of the nodes act as beacons, so the flooding overhead is reasonable.

Next, we discuss the path stretch, i.e. the length of the BVR routing path over the shortest path. The average path length in the above described simulations is 17.5, and the path stretch is 1.05. Furthermore, the authors argue, that in all their simulations the path stretch never exceeded 1.1, resulting in a low overhead over shortest path routing.

11.4.4 Concluding BVR

Beacon Vector Routing is a promising approach to provide point-to-point routing in wireless sensor networks. The efficient and low overhead communication as well as its simplicity, resulting in an easy implementation on sensor nodes with limited resources, are the main advantages of BVR. Furthermore, simulation as well as testbed implementations show that BVR achieves good performance and can even exceed the quality of geographic routing, as BVR uses the position of a node in the topology and not its geographic position.

11.5 Algorithmic View

After presenting virtual coordinate based routing, we take a step back and look at virtual coordinate based routing from a more algorithmic point of view. Especially, we discuss the number of beacon nodes necessary to provide unique names for each node (naming problem), and how to route based on the unique names.

11.5.1 Introducing the Network Model

Let the network consist of n nodes, and k of these nodes, denoted by $a_1, a_2, \ldots, a_k$, be beacon nodes. Furthermore, assume that there exists a unique order on the beacon nodes: $a_1 \prec a_2 \prec \cdots \prec a_k$. Then, each node is able to determine its graph distance (in hops) d_i to each of the k beacon nodes. Based on this hop distance each node can determine its virtual position in the network, similar to BVR. Thus, the network can be embedded into a k-dimensional space and the position of a node is $(d_1, d_2, \ldots, d_k)$.

11.5.2 Bounding

We are interested in four points:

1. A lower bound, which describes the minimum number of beacon nodes necessary. I.e. the number of beacon nodes necessary when these are placed optimally.
2. An upper bound, e.g. the worst case, for the number of beacon nodes.

3. Whether and with how many beacon nodes the naming problem can be solved locally. Locally in this case means using neighboring nodes only.
4. How can routing be done in such an environment?

In the following sections we discuss these points for various topologies, namely a line of nodes, a grid layout and the unit disk graph. The corresponding paper [389] also discusses ring, tree, butterfly, and hypercube graphs. Please refer to it for the corresponding graphs and proofs.

11.5.3 A Line of Nodes

For a line of nodes, we can observe:

1. **Lower bound:** Choosing a node with degree one solves the naming problem.
2. **Upper bound:** Two arbitrarily chosen nodes solve the naming problem
3. **Local naming:** Any two neighboring nodes solve the naming problem.
4. **Routing:** By comparing the position of the destination with its own position, the routing node can decide in which direction to route. Furthermore, the resulting route is the shortest path.

11.5.4 Grid

If we choose the anchors properly, we can solve the naming problem with two beacon nodes in the grid. For example, one node a in the upper left corner and a node b in the upper right corner of the grid solve the naming problem.

Once all nodes have unique names, data can be routed on the shortest path. A node a can determine the direction each beacon node lies in by comparing its own position to the position of its neighbors. By routing towards the neighbor that is the closest to the destination (greedy forwarding) the packet is routed on the shortest path.

1. **Lower bound:** Two properly chosen nodes solve the naming problem.
2. **Upper bound:** $(\sqrt{n} - 1)^2 + 2$ arbitrarily chosen nodes solve the naming problem.
3. **Local naming:** It is not possible to solve the naming problem locally.
4. **Routing:** By comparing the position of the destination with the position of the neighbors and its own position, the routing node can decide in which direction to route. Furthermore, the route is the shortest path.

11.5.5 Unit Disk Graph

Unit disk graphs are graphs were there is an edge between two nodes if and only if their Euclidean distance is less or equal one. Thus, compared to the grid layout, a unit disk graph is less regular and so the lower bound depends on the graph layout. For the unit disk discussion we therefor look at an maximum lower bound.

	lower bound	upper bound	local	shortest path routing
Line	1	2	√	√
Ring	2	3	√	√
Grid	2	$(\sqrt{n}-1)^2 + 2$	-	√
Tree	$mc(T)$	$n-1$	-	√
Unit Disk	$((\sqrt{n}+9)12)^2$	$n-1$	-	√
Butterfly	$\frac{n}{\log n}$	$n-1$	-	-
Hypercube	$\frac{\log n}{\log \log n}$	$n/4 + 1$	√	√

Table 11.5. Upper and lower bounds for various topologies.

1. **Lower bound:** There are unit disk graphs were at least $((\sqrt{n}+9)/12)^2$ beacon nodes are needed to provide unique node names.
2. **Upper bound:** There are unit disk graphs for which $n-1$ beacon nodes are needed.
3. **Local naming:** It is not possible to solve the naming problem locally.
4. **Routing:** When the nodes have unique names, e.g. unique coordinates, routing decisions can be made similar to the approach described in the Grid section. Furthermore, the route is the shortest path.

The upper bound is quite interesting as it is very large. As we expect radio propagation in sensor networks to behave similar to the unit disk graph, the bounds in this section give an interesting insight into the number of beacon nodes needed.

11.5.6 Conclusion

In this section we discussed geographic routing from a more algorithmic point of view. We presented upper and lower bounds on the number of beacon nodes necessary for unique naming and shortest path routing for various topologies. However, we expect a "real world" sensor network not to be as regular as, for example, a unit disk graph and so we expect the bounds to be even worse.

11.6 Related Work

Next to the described algorithms, we want to mention and shortly discuss three more routing algorithm based on virtual coordinate systems.

Logical Coordinate Routing [60] is similar to BVR as it uses a set of beacon nodes to assign coordinates to each node. Similar to BVR, the routing algorithm tries to minimize a distance function until the destination or a local minimum is reached. When a local minimum, e.g. dead end, is reached, the packet is tracked back along its path, until a valid route is found. As a result, no scoped floods as in BVR are needed, but each node needs to store the forwarded packets for back tracking. This increases the storage overhead.

The NoGeo algorithm [328] does not use landmarks, or beacons to build a virtual coordinate system. Instead it uses an iterative relaxation to find the nodes on the border of the network and embed all nodes into a Cartesian coordinate space. The Cartesian coordinate space reduces the computational overhead of the routing algorithm, as routing in a two dimensional space is less complex than routing in a multi-dimensional vector space as done in BVR. However, NoGeo requires $O(\sqrt{n})$ nodes to flood the network and all these flooding nodes are required to store the hop matrix between all flooding nodes, resulting in $O(\sqrt{n} * \sqrt{n}) = O(n)$ states at $O(\sqrt{n})$ nodes. For today's sensor nodes such a storage requirement seems to be impractical.

In [290] a more algorithmic approach is taken. It presents an embedding of a multidimensional beacon space into 2-dimensional space and discusses its quality. Although the algorithm provides interesting insight into the quality of the embedding, the it requires a global view on the network and therefore its application is limited.

11.7 Chapter Notes

It is necessary that basic communication patterns as any-to-any communication work reliably and scalably in the resource constrained sensor network environment. Recently proposed applications for sensor networks (see section 11.1) require such routing paradigms. Thus, without any scalable any-to-any routing the new applications and the ones still to come, cannot be deployed and evaluated.

In this chapter we presented and discussed a routing scheme, namely "Beacon Vector Routing" [132], which provides scaleable any-to-any routing. It works in the resource constrained sensor network environment and does not make assumptions on the availability of positioning systems such as GPS. Furthermore, the evaluation shows, that the presented algorithm performs as well as geographic routing and under some conditions can even outperform it. Additionally, it proposes little bandwidth and memory overhead compared to ideal scaling properties of geographic routing schemes. Other algorithms taking a similar approach are Logical Coordinate Routing [60], the NoGeo algorithm [328], and [290].

Additionally, this chapter discussed algorithmic bounds for naming and routing in various topologies [389]. Such algorithmic bounds are important as they are the limits for implementing routing protocols in the "real world".

12

Minimal Range Assignments for Broadcasts

Christian Gunia

12.1 Introduction

Providing people with up-to-date information has become more and more crucial over the past centuries. Right now, this is mostly done via newspapers or television. While this is no big issue in densely populated areas like cities, where a reliable infrastructure exists, it is rather challenging in other regions like deserts or jungles. Distributing daily newspapers could involve enormous cost. In order to solve this problem for a region without a highly connected infrastructure, one could imagine a digital newspaper broadcasted via wireless links between houses—in the following called stations. This process starts at one station s acting as a server. To keep operation expenses as low as possible multi-hop strategies are used. In this chapter we consider a single-hop strategy to be in particular a multi-hop strategy; in that way this solution is implicitly regarded as well. The resulting question is then how to choose the transmission range of each station. To measure the quality of such a range assignment we use the sum of the energy needed to provide it: based upon experiences and theoretical research from communications engineering (see Section 2.3), the power P_u required by a station u to correctly transmit data to another station v must satisfy

$$\frac{P_u}{d(u,v)^\alpha} > \gamma, \qquad (12.1)$$

where $d(u, v)$ is the Euclidean distance between u and v, $\alpha \geq 1$ is the path-loss exponent and $\gamma \geq 1$ describes properties of the used technology (including antenna characteristics, transmission quality, etc.). Fixing the range assignment induces the transmission graph: edge (u, v) represents the ability of u to communicate with v. Note that this graph is directed since the transmission ranges of u and v may differ. The task is to find a range assignment that keeps the sum of the energy used by all stations as low as possible while guaranteeing s the ability to send messages to all other stations. We assume all stations to be embedded in the plane, i. e., the problem is 2-dimensional, and refer to [77] for a generalization of this Range Assignment Problem to d-dimensional space.

D. Wagner and R. Wattenhofer (Eds.): Algorithms for Sensor and Ad Hoc Networks, LNCS 4621, pp. 215–235, 2007.

At first glance, the choice of a station's transmission range may seem to be inherently continuous. However, one can easily verify that this range can be chosen in discrete steps: have a look at an arbitrary station u. Let v, v' be two other stations fulfilling $\not\exists$: $d(u, v)$ $d(u,)$ $d(u, v')$. Increasing u's transmission range continuously from $d(u, v)$ to $d(u, v')$ will not derive in a change of the transmission graph until reaching $d(u, v')$. Hence, the problem can be seen as a discrete optimization problem. Another viewpoint could be useful. Let the complete graph G consist of all stations and the weights of the edges represent the energy needed to establish a connection between them. Is the original problem equivalent to choose a subset of edges providing an minimum spanning tree (MST) rooted at s?—Not exactly, choosing an edge can render other adjacent edges obsolete as they are chosen implicitly without causing additional cost (see Figure 12.1(a)). To be more precise, only the most expensive of all selected edges at each node will raise cost. Nevertheless, the connection between the original problem—which has been proven to be $\mathcal{NP}$-hard for any dimension greater 1 and each $\alpha > 1$ by Clementi et al. [77]—and MST is strong enough to be exploited.

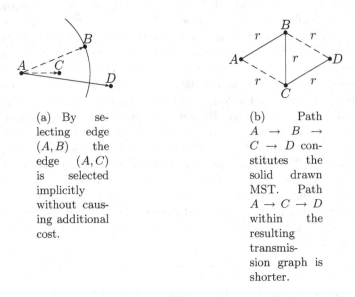

(a) By se-lecting edge (A, B) the edge (A, C) is selected implicitly without caus-ing additional cost.

(b)   Path $A \rightarrow B \rightarrow C \rightarrow D$ con-stitutes the solid drawn MST. Path $A \rightarrow C \rightarrow D$ within the resulting transmis-sion graph is shorter.

Fig. 12.1. Example for 'induced' edges and multiple paths.

After this short introduction we will continue to present an algorithm using this connection in the next section. Since this algorithm is based upon the computation of an MST, details on its distributed computation are presented in Section 12.3. Section 12.4 will show up new achievements on the Range Assignment Problem. Finally, we will finish with conclusions, open questions and chapter notes in Section 12.5.

12.2 The Algorithm RAPMST and Its Analysis

Given the problem instance by the location of its n stations, the source s, $\alpha \geq 1$ and $\gamma \geq 1$ we consider the algorithm RAPMST (see Algorithm 29), that computes a range assignment by employing the above mentioned considerings about the connection to MSTs. Observe that the unique route from s to another station u induced by T will also be available within the resulting transmission graph due to the choice of the transmission energy. However, it neither has to be the only existing route nor the shortest route from s to u (see Figure 12.1(b)).

---

1 Compute the undirected version $G = (V, E_G)$ of G_D.
2 Compute an MST $T = (V, E_T)$ on graph G.
3 Make T downward oriented by rooting it at s.
4 To create the range assignment r choose the transmission energy of station v
   to be the maximum of the weights of v's outgoing edges.

Algorithm 29: The algorithm RAPMST.

We will bound algorithm RAPMST's approximation ratio from above. The key idea is to find a connection between the size of an MST and the cost caused by an optimal range assignment. Therefore, for the moment we consider the case $\alpha = 2$. The energy needed to establish a link between two nodes is proportional to the square of their Euclidean distance. Hence, we associate each edge of the MST with a disk whose diameter depends on the edge's length. By bounding the area covered by them from above and showing that these disks are disjoint, we will obtain the following theorem [77].

Theorem 12.2.1. *For* $\alpha \geq 1$ *the cost induced by the solution of algorithm* RAPMST *is at most* $5^{\alpha/2} \cdot 2^\alpha$ *times the optimal cost.*

Before proving Theorem 12.2.1 we want to develop some tools making the proof easier and more straightforward. For any fixed ordering of T's edges, let $e_i = (u_i, v_i)$[11] be the i-th edge in T and c_i its center. We denote by D_i the diametrical open disk of e_i, i.e., $D_i := \{x \in \mathbb{R}^2 \mid d(x, c_i) < d(u_i, v_i)/2\}$. Finally, D_i^* and $\overline{D_i^*}$ represent the open and closed disk, respectively, with the same center but half the radius.

Lemma 12.2.2. *For any edge* $e_i = (u_i, v_i) \in V$, *no other node than* u_i *and* v_i *is contained in disk* D_i.

---

[11] We use (u, v) to denote both a directed and an undirected edge between nodes u and v; although this usage is polysemous the context pinpoints its exact meaning.

Proof. Suppose there is a node $x \in V$ that is contained within disk δ_i. Since T is an MST, it contains either a path from x to u_i without touching v_i or vice versa. Without loss of generality, we assume this path to end in u_i (see Figure 12.2). By adding (x, v_i) to T a cycle containing edge e_i is closed. Since δ_i is an open disk, $d(x, v_i) < d(u_i, v_i)$ holds and removing e_i leads to a spanning tree T' of less weight. This is a contradiction to the optimality of T. $\square$

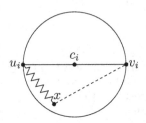

Fig. 12.2. Visualization of the situation of Lemma 12.2.2: path (x, u_i) is cheaper than (u_i, v_i).

With this lemma at hand, we can show an even stronger property: δ_j not only contains no other node u_i or v_i, but even no center c_i.

Lemma 12.2.3. *For any $i, j \in \{1, \ldots, |V| - 1\}$ with $i \ne j$, center c_i is not contained in disk δ_j.*

Proof. To obtain a contradiction assume there are two disks δ_i and δ_j such that c_i is contained in δ_j. Without loss of generality, we further assume $d(u_j, v_j) \ge d(u_i, v_i)$; otherwise we switch roles of e_i and e_j. Partition the plane along the line containing centers c_i and c_j into half-planes H^+ and H^-. We assume nodes v_i and v_j to be contained in H^+ and nodes u_i and u_j in H^-, respectively. Let $z \in H^-$ be the intersection point of the two circumferences determined by disks δ_i and δ_j. The diametrically opposite points of z with respect to δ_i and δ_j are denoted by z_i and z_j, respectively. We want to show that

$$\max\{d(u_i, u_j), d(v_i, v_j)\} \overset{(I)}{\le} d(z_i, z_j) \le d(u_j, v_j)$$

holds. Just for the moment, we assume $d(v_i, v_j) \ge d(u_i, u_j)$. Figure 12.3 summarizes the situation obtained so far.

According to Lemma 12.2.2 node u_i is not contained in disk δ_j and node u_j is not contained in disk δ_i, $d(v_i, v_j) \le d(z_i, z_j)$. Therefore inequality (I) holds. One can easily verify that the triangles $(z\, z_i\, z_j)$ and $(z\, c_i\, c_j)$ are similar. As center c_i was assumed to be inside of disk δ_j, clearly, $d(z_i, z_j) \le d(z, z_j)$ holds. Hence, by similarity, $d(z_i, z_j) \le d(z, z_j) = d(u_j, v_j)$ is also fulfilled. By redefining z to be the intersection point within the half-plane H^+, we

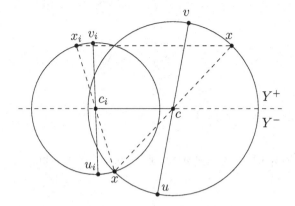

Fig. 12.3. Visualization of the construction used in Lemma 12.2.3.

obtain the same result in the case $d(v_i, v_j) \quad d(u_i, u_j)$. Therefore, the claimed inequality is fulfilled:

$$\max\{d(u_i, u_j), d(v_i, v_j)\} \quad d(u_j, u_j).$$

Depending on the actual topology of the MST T, inserting either edge (u_i, u_j) or (v_i, v_j) into T closes a cycle within T which contains the edge $e_j = (u_j, v_j)$. Thus, removing this edge results in a spanning tree T' with $(T') \quad (T)$. This contradicts the minimality of T. □

Lemma 12.2.4. *For any* $i, \in \{1, \ldots, |V| - 1\}$ *with* $i \neq$, *disks* $_i^*$ *and* $_j^*$ *are disjoint.*

Proof. We adopt the former notations and, additionally, denote the radii of disks $_l$ and $_l^*$ by r_l and r_l^*, respectively. Due to Lemma 12.2.3

$$\frac{d(_i, _j)}{2} \geq \frac{\max\{r_i, r_j\}}{2} = \max\{r_i^*, r_j^*\}$$

holds which implies our claim. □

Lemma 12.2.5. *Let* *be the smallest closed disk containing all the points in* V *and let* diam *be its diameter. Furthermore, let* $'$ *be* *whose radius was increased by a factor* $\sqrt{5}/4$ *and* $U := \cup_{e_i \in T} \overline{_i^*}$. *Then,* U *is contained within* $'$.

Proof. We will show that each disk $\overline{_i^*}$ is contained within $'$. Consider an arbitrary $e_i = (u_i, v_i)$ and its diametral disk $_i$. Since the diametral disk of all points $'$ and $'$ laying on $\overline{}$ is contained within $_i$, we assume, without loss of generality, nodes u_i and v_i to lie on the boundary of . Let z be an arbitrary point within disk $\overline{_i^*}$. Although this is an open disk, we assume z to lie on its 'boundary', i.e., $d(z, _i) = d(u_i, v_i)/4$ (see Figure 12.4).

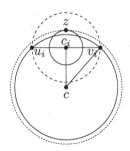

Fig. 12.4. Worst-Case position of disk D_i^* within D'.

By applying the triangular inequality and the Pythagorean Theorem we get

$$d(\ ,z) \le d(\ ,\ _i) + d(\ _i, z) = d(\ ,\ _i) + d(u_i, v_i)/4$$

and

$$d(\ ,\ _i)^2 + d(\ _i, v_i)^2 = d(\ , v_i)^2 = \mathrm{diam}^2/4.$$

Hence,

$$d(\ , z) \le \sqrt{\frac{\mathrm{diam}^2 - d(u_i, v_i)^2}{4}} + d(u_i, v_i)/4$$

is fulfilled, reaching its maximum of $\sqrt{5}/4 \cdot \mathrm{diam}$ for $d(u_i, v_i) = \mathrm{diam}/\sqrt{5}$. $\square$

Theorem 12.2.6. *Let* diam *denote the diameter of the smallest disk containing all stations. Then, for any MST T of G,* $(T) \le 5^{\ 2} \cdot \mathrm{diam}$ *holds.*

Proof. We first prove our claim for $\alpha = 2$ and afterwards extend it to general $\alpha \ge 2$. For $\alpha = 2$ we have to prove

$$(T) = \sum_{i=1}^{|\ |-1} d(u_i, v_i)^2 \le 5 \cdot \mathrm{diam}^2.$$

Let Area$(\ _i^*)$ denote the area of disk $\overline{\ _i^*}$. Since the radius of $\ _i^*$ equals $d(u_i, v_i)/4$, we obtain

$$\sum_{i=1}^{|\ |-1} d(u_i, v_i)^2 = \frac{16}{\pi} \underbrace{\sum_{i=1}^{|\ |-1} \mathrm{Area}(\ _i^*)}_{(*)}.$$

Due to Lemma 12.2.5 the size of the union of $\ _i^*$'s area sums up to at most $\pi \left(\sqrt{5}/4 \cdot \mathrm{diam} \right)^2$. From Lemma 12.2.4 we learn that the $\ _i^*$ are pairwise disjoint. Hence, the size of this union equals ( ). This induces the theorem for the case $\alpha = 2$. For $\alpha > 2$ we obtain by using Minkowski's inequality

$$w(T) = \sum_{i=1}^{|V|-1} d(u_i, v_i)^\alpha = \sum_{i=1}^{|V|-1} \left(d(u_i, v_i)^2\right)^{\alpha/2}$$

$$\leq \left(\sum_{i=1}^{|V|-1} d(u_i, v_i)^2\right)^{\alpha/2} \leq 5^{\alpha/2} \cdot \mathrm{diam}^\alpha$$

which completes the proof of this theorem. □

Finally, with this collection of tools Theorem 12.2.1's proof is straightforward.

Proof (of Theorem 12.2.1). Since the MST T computed by the algorithm RapMst is obviously in particular a spanning subgraph, it is sufficient to show that there exists a spanning subgraph $G' \subseteq G = (V, E_G)$ such that $w(G') \leq 5^{\alpha/2} \cdot 2^\alpha \cdot$ opt. Let r_{opt} be an optimal range assignment, i.e., opt $= \sum_{v \in G} r_{\mathrm{opt}}(v)$ holds. For any $v \in V$ define

$$S(v) := \{u \in V \mid w(v, u) \leq r_{\mathrm{opt}}(v)\}$$

and let $T(v)$ be an MST of the subgraph induced by $S(v)$. Recall that the weight of each edge in G equals its Euclidean length to the α-th power. Hence, the diameter of $T(v)$ is as most $2 \cdot (r_{\mathrm{opt}}(v))^{1/\alpha}$ which results in $w(T(v)) \leq 5^{\alpha/2} \cdot 2^\alpha \cdot r_{\mathrm{opt}}(v)$ due to Theorem 12.2.6.

Consider the spanning subgraph $G' := (V, E') \subseteq G$ such that

$$E' := \bigcup_{v \in V} \{e \in E \cap T(v)\}.$$

We conclude that

$$w(G') \leq \sum_{v \in V} w(T(v)) \leq 5^{\alpha/2} \cdot 2^\alpha \cdot \sum_{v \in V} r_{\mathrm{opt}}(v) = 5^{\alpha/2} \cdot 2^\alpha \cdot \mathrm{opt}.$$

□

We want to point out that the algorithm RapMst is solely based upon the computation of an MST. Therefore, all known techniques for computation and maintenance of MSTs in static and dynamical systems are applicable.

In most ad hoc networks there is no infrastructure known a priori and no dedicated station that takes care of establishing links between the stations. Therefore, at the beginning it is (nearly) impossible to guarantee an organized communication. Often, to overcome this issue as fast and efficient as possible a routing plan is developed in a distributed manner. Hence, a protocol is invoked and the stations following this protocol build up an organized network to guarantee a certain Quality of Service regarding the communication. In our case this Quality of Service is to establish a structure capable of relaying the broadcast from station s to all stations. In the next section we present a way to compute an MST in a distributed fashion. We observe that this MST can be converted into a routing plan locally by following Step 4 of the algorithm RapMst.

12.3 Distributed Computation of an MST

In the last section we have seen that the computation of a "'good"' range assignment in a distributed fashion can be done provided that an MST is given. In this section we want to present a technique of computing the MST in a distributed manner as well. It is based upon a work of Kutten and Peleg [149]. First of all we have to specify the model in more detail. In the model we use, each node of the graph $G = (V, E_G)$ has the ability to process information and communicate with its adjacent nodes. This communication is done over G's edges and synchronously in rounds. Thereby, each edge can transfer $O(\log n)$ bits per round. The weight of an edge corresponds to the energy needed to establish a link between the two stations at its endpoints. We use the common assumption that nodes have unique identifiers and the weight $w(e)$ of each edge e is known to all adjacent nodes. At first glance, the last assumption seems to be very restrictive: as mentioned before, in an ad hoc network typically no station is aware of any other station a priori. Nevertheless, this information can be gathered via handshakes. Probably one of the easiest ways is achieve this to exchange the stations' IDs in a first step. Afterwards, assigning each station an unique time slot and performing the handshake in a round robin fashion. Therefore each station sends a "hello-world" message within the time slot assigned to her. During this procedure stations have to send at their maximal power level to guarantee reaching each other station within their range. As this method is crude, its energy consumption is likely far from being optimal. Nevertheless, in situations where the system's topology is static for a "long" period of time this method suffices as the energy spent during the operation phase compensates for the relatively high cost induced by this initialization phase.

The task is to compute an MST T, i.e., after the computation is done each node has to know the subset of its adjacent edges belonging to T. Hence, collecting this information over all nodes would admit constructing T. To point out that this MST T is consistent over all nodes, we henceforth use the term *the* MST; well aware that it does not have to be unique. Since the time bound we are going to show will allow us to synchronize the computation between all nodes, we assume it to be already synchronized in the first place. As the bottleneck of most distributed protocols is the limited communication between the nodes—instead of the computational power in each node—we are interested in the number of rounds until the MST is computed. The actual size of these rounds will not concern us, as we assume that the nodes have sufficiently large computational power to perform their computation within an arbitrarily small amount of time.

Unfortunately, the protocol used here is somewhat more involved than Kruskal's well-known algorithm for computing an MST as it tries to keep the computation "local". Hence, we try to find a trade-off between a detailed and a comprehensible formalization of the protocol. For a more detailed description we refer to Kutten and Peleg [246]. The key idea is to compute "easy parts" of the MST T first. Afterwards, compose these parts to finally form T. The following definition is useful for specifying these parts.

Definition 12.3.1. *A* (σ, ρ) *spanning forest of a graph* $G = (V, E)$ *is a collection of node disjoint trees* $\{T_1, \ldots, T_m\}$ *with the following properties:*

1. *The trees consist of edges of* E *and span all the nodes of* V.
2. *Each tree contains at least* σ *nodes.*
3. *The radius of each tree is at most* ρ.

We compute a $(k + 1, n)$ spanning forest of G with the additional property that each tree forms an MST-fragment, i. e., each tree is a subtree of T. The next section explains how this goal can be achieved by applying the protocol SIMPLEMST. The parameter k is chosen appropriately later on.

12.3.1 Protocol SIMPLEMST

In the following, we use the expressions "fragment" and "tree" interchangeably since each MST-fragment consists of a tree. In the beginning, each node forms its own fragment. The synchronous protocol operates in phases $1, 2, \ldots, \lceil \log(k + 1) \rceil$ where phase i lasts $O(2^i)$ rounds[12]. At any point of time, each fragment contains one specified root node which aggregates the information collected inside the tree. Furthermore, each node knows which of its adjacent edges belongs to T restrained to its fragment. We describe the behaviour of an arbitrary fragment A in phase i.

In phase i the root of fragment A first finds out whether the depth of its tree is larger than 2^i. All trees with a depth of more than 2^i are inactive, the remaining ones are active within the current phase. Obviously, this can be done by broadcasting a counter down the tree: if this counter exceeds 2^i, a failure message is routed back to the root. Whenever we send data over a given structure to collect information in a tagged node, we call this process a convergecast. Hence, the failure message is convergecasted back to the root. This operation can be finished within $2 \cdot 2^i$ rounds. Afterwards, in each active fragment, the root broadcasts its identity over the tree. Hence, after 2^i more rounds each node knows whether it is in an active or inactive fragment. At this point, each node belonging to an active fragment transmits the identity of its root over all adjacent edges. By the next round each node v receives over some of its edges identities of the corresponding roots. Those edges over which it receives an identity equal to that of its own root are recognized as *internal* edges leading to other nodes inside the same fragment. The remaining edges are *outgoing* edges.

Since each node knows its adjacent edges belonging to the same fragment, it can decide weather it is a leaf or not. Each leaf of an active fragment selects

---

[12] The exact number of rounds can be determined later on. As it is only technical, we will omit it here.

its minimum weight outgoing edge and convergecasts this weight to the root. Thereby, each inner node waits until it has collected all convergecasted weights of its children. Then it forwards the minimum of these weights and its own outgoing edges to the root. Each node remembers the child—or itself—whose weight was transmitted. Hence, the route to the minimum weight outgoing edge e can be reconstructed.

As the root receives the weight of this edge, it stops being root and sends a message declaring the endpoint of e that belongs to the root's fragment to be the new root. This new root waits till the time $T = O(2^i)^{13}$. The new root then notifies the other endpoint of e that it wishes to connect. If that endpoint does not answer within the next round—by sending a connection-message as well — the root stops being a root. In that case the other fragment is responsible for selecting a (new) root for the combined fragment. However, if the other endpoint answers, the one with the larger identity number becomes the new root. The edge e is marked as an MST edge by both endpoints.

Lemma 12.3.2. *The* SimpleMst *protocol terminates after $O(k)$ rounds. When it terminates, the collection of resulting fragments forms a $(k + 1, n)$ spanning forest $\mathcal{F}$ for the graph. Moreover, each tree of this forest is a fragment of the same MST of the graph.*

Proof. The time bound results from the length of each phase: since each phase lasts $O(2^i)$ rounds, all $\lceil \log(k + 1) \rceil$ phases are completed within $O(k)$ rounds.

At the beginning of the protocol each fragment is of size 1. Have a look at a smallest fragment. Assume that at the beginning of phase i its size is at least 2^{i-1}. Since each active fragment merged with some other fragment, the size of a smallest active fragment at least doubles, i.e., is at least 2^i at the end of this phase. Inactive fragments already have at least this size. Hence, at the end of phase $\lceil \log(k + 1) \rceil$ each (smallest) fragment is of size at least $k + 1$. Due to the pigeonhole principle the number of fragments is bounded above by $n/(k + 1)$. The depth of each one is trivially bounded above by n.

Clearly, an MST can be constructed by starting with an empty set and step by step adding an edge with the lowest weight that is not closing a cycle. Exactly the same operation is done here, but in a parallel manner: each fragment tries to connect to another fragment using its minimal weight outgoing edge. However, we do not obtain a complete MST since the operation is aborted after $\lceil \log(k + 1) \rceil$ rounds. There is no guarantee that all fragments are connected by that time. $\qquad\square$

Since the protocol SimpleMst terminates after $O(k)$ rounds we only obtain a forest of fragements of the MST T. Hence, it remains to choose the right subset of inter-fragment edges to complete the MST. We will present a procedure for this task later on, but first limit the depth of the given fragments in order to reduce the running time of the remaining computation.

---

[13] Details on the choice are omitted here; however, note that all fragments are henceforth synchronized.

12.3.2 Constructing a $(k+1, O(k))$ Spanning Forest

By applying the protocol SIMPLEMST on G, we have already obtained a partition into trees T_i that are subtrees of T. Hence, it is sufficient to describe how each one of these trees is split up into trees of size at least $k + 1$ and radius at most $O(k)$. This is done in detail by Kutten and Peleg [246]. Here we will only present their key idea and obtain a forest consisting of trees of size at least $k + 1$ and radius $O(k^2)$. Furthermore, the running time of our protocol will not satisfy the needs of $O(k \log^* n)$. We will make use of the following definition.

Definition 12.3.3. *A balanced dominating set of a graph $G = (V, E)$ with n nodes is a set of nodes $D \subseteq V$ and a partition P of G's nodes if and only if*

a) $|D| \le \lceil n/2 \rceil$.
b) D is a dominating set, i. e., each $v \in D \backslash V$ has a neighbor inside D.
c) Each cluster of the partition P has at least two nodes (including the nodes in D).

We focus on an arbitrary tree T_i. We assign each node v its distance $L(v)$ from the next leaf according to

$$L(v) := \begin{cases} 0, & v \text{ is a leaf of } T_i \\ 1 + \min_{u \in \text{child}(v)} L(u), & \text{otherwise} \end{cases}$$

and partition the tree into levels $L_j := \{v | v$ is node of T_i and $L(v) = j\}$. The protocol BALANCEDDOM (see Algorithm 30) shows how to compute a balanced dominating set of tree T_i in time $O(log^* n)$.

---

1 Mark nodes on Levels L_0, L_1 and L_2.
2 Select Maximal-Independent-Set Q on unmarked nodes R.
3 $D := Q \cup L_1$.
4 For every singleton $\{v\} \in P$, v quits the set D and selects an arbitrary neighbor $u \notin D$ as its dominator.
5 Each node $u \notin D$ that was selected as a dominator in step (4) adds itself to D, quits its old cluster in P, and forms a new cluster, consisting of itself and all the nodes that chose it as their dominator.
6 Consider a node $v \in D$ whose cluster in the modified partition P is now a singleton. Then v chooses arbitrarily one node U that was in v's cluster in the original partition and left in step (4). Also, v quits the dominating set.

Algorithm 30: The protocol BALANCEDDOM.

Lemma 12.3.4. *Protocol BALANCEDDOM computes a balanced dominating set of tree T_i in time $O(log^* n)$.*

Proof. For the proof of the time bound we refer to Goldberg et al. [163], but we want to point out that the computation of the Maximal-Independent-Set dominates the running time. The remaining steps can be done in constant time.

In order to show the protocols's correctness we first observe that D actually is a dominating set since $V = L_0 \cup L_1 \cup L_2 \cup R$ holds. The set deriving of Step 6 is still a dominating set since each node leaving D selects a valid dominator. It remains to show that in Step 6 there exists a node u which v can choose. Node v has to be contained within the original partition since all nodes joining D in Step 5 are not within a singleton. Furthermore, v was not contained within a singleton in the original partition since it would have quit it in Step 4. Hence, there has to be an adequate neighbor u in Step 6. Therefore, properties (b) and (c) hold, and (a) follows from (c).   □

By using the protocol BALANCEDDOM we define the following protocol DOM-PARTITION(k) (see Algorithm 31) that is applied to each tree T_i.

---

1 **for** $\lceil \log(k+1) \rceil$ times **do**
2   Perform protocol BALANCEDDOM, assigning each node not in the dominating set to a cluster with an arbitrary neighbor in the dominating set.
3   Contract each cluster to one node.
4 **end**

Algorithm 31: The protocol DOMPARTITION(k).

Lemma 12.3.5. *Protocol* DOMPARTITION(k) *constructs in time* $O(k^2 \log^* n)$ *a partition* P *in which each cluster* C *consists of at least* $k + 1$ *nodes and its radius is bounded above by* $2k^2$.

Proof. Let us omit the precise details on how to implement this in a distributed manner momentarily. For the time being, we focus on the size and the depth of the constructed clusters. In each iteration of the algorithm the size of the smallest clusters at least doubles since it merges with a cluster whose size is at least equally large. As the minimal cluster has the size 1 in the beginning of the algorithm, it reaches at least $k + 1$ after $\lceil \log(k+1) \rceil$ iterations. Let $d(i)$ denote the maximal cluster radius after the i-th iteration. Obviously, $d(0) = 0$ holds since we start with n single clusters. To figure out what happens to the cluster's radius within one iteration we observe that each newly formed cluster consists of old clusters composed in a star-like manner. Therefore, we gain a situation depicted in Figure 12.5. Let v be the root of the old cluster in the middle used in the definition of its radius. Note that the diameter of each cluster is at most twice its radius. Since the distance from v to each cluster is at most $d(i) + 1$ we obtain $d(i) \leq 3 \cdot d(i-1) + 1$.

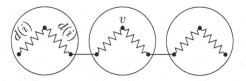

Fig. 12.5. The star-like structure occurring by building new clusters.

Solving this with the constraint $d(0) = 0$ leads to $d(i) = \sum_{j=0}^{i-1} 3^j = (3^i - 1)/2$. Substituting i by $\lceil \log(k+1) \rceil$ leads to $2 \cdot (k+1)^{1\,6}$ as an upper bound for the radius of each cluster. This is at most $2k^2$ for each $k \geq 4$. Hence, we proved the clusters' claimed attributes and focus on the distributed implementation. In each cluster we appoint one of its nodes as a representative of this cluster. Whenever a cluster is requested by BALANCEDDOM to perform an operation this node does it, while the remaining nodes merely serve as communication links between the clusters. Since the radius of each cluster is at most $2(k+1)^{1\,6}$, $O(\log(k+1))$ iterations are sufficient and each appliance of BALANCEDDOM costs $O(\log^* n)$ rounds; the time bound follows.   □

As mentioned above, the radius of the clusters is too large for our purposes. We need clusters with a radius of $O(k)$. The following lemma states that such clusters can be computed.

Lemma 12.3.6. *A modification of algorithm* DOMPARTITION*(k) constructs in time $O(k \log^* n)$ a partition P in which each cluster C consists of at least $k + 1$ nodes and its radius is bounded above by $O(k)$.*

Proof (Sketch). The main idea is to remove a cluster that reaches the radius k from the tree. Hence, this cluster stops growing. There are two major difficulties with this approach. First, the original tree may be split up into trees. This can be handled by generalizing BALANCEDDOM canonically to work on trees. The second problem is that it may occur that "shallow" clusters, i. e., clusters not having reached the desired number of $k+1$ nodes, may not be able to join neighboring clusters, since all of them have been removed. We overcome this difficulty by also removing them from the tree and storing them separately. After all merging steps are finished, we will merge these shallow trees once with their—sufficiently large—neighboring clusters. Hence, we can guarantee a minimal size of $k + 1$ and an upper bound of $O(k)$ for the radius of each cluster.   □

12.3.3 Selecting the Missing Edges

If we first use the protocol SIMPLEMST to compute the trees T_i and afterwards apply the protocol cited in Lemma 12.3.6 to each tree T_i, we obtain a forest $\mathcal{F}$ that consists of tree that are fragments of the MST T, consist of at least $k+1$

nodes and whose radius is bounded above by $O(k)$. It remains to determine the "correct" subset of inter-fragment edges to complete the MST T. Assume for the moment, there is no limit for the number of bits broadcasted over an edge in a single round. Hence, we could compute the MST in the following way: appoint an arbitrary node as the root $r(B)$ and compute a breadth-first-search tree B by starting in $r(B)$. Afterwards, beginning at the leaves, convergecast all inter-fragment edges to the root. Now the root $r(B)$ holds all needed information to apply Kruskal's algorithm and broadcast the resulting MST down the tree B. Let $diam(G)$ denote the diameter of G measured in terms of hops. Therefore, after $O(diam(G))$ rounds T is computed. However, as there is a limit on the edges, we have to modify this approach to reduce the number of convergecasted edges.

The fragments constructed by the execution of SIMPLEMST have at least size k and are disjoint. Hence, their number is at most n/k. Let us have a look at the protocol PIPELINE (see Algorithm 32).

Clearly, the algorithm's first step can be done in a distributed manner within $O(diam(G))$ rounds since the convergecast of a counter is sufficient: if a station receives the counter for the first time, it increases the counter by one and forwards it over its edges. Similarly, the last step can also be done in $O(diam(G))$ rounds.

The next Lemma states some properties of the execution of PIPELINE that help us to show its correctness and running time. We omit the technical proof of the lemma here and refer to Garay et al. [149].

Lemma 12.3.7. *If procotol PIPELINE is applied to the spanning forest $\mathcal{F}$, it holds:*

1. *Every node v starts sending messages upwards at round*

$$\hat{H}(v) := \begin{cases} 0 & \text{if } v \text{ is a leaf.} \\ 1 + \max_{u \in child(v)}\{\hat{H}(u)\} & \text{otherwise} \end{cases}$$

2. *Node v upcasts edges in nondecreasing order.*
3. *After a node v has terminated its participation in the algorithm, it will learn of no more edges reportable to the root $r(B)$.*

Since there are at most $O(n/k)$ fragments, each node v can send at most $O(n/k)$ inter-fragment edges to root $r(B)$ without closing a cycle. According to the first statement of the previous lemma, $r(B)$ starts receiving the first edges after at most $diam(G)$ rounds. Hence, it has collected and convergecasted back all information after $O(n/k + diam(G))$ rounds.

According to the second and third statement of Lemma 12.3.7 and the "red rule" for MSTs (see Tarjan [358], p. 71), the selection of edges convergecasted to $r(b)$ enable this node to select the edges belonging to the MST T and connecting the fragments. Hence, each node knows its adjacent edges belonging to T. We conclude that the successive appliance of protocols SIMPLEMST, DOMPARTITION and PIPELINE results in the MST T. Choosing k as $\sqrt{n}$ yields the following theorem.

Theorem 12.3.8. *There exists a distributed Minimum-weight Spanning Tree algorithm with time complexity* $O(\sqrt{n}\log^* n + \mathrm{diam}(G))$.

---

1 Build a breadth-first-search tree B on G with root $r(B)$.
2 Let $Q(v)$ denote the set of inter-fragment edges currently known to node v ordered by nondecreasing weights. Initialize Q with the set of inter-fragment edges adjacent to v. Let U be the set of edges already sent up to $r(B)$.
3 Each node starts sending edges up to $r(B)$ one round after it has received messages from all its children.
4 By denoting all edges of Q closing cycles in U by $\mathrm{Cyc}(U, Q)$, compute the set $RC := Q\backslash(U \cup \mathrm{Cyc}(U, Q))$. If RC is empty send a "terminating" message to parent. Otherwise, send up the lightest edge of RC.
5 The root computes locally the set S of the at most n/k edges connecting the fragments within MST T and broadcasts this information back to all nodes within G.

Algorithm 32: The protocol PIPELINE.

12.4 Further Advances

In this section we are going to present recent advances on the Range Assignment Problem and the distributed computing of an MST. Thereby we begin with the Range Assignment Problem.

12.4.1 The Range Assignment Problem

The problem introduced in the first section can be generalized: given a graph $G = (V, E)$ embedded in the d-dimensional plane, the task is to compute a range assignment such that the corresponding transmission graph fulfills a given property π. Thereby, the energy needed to establish a link between two nodes follows from equation (12.1). We already know the property "there have to exist paths from a given node s to all others" (π_1). For $\alpha = 1$ the optimal range assignment is trivially obtained by setting the transmission range of s sufficiently large to establish a link between s and the node that has the greatest Euclidean distance to s. This holds due to the Triangle Inequality. However, Clementi et al. [77] showed the $\mathcal{NP}$-completeness of this problem for $\alpha > 1$ and $d \geq 2$. The one-dimensional problem for $\alpha \geq 2$ is solved by Clementi et al. [78] by presenting an algorithm based upon dynamic programming.

The analysis of algorithm RAPMST presented in Theorem 12.2.1 is far from being tight. For each $\alpha \geq 2$ Figure 12.6 depicts an input graph on which RAPMST obtains an approximation ratio of 6. The node in the middle is s and each of the outer ones has a distance of exactly 1 to s. The distance from s to the inner nodes is $\varepsilon > 0$. Therefore, the cost of the optimal range assignment sums up to 1. However, since the distance between two outer nodes is at least 1, each MST will contain 6 edges of length $1 - \varepsilon$ and 6 edges of length ε. Converting this solution into a range assignment yields cost of $\varepsilon^2 + 6 \cdot (1 - \varepsilon)^2$ which converges to 6 as ε tends to 0. Hence, the algorithm RapMst cannot perform better than 6 in the worst case.

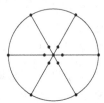

Fig. 12.6. A worst case example for algorithm RapMst.

By replacing the disks used in the presented proof of RapMst's approximation ratio by diamonds, i. e., two isosceles triangles with an angle of $2\pi/3$, Wan et al. [383] proved an approximation ratio of 12.15. Thereby, they first focused on bounding the cost of an MST of a set of points with radius 1. Afterwards, they showed that this cost can be used to obtain the claimed approximation ratio of RapMst. An important attribute of these diamonds is that they do not intersect with each other—similar to the disks used here. Ambühl [13] refined this proof by using pairs of equiliteral triangles instead of diamonds. In contrast to the diamonds, the equiliteral triangles may intersect. Using them Ambühl showed the following theorem and lemma.

Theorem 12.4.1. *Let S be a set of points from the unit disk around the origin, with the additional property that the origin is in S. Let $e_1, e_2, \ldots, e_{|\ |-1}$ be the edges of an Euclidean minimum spanning tree of S. Then*

$$\mu(S) := \sum_{i=1}^{|\ |-1} |e_i|^2 \leq 6$$

holds.

Lemma 12.4.2. *A bound on $\mu(S)$ automatically implies the same bound on the approximation ratio of RapMst for $\alpha \geq 2$.*

Hence, Ambühl was able to bound the approximation ratio of algorithm RapMst from above by 6. Therefore, this analysis is tight regarding the worst case approximation ratio. We want to point out that minor modifications to the proof presented in this paper and the usage of the last lemma lead to an approximation ratio of 20, i. e., an ratio independent of α. The table of Figure 12.7 summarizes the cited results.

	$d = 1$	$d = 2$	$d > 2$
$\alpha = 1$	$\in P$	$\in P$	$\in P$
$\alpha \geq 2$	$\in P$ [78]	$\mathcal{NP}$-hard [77] 6-approximation [13]	$\mathcal{NP}$-hard [77]
$2 < \alpha < d$	×	×	$\mathcal{NP}$-hard [77]
$\alpha \geq d$	×	×	$\mathcal{NP}$-hard [77] $2^{O(d)}$-approximation [77]

Fig. 12.7. Complexity of the range assignment problem for π_1 based on the dimension d and the distance power gradient α.

Another well studied graph property is "the transmission graph has to be strongly connected" (π_2). Regarding $d = 1$, Kirousis et al. [218] presented an algorithm computing the optimal range assignment within polynomial time based upon dynamic programming. For any $\alpha > 1$ and $d \geq 2$ the problem becomes $\mathcal{NP}$-hard ([80], [218]) and even $\mathcal{APX}$-hard for $d \geq 3$ [218]. However, Kirousis et al. [218] obtained a 2-approximation by using an algorithm that is quite similar to the one presented in the last section. First, they also compute an MST T of G. Since the range assignment problem at hand is somewhat more "symmetrical", they convert T into a range assignment according to

$$r(v) := \max_{u:(u,v)\in T} \{d(u,v)^\alpha\}.$$

The table in Figure 12.8 summarizes the known results.

	$d = 1$	$d = 2$	$d > 2$
$\alpha = 1$	$\in P$ [218]	2-approximation [218]	$\mathcal{NP}$-hard [218] 2-approximation [218]
$\alpha \geq 2$	$\in P$ [218]	$\mathcal{NP}$-hard [80] 2-approximation [218]	$\mathcal{APX}$-hard [80] 2-approximation [218]

Fig. 12.8. Complexity of the range assignment problem for π_2 based on the dimension d and the distance power gradient α.

12.4.2 Distributed Computation of an MST

The given algorithm for computing an MST in a distributed manner marks an important step in the quest for the "correct" the graph parameter that reflects the problem's complexity to as large extend as possible. Since the running time of the presented algorithm can be significantly smaller than the number of stations within G, the graph's diameter seems to be more suitable than the size of the graph—which limited the running time of the algorithm introduced by Awerbuch [25], for example. Peleg [313] suggested that the cyclic radius could be the parameter sought-after. Thereby, the cyclic radius $\gamma(G)$ is defined by

$$\gamma(G) := \max_{\subseteq} \min_{\in} \max_{\in} \{\mathrm{dist}(u,v)\},$$

whereas C is a cycle of G and $\mathrm{dist}(u,v)$ denotes the number of edges on the shortest path between stations u and v.

However, Elkin [109] showed that the so-called MST-radius $\mu(G,\omega)$ represents this complexity even better. For a given undirected graph $G = (V, E)$ with a weight function ω on its edges let $z(e)$ denote all cycles in which edge e is the heaviest edge. Furthermore, let $N(u)$ define all edges adjacent to node u. Now we can define the MST-radius as

$$\mu(G,\omega) := \max_{\in} \max_{e \in (\ )} \underbrace{\min_{\in (e)} \max_{e' \in} \min_{' \in e'} \{\mathrm{dist}(u,v')\}}_{\text{VxEdgElimRad}(u\ e)},$$

where VxEdgElimRad denotes the Vertex-Edge-Elimination-Radius, that is set to 0 if $z(e)$ contains no cycle. Broadly speaking, VxEdgElimRad(u, v) reflects the number of rounds u requires to discover that e does not belong to the MST—this is due to the "red rule" and the definition of $z(e)$. As graph G and its weight function ω are fixed, we use μ equivalently to $\mu(G,\omega)$. Figure 12.9 shows an example in which $\gamma(G) = n/2$ and $\mu(G,\omega) = 1$ hold. Have a look at some arbitrary node u. For each edge e with $\omega(e) = 0$ obviously $z(e) = 0$ holds. Therefore, all interesting adjacent edges e are going top-down. The dashed box in the figure depicts the cycle C minimizing the expression above. Hence, the MST-radius equals 1. It is not hard to verify that the cyclic radius is $k = n/2$. Hence, there can be a significant discrepancy between these two values.

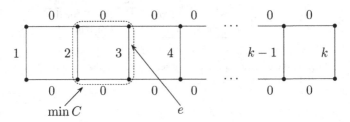

Fig. 12.9. For $k \in \mathbb{N}$, there can be a discrepancy of $\Omega(n)$ between $\gamma(G)$ and $\mu(G,\omega))$.

Computing an $(\Omega(k), O(k))$ Spanning Forest.

Using the parameter $\mu(G,\omega)$, Elkin [109] showed that there exists a (randomized) protocol constructing the MST in $\tilde{O}(\mu(G,\omega) + \sqrt{n})$ rounds[14] with

---

[14] Note, that polylogarithmic factors are omitted due to the usage of $\tilde{O}$.

sucess-probability $1 - O(1/\text{poly}(n))$. He presented a lower bound of $\Omega(\mu(G, \omega))$ as well to emphasize the suitability of the MST-radius to reflect the problem's complexity. In the remaining part we want to outline the key ideas of the proof of the upper bound. Let us start by giving an overview of the protocol.

First of all, it creates an $(\Omega(k), O(k))$ spanning forest $\mathcal{F}$ whose trees are fragments of the MST T. As we will see later on, this can be achieved within $O(k \log^* n)$ rounds. Afterwards, it computes a $(\log n, \mu)$-neighborhood cover C. We will soon present its definition and state right now that this step takes $O(\log^3 n \cdot \mu)$ rounds. For each tree $\tau \in C$ and $T \in \mathcal{F}$ let T^τ denote the forest with the vertex set $V(T) \cap V(\tau)$ and edge set $\{e = (u, w) \in E(T) | u, w \in V(\tau)\}$. The protocol constructs the multigraph $\hat{G}^\tau$ by contracting for each T the vertices of T^τ within G to one supervertex. The multigraph's vertex set consists of all inter-fragment edges, i. e., for each edge $e = (u, w)$ with $u \in T_1^\tau$ and $w \in T_2^\tau$ an edge between T_1^τ and T_2^τ is inserted into $\hat{G}^\tau$. Next, for each $\tau \in C$ the MST of $\hat{G}^\tau$ is computed in parallel by using the protocol PIPELINE. Due to the size of $\hat{G}^\tau$ and the parallelism this is done within $O(\mu \cdot \log^3 n + n \log^2 n/k)$ rounds. Finally, a quite simple property is exploited to select the edges belonging to T out of the ones contained within $\mathcal{F}$ and the MSTs for $\hat{G}^\tau$.

Now that we know the protocol's fundamental structure, we specify the missing parts. Kutten and Peleg [246] showed that the forest $\mathcal{F}$ can be computed in $O(k \log^* n)$ rounds and we presented the corresponding protocol in section 12.3.2. Therefore, by sequentially performing SIMPLEMST and BALANCEDDOM on graph G, we obtain a $(\Omega(k), O(k))$ spanning forest whose trees still consist of edges of MST T in $O(k \log^* n)$ rounds.

Computing a $(\log n, \mu)$-Neighborhood-Cover.

Following Elkin's protocol we want to apply the already known PIPELINE protocol in a parallel manner. To this end we partition G by using a $(\log n, \mu)$-neighborhood-cover.

Definition 12.4.3. *For an undirected graph $G = (V, E)$ a collection C of trees is called a (κ, W)-neighborhood-cover iff*

1. *For each tree $\tau \in C$, $\text{depth}(\tau) = O(W \cdot \kappa)$.*
2. *Each vertex $v \in V$ appears in at most $\tilde{O}(\kappa \cdot n^{1/\kappa})$ different trees $\tau \in C$.*
3. *For each vertex $v \in V$ there exists a tree $\tau \in C$ that contains the entire W-neighborhood of the vertex v, i. e., for each vertex $w \notin C$ there exists a vertex $w' \in C$ with $\text{dist}(w, w') \leq W$.*

We omit the details of constructing the protocol and refer to Elkin [109]. It obtains $(\log n, \mu)$-neighborhood cover with a probability of $1 - O(1/\text{poly}(n))$. By defining $\kappa = O(\log n)$ and $W = \mu = \mu(G, \omega)$ the protocol yields a running time of $O(\log^3 n \cdot \mu)$ rounds.

Afterwards, the algorithm PIPELINE is applied to all graphs $\hat{G}^\tau$ simultaneously. Each $\hat{G}^\tau$ contains one node for each fragment $T \in \mathcal{F}$, resulting in at most $O(n/k)$ nodes. Since each vertex appears in at most $\tilde{O}(\log^2 n)$ trees of the cover, at most $\tilde{O}(\log^2 n)$ executions of the protocol PIPELINE can utilize the same edge. This increases the running time of PIPELINE by $\tilde{O}(\log^2 n)$ to $O(\mu \cdot \log^3 n + n \log^2 n/\kappa)$.

Selecting the Missing Edges.
After all executions of PIPELINE have finished, the edges of the $(\Omega(k), O(k))$ spanning forest $\mathcal{F}$ and the MST of each $\hat{G}^\tau$ are known. The MST T is composed of the following edges:

1. The set of edges of $\mathcal{F}$.
2. $\{e = (v, w) \mid \forall \tau \in C \text{ s. t. } v, w \in V(\tau), e \in \text{MST}(\hat{G}^\tau)\}$.

The following two lemmata [109], whose proofs are omited, state that this selection is correct.

Lemma 12.4.4. *If $e = (u, w)$ is an inter-fragment edge that belongs to the MST of G, then it belongs to the MST of each multigraph $\hat{G}^\tau$ such that both its endpoints u and w belong to the vertex set $V(\tau)$.*

Lemma 12.4.5. *Let $e = (u, w)$ be an inter-fragment edge that does not belong to the MST of G. Let C be a cover that coarsens a (κ, μ)-neighborhood cover. Then there exists a tree $\tau \in C$ with $u, w \in V(\tau)$, such that e does not belong to the MST of the multigraph $\hat{G}^\tau$.*

This completes the description of this protocol and, consequently, our discussion of computing the MST of an undirected graph G in a distributed fashion.

12.5 Conclusion and Open Questions

We considered the problem of finding a range assignment for an arbitrary network given by a transmission graph G such that (a) a (multi-hop) broadcasts from a tagged base station to all other stations is possible and (b) the energy needed to provide this broadcast is minimized. This problem is $\mathcal{NP}$-hard and, therefore, most likely not solvable with acceptable computational effort. However, we have seen that a 6-approximation can be computed by a central station knowing the whole graph. Since this is not given in an ad hoc network we focused on its distributed computation and showed that the computation can also be achieved in a distributed manner within $O(\mu(G, \omega) + \sqrt{|G|})$ rounds, where $\mu(G, \omega)$ denotes the MST-Radius of the graph G.

However, a lot of questions are left open. See [79] for an overview. We will only cite a few of them. The maximal number of hops a broadcast message from the base station passes until it reaches a station, that has not recieved this message yet, is only trivially bounded above by n. Each forwarding takes a certain amount of time—consumned by decoding and encoding, collisions on the medium, etc.—Hence, one could limit the maximal number of hops by an natural number h to restrict the transmission delay. Up to now there is only very little known about this version of the range assignment problem.

Even for the case $h = 2$ the question whether it is $\mathcal{NP}$-hard or polynomially solvable is left open.

Instead of considering the goal to create a structure able to broadcast from one station to all others, one could try to create a structure allowing each pair of stations to communicate with each other; again the energy needed to provide this communication is to be minimized. This is identic to the property π_2 introduced in the last section. Although it is known that this problem is $\mathcal{NP}$-hard for all dimensions $d \geq 2$ and $\mathcal{APX}$-hard for $d \geq 3$, the question whether it is also $\mathcal{APX}$-hard for $d \geq 2$ or admits a PTAS, i.e., an $(1 + \varepsilon)$-approximation within polynomial time for an arbitrary constant $\varepsilon > 0$, is left open.

In the last section we stated that the algorithm presented by Kirousis et al. [218] provides a 2-approximation. However, it is not known whether the approximation ratio of this algorithm is really 2 or the analyis can be improved. Even if the analysis cannot be improved, maybe for some restriction on the input, e. g., euclidean input instances, a better algorithm can be found.

The field of range assignment problems seems to be a promising one for future research; it is challenging and can provide interesting results applicable to real life applications.

12.6 Chapter Notes

This chapter gives a brief overview on the topic, for detailed information we refer to cited references. Clementi et al. [79] give an overview on the results dealing with broadcast range computation. The first section of this chapter is solely based upon a work by Clementi et al. [77]. The results presented here have been improved by Wan et al. [383] and recently by Ambühl [13].

To the author's best knowledge Awerbuch [25] presented the first algorithm to compute an MST in a parallel manner. He shows that his algorithm is worst case optimal when merely using the number of vertices in the graph to bound the running time. The second section of this chapter is mainly based upon a work on Garay et al. [149] and Kutten et al. [246]. They proposed to additionally use the graph's diameter to bound the running time and developed more elaborate protocols for the computation that are faster for sublinear diameters. The quest for the correct parameter to reflect the problem's complexity to as large extend as possible resulted amongst others in works by Kirousis et al. [218], Peleg [314] and Elkin [109].

13

Data Gathering in Sensor Networks

Ludmila Scharf

13.1 Introduction

In this chapter we consider sensor networks, in which all nodes periodically produce some amount of information by sensing an extended geographic area. In each round all data from all nodes needs to be transmitted to an information sink for processing. Because of limited transmission ranges of the nodes, multi-hop routing techniques are applied to transmit the data from all nodes to the sink. Since different nodes partially monitor the same spacial region, data is often correlated. In this case and in order to save energy, data can be processed on its way to the sink. This technique is referred to as in-network data aggregation. For most algorithms presented here, we will assume that the energy cost required by data aggregation is negligible compared to transmission costs, though [270] presents a network model, which incorporates data aggregation and communication costs.

Several coding strategies have been proposed in recent research. Most are based on explicit communication, i.e. nodes can exploit the data correlation only by receiving explicit side information form other nodes. For example, if a node relays information for one or more nodes it can use a so called aggregation function to encode the data available at that node before forwarding it to the sink.

Another approach is based on signal processing techniques and allows nodes to use joint coding without explicit communication, using Slepian-Wolf coding [88, 89]. Consider two correlated data sources x_1, x_2 and assume that their data can be jointly coded with a rate r_{x_1,x_2}. Slepian and Wolf [350] showed that under certain conditions the data from x_1 and x_2 can be coded with a total rate r_{x_1,x_2} even if the two sources are not able to communicate with each other. This result can be extended to n sources. The correlation structure of the data has to be known in advance.

The explicit communication coding strategies can be divided into multi-input [162, 401, 205, 270] and single-input [88, 379, 90] coding strategies. Multi-input coding means, that the data from multiple sources is gathered

D. Wagner and R. Wattenhofer (Eds.): Algorithms for Sensor and Ad Hoc Networks, LNCS 4621, pp. 237–263, 2007.
© Springer-Verlag Berlin Heidelberg 2007

and aggregated in a single node. The aggregation can be performed at a node only if all the input information is available. Single-input coding means that the encoding of data collected by one node depends only on information of one other node. Single-input strategies can be further subdivided into self-coding and foreign coding strategies. In a self-coding scheme, a node can only encode its own data and only in presence of side information, i.e., when data generated by another node is relayed through this node. In a foreign coding scheme, a relay node is only able to encode information originating from another node using its own data. The major advantage of single-input coding strategy is that it can be applied in asynchronous networks where no timing assumptions can be made, and a node performing the aggregation does not have to delay data waiting for more information to arrive. On the other hand, in the multi-input case a node can achieve higher compression rate aggregating joint information from several nodes. Although, certain timing assumptions have to be made in order to avoid indefinite information packet delays and some data freshness-aggregation trade-offs have to be met.

Once a network model is established and a coding strategy is selected, the problem of correlated data gathering is to find a coding scheme and a routing scheme, which minimize the overall energy consumption. This problem statement is typical for the single-input coding strategies and Slepian-Wolf coding. In case of multi-input coding strategies the latency should also be kept low [352, 401, 259]. An alternative problem statement is to find a coding and a routing strategy which maximize the lifetime of a sensor network. The lifetime of a network is defined either as the number of data collecting rounds until the first node is drained out of its energy [205], or the time elapsed until some specified portion of nodes die [356]. The lifetime problem is closely related to the problem of minimizing overall energy consumption, but the optimal routing and coding schemes may vary. Whereas most research on the data gathering problem operates on a network model, where positions of the nodes are fixed and are part of the input, the work in [144] and [91] deals with finding optimal placement and density of sensor nodes while minimizing overall energy consumption.

We will primarily address the problem of minimizing overall energy consumption. Obviously, if the information collected by the nodes is statistically independent, i.e., uncorrelated, then the aggregation function does not result in any data compression. If the energy costs for transmitting data over a link are fixed, then the problem becomes trivial, and the best strategy for a node for sending its data to the sink is to transmit it over the path with lowest energy costs without any data aggregation. Consider a simple network example depicted in Figure 13.1 with three sensor nodes u, v, and w, each producing a data packet of size s, and a sink node t. Communication links exist between nodes u and v, u and w, v and w, and between w and t, and all links have energy cost 1. The best routing strategy for uncorrelated data is the shortest path tree, Fig. 13.1(a), with total cost $5s$. Now let's assume the data is correlated and an information packet can be encoded in presence of data from one other node, i.e., single-input coding, and the size of an encoded data packet is $s' \quad s$, independent of which two nodes are involved. In this example we use the self-coding strategy. Now with self-coding strategy and the routing scheme given by the shortest paths tree, the node , which relays data for nodes u and v, can use data generated by either one of those to encode its own information. The total cost is now $4s + s'$, which is less than the total cost of gathering uncorrelated data. However, if the size of the encoded packet is at most half of the raw data size, the shortest paths are no longer the optimal routing strategy even for this simple example: If we let the node u send raw data to node v, the node v can encode its own information utilizing the data from node u. The total cost of the gathering tree in Figure 13.1(c) is then $3s + 3s'$, which is smaller than the cost of shortest path tree under the correlation assumption $s' \leq \frac{1}{2}s$. From this simple example it can be seen, that correlated data gathering is a non-trivial optimization problem.

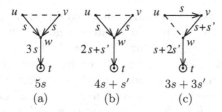

Fig. 13.1. Simple network example: (a) uncorrelated data gathering with shortest paths tree; (b) correlated data gathering with self-coding strategy and shortest paths tree as routing scheme; (c) self-coding strategy and alternative routing scheme.

Content Structure. The next section defines a theoretical model of sensor networks. This model is used in all presented algorithms unless otherwise stated. In Section 13.3 we address the problem of minimum energy data gathering. Two single input coding strategies are presented together with a detailed description of the algorithms and an $\mathcal{NP}$-hardness proof for one of the strategies. We also describe an algorithm which finds a good routing structure for a given graph for the class of canonical aggregation functions. Section 13.4 gives a brief overview of some heuristic approaches to the maximum lifetime data gathering problem.

13.2 Network Model

We model a sensor network as an undirected graph $G = (V, E)$, where V denotes the set of n nodes, and $E \subset V^2$ a set of   edges. Nodes in V correspond to sensor nodes, and edges represent the communication links between node-pairs. We assume that communication links can be used in both directions, and therefore the edges of the graph are undirected. The edges are usually determined by either the transmission range of nodes, or by the k-nearest neighborhood. There is a special node $t \in V$ called the *sink* node where the data from all nodes in V should be gathered. We refer to the process of gathering information during a certain time interval from each sensor node in V to sink t as a *round*.

We consider nodes to be able to adjust their radio signal strength in order to save energy. The weight $w(e)$ for an edge $e = (v_i, v_j) \in E$ is defined to be the cost for transmitting a unit of information (i.e., one bit of data) from node v_i to node v_j or vice versa. Therefore, the radio model is abstracted in the edge weights. The *transmission cost* for sending s units of data along the edge e is given by $s \cdot w(e)$.

13.3 Minimum Energy Data Gathering

In this section we introduce two single-input coding strategies based on work in [379] and an algorithm which guarantees a good approximation for any concave multi-input aggregation function. For the foreign coding correlation model we describe an algorithm, which results in a minimum-energy data gathering topology under the assumption, that topology of the network and data correlation is known in advance. Further, a distributed version of the algorithm is presented for a restricted network model. In case of self-coding, we show that the problem of finding a minimum-energy data-gathering tree is $\mathcal{NP}$-complete, the datailed proof is given in [88]. For this problem we present an approximation algorithm with the approximation ratio $2(1 + \sqrt{2})$, the algorithm is due to [379].

13.3.1 Foreign Coding

Correlation Model. We assume that each node $v_i \in V$ generates a data packet p_i of size s_i bits per round that contains information measured by v_i. We refer to the packet p_i as *raw data*. Whenever a data packet is routed tward the sink node, data aggregation can potentially take place at any intermediate node. Once a packet is encoded at a node, it is not possible for other intermediate nodes to decode or further encode that packet. Hence, an encoded data packet can not be altered. A packet from node v_i, that is encoded at a node v_j is denoted by p_i^j, and its corresponding size is s_i^j. The compression rate depends on the data correlation between the involved nodes v_i and v_j, denoted by the correlation coefficient $\rho_{ij} = 1 - s_i^j/s_i$. Encoding at a node v_j depends only on data collected by the node v_j and no other data routed through that node. Since decoding of a data packet p_i^j at the sink node requires a raw data packet p_j, it must be guaranteed that encoding does not result in cyclic dependences, that is, the sink node t should be able to decode data packets from all nodes in V.

The *minimum-energy data gathering* problem for a given graph $G = (V, E)$ with positive edge weights $w : E \to \mathbb{R}^+$ and a sink node $t \in V$ is defined as follows. Find a routing scheme and coding scheme to deliver data packets from all nodes in V to the sink node $t \in V$, such that the overall energy consumption is minimal. Let $f(e)$ denote the *load* of an edge $e \in E$, that is, the number of bits transmitted over e in one round. Minimizing overall energy consumption means then to minimize the cost function

$$C(G) = \sum_{e \in E} f(e)w(e) \; . \tag{13.1}$$

Algorithm MEGA. In the following we present the *Minimum-Energy Gathering Algorithm* (MEGA). The resulting data gathering topology of the algorithm is a superimposition of two tree constructions. A directed minimum spanning tree, which will be specified later, is used to determine the encoding relay for all data packets. A directed edge (v_i, v_j) in this tree means, that data packets generated by the node v_i will be encoded at node v_j. Once a packet is encoded, it is routed on the shortest path towards the sink node, in order to save energy. The shortest paths are combined in the shortest path tree structure.

Given a graph $G = (V, E)$ as described in Section 13.2 with a sink node $t \in V$, MEGA first computes a *shortest path tree* (SPT) rooted at t, using, for example, Dijkstra algorithm [83]. The SPT comprises shortest paths from all nodes in V to the sink node t, and thus energy-minimal paths from all nodes to t.

In the second step, the algorithm computes for each node $v_i \in V$ a corresponding node $v_j \in V$, which encodes the information packet p_i using its own data. A pair (v_i, v_j) corresponds then to a directed edge in a coding graph. Since no cyclic dependences are allowed in order to guarantee decoding, the coding graph should be a *coding tree*. MEGA computes the coding tree using an algorithm for solving the *directed minimum spanning tree* (DMST) problem. The DMST problem, also referred to as *minimum weight arborescence* problem is defined as: given a directed graph $\tilde{G} = (V, \tilde{E})$ with edge weights and a root node $t \in V$, find a rooted directed spanning tree, such that the sum of weights over all tree edges is minimized, and every node in V is reachable from the root node. Chu and Lin [72], Edmonds [107], and Bock [42] have independently given efficient algorithms for solving the DMST problem. Tarjan [357] and Camerini et al. [58] give an efficient implementation. Furthermore, a distributed algorithm is given by Humblet [190].

MEGA builds first a complete directed graph $\tilde{G} = (V, \tilde{E})$, which has the same set of vertices as G. The weight of a directed edge $e = (v_i, v_j) \in \tilde{E}$ is defined as

$$\begin{aligned} \tilde{w}(e) &= s_i(w_{\mathrm{SP}}(v_i, v_j) + w_{\mathrm{SP}}(v_j, t)(1 - \rho_{ij})) \\ &= s_i \cdot w_{\mathrm{SP}}(v_i, v_j) + s_i^j \cdot w_{\mathrm{SP}}(v_j, t) \; , \end{aligned} \tag{13.2}$$

where $w_{\mathrm{SP}}(v_i, v_j)$ denotes the weight of a shortest path from v_i to v_j in graph G. The weight $\tilde{w}(e)$ of a directed edge $e = (v_i, v_j)$ represents the total cost for routing a data packet collected by v_i to the sink using v_j as encoding relay. It depends on the edge weights, i.e., the shortest paths weights, in the original graph and on the correlation coefficients. In order to compute the edge weights in the directed graph $\tilde{G}$ we need to know weights of shortest paths in G for all pairs of nodes. We can do this either by applying Dijkstra algorithm n times, or using Floyd-Warshall algorithm [83] for solving all-pairs shortest paths problem. Both alternatives have the same computational complexity.

A DMST is by definition a minimum spanning tree, where the edges are directed off the root node. Here, we aim at a directed tree with the edges pointing towards the root node, which is the sink node t. Therefore, MEGA applies a DMST algorithm to a transposed graph $\tilde{G}^{\mathrm{T}}$, i.e., a graph, where the direction of each edge is reversed. The resulting tree $\tilde{T}$ is again transposed, that is, the tree $T = \tilde{T}^{\mathrm{T}}$ contains original edges from $\tilde{G}$ and defines encoding relays for all nodes in V. If a node has a direct edge in T to the sink node, that node transmits its raw data packets to the sink.

The resulting data gathering topology comprises for each node v_i edges of a shortest path to its coding relay v_j found by the DMST algorithm and edges of the shortest path from v_j to the sink node t. If all data is pairwise independent, one can see that the topology produced by MEGA is a shortest path tree, which corresponds to the observation made in the introduction.

Input: Graph $G = (V, E)$, sink node $t \in V$, edge weights $w : E \to \mathbb{R}^+$,
correlation ratios ρ_{ij}.
Output: Minimum-energy data gathering topology for G
1 $T_{\mathrm{SP}}(G, t) =$ shortest path tree for G routed at t
2 $\tilde{G} = (V, \tilde{E}) =$ complete directed graph
3 **forall** $(v_i, v_j) \in \tilde{E}$ **do**
4 $\quad|\quad \tilde{w}(v_i, v_j) = s_i(w_{\mathrm{SP}}(v_i, v_j) + w_{\mathrm{SP}}(v_j, t)(1 - \rho_{ij}))$
5 **end**
6 $\tilde{T} = $ DMST on $\tilde{G}^{\mathrm{T}}$
7 $T = (V, E_T) = \tilde{T}^{\mathrm{T}}$
8 **forall** $v_i \in V$ **do**
9 $\quad|\quad$ consider v_j such that $(v_i, v_j) \in E_T$
10 $\quad|\quad$ set v_j as encoding relay for p_i
11 **end**

Algorithm 33: MEGA

An example in Figure 13.2(a) depicts a simple network, where data gathered by every node has to be sent to the sink node t. For simplicity we assume that all nodes produce raw data packets of equal size s. The edge weights correspond to the energy cost of sending one bit of information along that edge.

Data correlation for this example is defined to be inverse proportional to the Euclidean distance d between the nodes, that is, $\rho_{ij} = \frac{1}{1+d(v_i,v_j)}$. Furthermore, a path loss exponent of 2 is assumed and thus $d(v_i, v_j) = \sqrt{c(e)}$, where $e = (v_i, v_j)$. Algorithm MEGA first computes a shortest path tree rooted at the sink node – bold edges in Figure 13.2(a). The directed coding tree determined by MEGA is depicted by dashed arrows. Every node has one outgoing arrow to its encoding relay. After a data packet has been encoded, it is sent towards the sink node along a shortest path.

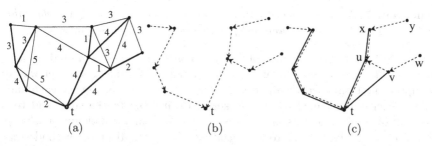

(a)   (b)   (c)

Fig. 13.2. A simple network G with a sink node t. Edges in (a) are labeled according to the unit energy consumption. Bold edges are the edges of a shortest path tree rooted at t. (b) depicts the directed minimum spanning tree rooted at t. DMST defines the coding relationships. (c) shows a minimum-energy data gathering topology for the given network: DMST combined with shortes paths tree. The dashed edges point from a node to its coding relay, encoded data is sent to the sink along the edges of the shortest path tree.

For an illustration of the data gathering process consider the right part of the graph in Fig. 13.2(b). The nodes x and w send raw data to their encoding relays y and v respectively. The node y encodes the data gathered by x and sends the coded package as well as its own raw data to the node u. The node v encodes the data of node w and transmits it direct to the sink node t. The raw data gathered by the node v is sent to v's coding relay u. The node u encodes the packages of y and v and sends them along with its own raw data and the forwarded encoded package of x to the sink. Now the sink node can recursively decode all gathered information.

Analysis. *Computational Complexity.* The running time of MEGA is dominated by solving the all-pairs shortest path problem on G, which takes $\mathcal{O}(n^3)$ time. The construction of the graph $\tilde{G}$ and solving the DMST problem on $\tilde{G}^{\mathrm{T}}$ have both running time of $\mathcal{O}(n^2)$. The total computational complexity of the algorithm is thus $\mathcal{O}(n^3)$.

Correctness. In order to show that MEGA produces an optimal data gathering topology we first analyze some properties of an optimal topology. In an optimal solution for the graph G each packet i is routed on a single path to the sink t, since multiple paths would unnecessarily increase energy consumption. By problem constraints, a packet p_i is encoded at most once on its way to the sink. It is also possible that p_i is sent unencoded to the sink, in this case t is considered to be the encoding relay for p_i. We can establish a directed coding graph $\tilde{G}_{\mathrm{opt}} = (V, \tilde{E}_{\mathrm{opt}})$, corresponding to the optimal solution, where each node v_i in V except the sink node t has exactly one outgoing edge pointing to its encoding relay in this solution. It follows that the coding graph $\tilde{G}_{\mathrm{opt}}$ has $n - 1$ edges. Since no cyclic coding dependences are allowed, $\tilde{G}_{\mathrm{opt}}$ must be cycle-free. These properties characterize a directed tree pointing into node t. Thus we obtain the following theorem:

Theorem 13.3.1. *Given a graph* $G = (V, E)$ *and a sink* $t \in V$, *algorithm MEGA computes a minimum-energy data gathering topology for* G.

Proof. In Equation (13.1) the total energy consumption is computed by summing up the load of each edge times the weight of the edge. Another way to compute the total cost of a data gathering topology is to charge to each node v_i the energy consumed by the packet p_i on its way to the sink, and then sum up over all nodes. As argued above, in an optimal topology a packet p_i is routed on a single path to its encoding relay v_j, and at the second stage the encoded packet p_i^j is routed from v_j to the sink t. Both sub-paths should be minimum-energy paths, that is, shortest paths in G. Therefore, the total energy consumed by the packet p_i in an optimal data gathering topology is exactly $\tilde{w}(v_i, v_j)$ as defined by Equation (13.2), where v_j is the optimal encoding relay for p_i. If we assign the energy consumption by the packet p_i on its way to the sink node to a single outgoing edge of the node v_i in the optimal coding graph $\tilde{G}_{\mathrm{opt}}$, then the total energy cost of the optimal solution is given by the sum of all edge weights in $\tilde{G}_{\mathrm{opt}}$. The algorithm MEGA computes exactly these account values for all possible coding relays of the node v_i and assigns them to the corresponding outgoing edges in the directed graph $\tilde{G}$. Applying a DMST algorithm on a transposed graph $\tilde{G}^{\mathrm{T}}$ a directed tree T pointing to the sink node t is obtained, that minimizes the sum of the edge weights. Since $\tilde{G}$ is a complete directed graph, it comprehends $\tilde{G}_{\mathrm{opt}}$ as a subgraph. In other words, $\tilde{G}_{\mathrm{opt}}$ is also a directed spanning tree in $\tilde{G}$ pointing to t. Therefore, MEGA minimizes the cost function $C(G)$. □

Distributed Computation. The presented algorithm MEGA requires total knowledge about data correlation among all nodes and the network topology. In the following a distributed algorithm is presented, which computes a minimum-energy data gathering topology for a restricted network model. We consider now a *unit disk graph* (UDG) model [75], where all nodes have the same limited transmission ranges. Additionally, all raw data packets are assumed to have equal size s. Data correlation in sensor networks is often assumed to be regional. Therefore, in the following the correlation between two nodes v_i and v_j is modeled to be inverse proportional to their Euclidean distance $d(v_i, v_j)$, e.g. $\rho_{ij} = 1/(1 + d(v_i, v_j))$.

In the first step of a distributed version of MEGA, the sink node t starts building a shortest paths tree rooted at t using, e.g., Dijkstra algorithm. Thus, each node in V is able to determine its energy cost for sending one bit of information to t. In the second step, nodes exploit the data correlation between the neighboring nodes: Each node v_i in the graph G broadcasts a sample packet p_i. Upon receipt of a packet p_j from a neighboring node v_j, the node v_i encodes p_i using p_j in order to compute the correlation coefficient ρ_{ij}. Additionally, the node v_i can determine the energy cost of transmitting to v_j by using a *Received Signal Strength Indicator* (RSSI) [329]. After that, the node v_i establishes a directed edge (v_i, v_j) with weight set according to Equation (13.2). The directed graph $\tilde{G}$ then consists of edges between direct neighbors in G only. The distributed algorithm described in [190] to compute DMST completes a distributed implementation of MEGA.

In order to show that distributed MEGA still results in an optimal topology we have to show that in an optimal solution each node and its coding relay are neighbors in graph G. Assume that in an optimal data gathering topology G_{opt} a node v_i and its encoding relay v_j are more than one hop away from each other. The packet p_i is then routed along a path $p(v_i, v_j) = (v_i, u_1, \ldots, u_k, u_l, u_m, \ldots, v_j)$ in G_{opt} to the encoding relay v_j. Since v_i and v_j are not neighbors and G is a UDG, the Euclidean distance between them is larger than the distance between v_i and its neighbor on the path: $d(v_i, v_j) < d(v_i, u_1)$. Consequently, $\rho_{v_i, v_j} < \rho_{v_i, u_1}$ because the data correlation is inverse proportional to the Euclidean distance. It follows that $s_i^{u_1} < s_i^j$. If v_i is not the encoding relay for u_1, that is (u_1, v_i) is not an edge in $\tilde{G}_{\mathrm{opt}}$, we choose u_1 to be the coding node of v_i and obtain a topology with lower energy consumption than G_{opt}, which contradicts the optimality assumption. If (u_1, v_i) is in $\tilde{G}_{\mathrm{opt}}$, we have a configuration depicted in Figure 13.3(a). Node v_i has an edge pointing to v_j in the coding graph $\tilde{G}_{\mathrm{opt}}$ and each node on the path $p(v_i, v_j)$ up to a certain node u_l has an edge to its predecessor on the path. Since $\tilde{G}_{\mathrm{opt}}$ must be cycle-free, there exists at least one node u_m on the path, which does not point to its predecessor. Raw data load caused by all nodes on the path is $2s$ for the edges of the sub-path $p(v_i, v_j)|_{v_i, u_l}$, and s for the edge (u_l, u_m). By changing the edges in $\tilde{G}_{\mathrm{opt}}$ subject to the configuration in Figure 13.3(b), and thus also the encoding relays, the edge (u_k, u_l) has load at most $s + s_{u_k}^{u_l}$ since raw packet p_{u_k} of size s is sent to u_l for encoding, and coded packet $p_{u_k}^{u_l}$ of size $s_{u_k}^{u_l}$ is possibly sent back along the same edge on its way to the sink. The same holds for all other edges on the sub-path where the coding relationship is reversed. Since $s_{u_k}^{u_l} < s$ for all direct neighbors in G, the transformation we made decreases the energy consumption of G_{opt}, which leads to a contradiction. Therefore, it is adequate to restrict the directed graph $\tilde{G}$ to comprise only edges connecting the direct neighbors in G. This allows for the distributed computation of the minimum-energy data gathering topology of G.

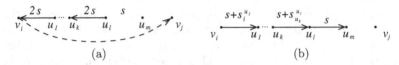

(a)   (b)

Fig. 13.3. Configuration of the coding graph $\tilde{G}$: (a) a node v_i is more than one hop away from its coding relay v_j; (b) changed configuration with one-hop relays.

13.3.2 Self-Coding

In this section we consider the problem of finding a minimum-energy data gathering topology in the model of explicit communication, where, in contrast to the previous section, nodes can only encode their own raw data in presence of other raw data routed through them. This problem is $\mathcal{NP}$-complete even in a simplified setting, the proof is due to [88]. Then, a lower bound for energy consumption of an optimal topology is given, and a constant factor approximation algorithm is presented, both as described in [379].

Correlation Model. We assume that all nodes $v_i \in V$ generate raw data packets of equal size s. In this model, if a raw data packet is routed on its way to the sink   through a node v_i, the node v_i can use that packet to encode its own raw data such that the size of $_i$ is reduced to s_e s. The correlation between data is assumed to be equal and $_{ij} = 1 - \frac{\cdot}{\cdot}$ for all $v_i, v_j \in V$ with $i \neq .$ If no side information is available at a node, that node sends its data unencoded.

We define the problem of *minimum cost correlated data gathering* as follows. Given a graph $G = (V, E)$ with positive edge weights $: E \to \mathbb{R}$ and a sink node $\in V$. Find a routing topology $T = (V, E')$, $E' \subset E$, and a partition of the node set $V = B \cup S$, $B \cap S = \emptyset$, such that the energy consumption is minimized. B is the set of nodes, which send their data to the sink without encoding, and S is a set of nodes using raw data from a node in B to encode its own packet. As in the previous section we assume, that sending k bits of information along an edge with unit energy cost   has energy consumption of $k \cdot .$ The cost function of a routing topology T is then given by

$$C(T) = \sum_{e \in '} \tau(e) \ (e) \ , \tag{13.3}$$

where $\tau(e)$ denotes the total load of an edge e.

$\mathcal{NP}$**-Completeness Proof.** In order to prove the $\mathcal{NP}$-hardness of the optimization problem, we show that its decision version is $\mathcal{NP}$-complete. The decision version of a *network data gathering tree cost* problem is then: given a graph $G = (V, E)$ with edge weights $: E \to \mathbb{R}$, a sink node $\in V$ and a positive integer   . Does G admit a spanning tree T such that, when the leaf nodes send their data without encoding to the sink, and the internal nodes code their data using the information routed through them, the total cost of T given by (13.3) is at most   ?

$\mathcal{NP}$-completeness is proven by reduction from the *set cover* problem [83], whose decision version is defined as follows: Problem instance is a finite set P, a collection C of subsets of P and an integer $0 < K < |C|$, with $|C|$ the cardinality of C. The question is whether there exists a subset $C' \subset C$ with $|C'| \leq K$, such that every element of P belongs to at least one of the subsets in C'. C' is then called a set cover for P.

Theorem 13.3.2. *Network data gathering tree cost problem with explicit communication and self-coding strategy is $\mathcal{NP}$-complete.*

Proof. First, the network data gathering tree cost problem is in $\mathcal{NP}$, since we can verify in polynomial time that a proposed routing tree has total cost less than the given value M.

Given an instance of the set cover problem, we construct an equivalent instance of a graph for our problem. Let $n = |P|$ be the cardinality of the set P, and $m = |C|$ the cardinality of C. The resulting graph is formed of three layers: a sink node t, a layer corresponding to the subsets $C_i \in C$, and a layer corresponding to the elements p_j of the set P, as pictured in Figure 13.4(a). Each subset $C_i \in C$ is represented by a structure formed of four nodes $x_1^i, x_2^i, x_3^i, x_4^i$ as shown in Figure 13.4(b). Each node x_3^i is linked to the sink node t, node x_4^i is connected only to the node x_1^i, the nodes x_1^i, x_2^i, x_3^i are all interconnected. There is an edge between a node p_j and a node x_1^i iff $p_j \in C_i$, that is, the nodes corresponding to the elements in P are connected to the structures corresponding to the subsets containing those elements. All edges between layers P and C have a weight $d > 0$, the edges (x_1^i, x_3^i) and (x_2^i, x_3^i) have a weight $a > 1$, all other edges are weighted with 1. Without loss of generality, we assume that the size of an encoded data packet is $s_e = 1$ and the size of a raw data packet is $s > 1$. We choose a positive value M to be $M = n(d+a+1)s + K(2as+3s+a+2) + (m-K)(as+3s+2a+4)$ and show that finding a spanning tree with cost at most M is equivalent to finding a set cover of cardinality at most K for the set P. Notice, that construction of the graph instance from the set cover instance can be done in polynomial time.

With a large enough value chosen for d, i.e., $d > m(2as + 3s + a + 2)/s$, a tree with total cost at most M would contain exactly n links between the layers P and C, since at least n such edges are needed to make sure that every node p_j is connected to the tree. If some node p_j is used as a relay, then more than n edges with weight d carry a packet of size s, together with the remaining tree load, it would result in a cost larger than M. That means, that no p_j node is used as a relay, so all nodes p_j send their data unencoded to the sink node. It also implies, that the structures corresponding to the subsets C_i can be connected to the sink only through their x_3^i nodes, so all x_3^i are internal nodes of the routing tree. All x_4^i nodes are necessarily leaf nodes and must be connected to the corresponding x_1^i nodes, thus x_1^i's are relay nodes. The only degree of freedom are the choices of two out of three edges between each triple x_1^i, x_2^i, x_3^i.

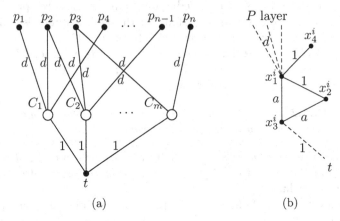

P layer

Fig. 13.4. Reduction from set cover problem. (a) A graph instance. (b) Structure of a block corresponding to a subset C_i.

Choose a value of > 1 such that $(\ +2)/\ \ \ s\ \ (\ +2)/(\ -1)$, since $s > 1$ it is always possible to find an appropriate value of . Now consider the structure corresponding to a subset C_i, the three possible configurations are depicted in Figure 13.5. The first pattern, see Figure 13.5(a), contains edges $(\ \frac{i}{4},\ \frac{i}{1}), (\ \frac{i}{1},\ \frac{i}{2}), (\ \frac{i}{2},\ \frac{i}{3}), (\ \frac{i}{3},\ )$ and has total cost caused by the nodes of this structure of $s(\ +3) + (\ +2) + (\ +1) + 1$. The second pattern, see Figure 13.5(b), contains links $(\ \frac{i}{4},\ \frac{i}{1}), (\ \frac{i}{1},\ \frac{i}{3}), (\ \frac{i}{2},\ \frac{i}{3}), (\ \frac{i}{3},\ )$ with cost $s(\ +2) + s(\ +1) + (\ +1) + 1$. The third one, see Figure 13.5(c), comprises edges $(\ \frac{i}{4},\ \frac{i}{1}), (\ \frac{i}{1},\ \frac{i}{3}), (\ \frac{i}{2},\ \frac{i}{1}), (\ \frac{i}{3},\ )$ with cost $s(\ +2) + s(\ +2) + (\ +1) + 1$. Since the second pattern is always better than the third one, we will only consider the first two configurations.

If no node from the P layer is connected to the structure corresponding to C_i, the optimal configuration is pattern one, remember that $s > (\ +2)/\ $. However, if the data packets from ≥ 1 nodes of the layer P are routed through $\frac{i}{1}$, the first pattern is no longer optimal, the total cost of its edges is then $s(\ +2) + s(\ +3) + (\ +2) + (\ +1) + 1$. The total cost of the edges of the second pattern is $s(\ +1) + s(\ +2) + s(\ +1) + (\ +1) + 1$. Since $s\ \ (\ +2)/(\ -1)$, the second pattern is more efficient.

Note, that in the optimal tree the cost of gathering data from each node $_j$ is the same, and is equal to $s(d +\ +1)$. Let k be the number of the C layer structures connected to the P layer in the optimal tree, then their total cost is k times the cost of the pattern two: $k(2s\ +3s +\ +2)$. The minimum cost of the remaining $\ -k$ structures is determined by the pattern one: $(\ -k)(s\ +3s + 2\ +4)$. The total energy consumption of the optimal solution is then $sn(d +\ +1) + k(2s\ +3s +\ +2) + (\ -k)(s\ +3s + 2\ +4)$, which is at most the chosen value exactly when $k \leq\ $.

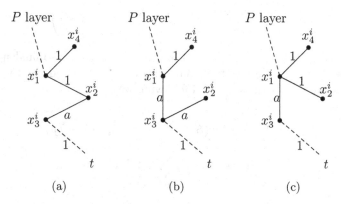

Fig. 13.5. Three possible configurations for a structure corresponding to the subset C_i.

That means that the problem of finding a tree with thr total cost at most
is equivalent to finding a set of at most   elements from the C layer to which
all nodes of the P layer are connected, which is, in turn, equivalent to finding
a set cover of the set P of cardinality at most   . Thus, the decision version
of the data gathering tree cost problem is $\mathcal{NP}$-complete.   □

Constant Factor Approximation Algorithm LEGA. Let us first bound
the energy consumption of the optimal solution from below.

Lemma 13.3.3 (Lower Bound). *The cost of an optimal topology C_{opt} is
bounded from below by $C_{\mathrm{opt}} \geq \max\{s_e \cdot C_{\mathrm{SSP}}, s \cdot C_{\mathrm{MST}}\}$, where C_{SSP} denotes
the sum of the costs of all shortest paths to the sink node   , C_{MST} is the cost
of a minimum spanning tree of the graph G, s and s_e denote the size of raw
and encoded data packages respectively.*

Proof. The sensor nodes can either send their raw data to the sink node or
use raw data from some other node to encode their data. Let B be the set of
nodes sending their data without encoding. Let S be the set of nodes which
encode their data using the data of a node $u \in B$. The set B and the sets S
for all u in B form a partition of the node set V.

After deciding how the nodes should be partitioned, the optimal solution
would send the encoded data of the nodes in $V \setminus B$ along the shortest paths
(SP) to the sink node since this minimizes the total energy consumption. The
data from the node $u \in B$ needs to be distributed to all nodes in S as well
as to the sink node since the sink   should be able to decode the packets of all
nodes in S . The optimal topology fpr this task is the minimum Steiner Tree
(ST) with all nodes in S and the nodes u,   being terminal nodes. The total
cost of the optimal solution is therefore

$$C_{\text{opt}} = \sum_{u \in B} \left(s \cdot \text{ST}(S_u, u, t) + \sum_{v \in S_u} s_e \cdot \text{SP}(v, t) \right) .$$

By definition, $\text{SP}(v_i, v_j) = \text{ST}(v_i, v_j)$ and any additional terminal node increases the cost of the Steiner tree. With this observation and due to the fact that $s_e < s$, we can bound C_{opt} in a following way:

$$
\begin{aligned}
C_{\text{opt}} &= \sum_{u \in B} \left(s \cdot \text{ST}(S_u, u, t) + \sum_{v \in S_u} s_e \cdot \text{SP}(v, t) \right) \\
&\geq \sum_{u \in B} s_e \cdot \text{SP}(u, t) + \sum_{u \in B} \sum_{v \in S_u} s_e \cdot \text{SP}(v, t) \\
&= \sum_{u \in V} s_e \cdot \text{SP}(u, t) = s_e \cdot C_{\text{SSP}} .
\end{aligned}
$$

On the other hand, every node in B sends a raw data packet and every node in $V \setminus B$ receives an unencoded packet. Therefore, raw data is distributed at least on a spanning tree. Since the minimum spanning tree (MST) is the cheapest spanning tree possible the second bound follows: $C_{\text{opt}} \geq s \cdot C_{\text{MST}}$. □

The constant factor approximation algorithm presented here is based on the *shallow light tree* (SLT), a spanning tree that approximates both minimum spanning tree and shortest path tree for a given root node. Given a graph $G = (V, E)$, a root node t and a positive number γ, the SLT has two properties:

- Its total cost is at most $(1 + \sqrt{2}/\gamma)$ times the cost of a minimum spanning tree of the graph G, $C_{\text{SLT}} \leq (1 + \sqrt{2}/\gamma)C_{\text{MST}}$.
- The distance on the SLT between any node v in V and the root node t is at most $(1 + \sqrt{2}\gamma)$ times the cost of a corresponding shortest path in G, $d_{\text{SLT}}(v, t) \leq (1 + \sqrt{2}\gamma)w_{\text{SP}}(v, t)$.

Construction of a shallow light tree is described in [214].

The *Low Energy Gathering Algorithm* (LEGA) first computes a spanning shallow light tree with the sink t as a root node, see Figure 13.6(a) for an example. At the data gathering stage the sink t broadcasts its packet p_t to all its children on SLT. When a node v_i receives a packet p_j from its parent on SLT, it encodes its own data using the parent raw dara and sends its information packet p_i^j along the path given by SLT to the sink node. Then v_i broadcasts its raw data packet p_i to all its children on SLT, see Figure 13.6(b). The sink node has its own raw data available locally and can recursively decode the data.

Theorem 13.3.4. *LEGA achieves a $2(1 + \sqrt{2})$-approximation of the optimal data gathering topology.*

Proof. Every node in G sends its raw data to its children on SLT and encoded data along the shortest path in SLT to the sink node. Therefore, the total data

---

Input: Graph $G = (V, E)$, sink node $t \in V$, edge weights $w : E \rightarrow \mathbb{R}^+$.
1 **Preprocessing:** Compute $\text{SLT}(t, G, w)$
2 **Data gathering:**
3   t broadcasts its packet p_t to all its children on SLT
4   **foreach** *node v_i upon receiving a packet p_j from its parent on SLT* **do**
5   encode p_i using p_j
6   send p_i^j on SLT to the sink t
7   broadcast p_i to all children on SLT
8   **end**

Algorithm 34: LEGA

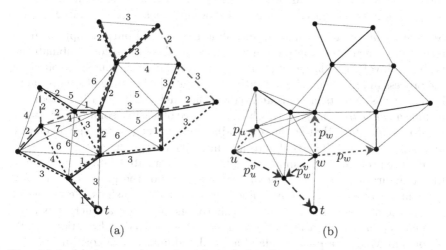

(a)   (b)

Fig. 13.6. (a) A network graph with dashed MST edges, dotted SPT edges and bold SLT edges. (b) A step of the LEGA data gathering stage: Nodes u and w send their packets coded with the information from the parent node to the sink, and broadcast their raw packets to the children on the SLT.

gathering cost achieved by LEGA is $C_{\text{LEGA}} = s \cdot C_{\text{SLT}} + \sum_{\in} s_e d_{\text{SLT}}(v,)$. Using the properties of SLT and setting $\gamma = 1$ we obtain

$$C_{\text{LEGA}} \leq s \cdot (1 + \sqrt{2}) C_{\text{MST}} + s_e \cdot (1 + \sqrt{2}) C_{\text{SSP}}$$
$$\leq 2(1 + \sqrt{2}) C_{\text{opt}} \ .$$

The second inequality follows from Lemma 13.3.3.   □

The computational complexity of the LEGA depends on the complexity of the construction of a shallow light tree. The algorithm in [214] runs in $\mathcal{O}(+ n \log n)$ time. Therefore the running time of LEGA is also $\mathcal{O}(+ n \log n)$.

Further Heuristics. A few heuristic approaches are described in [90]. They all construct a spanning tree, where during the data gathering process the leaf nodes send their raw data to the sink along the paths given by the tree. The internal nodes use relayed data to encode their own information and send the encoded data packets along the tree edges.

Shortest Path Tree. The first approximation suggests a shortest path tree (SPT) for the data gathering topology. The SPT is optimal for uncorrelated data gathering, but can be arbitrary bad as the data correlation gets high, as shown in the same paper.

Greedy Algorithm. This is an incremental algorithm, where the initial subtree comprises only a sink node. In each iteration step, a node whose addition results in the minimum cost increment is added to the tree together with the corresponding edge. This algorithm performs poor as well, since far nodes are being left out, so they need to connect to the sink via a path with large weight.

Simulated Annealing. Simulated annealing is a general purpose optimization technique, which is known to provide near optimal solutions for combinatorial problems involving a large number of variables. The technique is inspired by the process in physics involving heating and controlled cooling of a material to increase the size of its crystals and reduce their defects. The basic ingredients include a *configuration space* Z, which is a space of feasible solutions or states, *neighbors of a state*, that is, the set of adjacencies among states, and an energy function $f : Z \to \mathbb{R}$. The goal is to minimize the energy function.

For the problem of minimum-energy network correlated data gathering tree the configuration space is a set of spanning trees of the graph $G = (V, E)$. A spanning tree is completely defined by parent relationships. Neighbors of a state are all spanning trees, where exactly one node has a different parent. The energy function is the total cost function as defined in Equation (13.3). Algorithm 35 gives a pseudocode of a simulated annealing algorithm.

```
 1  Initialize temperature t = t_max
 2  Initialize a spanning tree T
 3  while t > t_min do
 4  |    while equilibrium not reached do
 5  |    |    T' = random neighbor of T
 6  |    |    E = C(T') - C(T)
 7  |    |    assign T = T' with probability p(E)
 8  |    end
 9  |    decrease t
10  end
```

Algorithm 35: Simulated annealing

A simple and typical way to define an equilibrium condition is just to perform a constant number of iterations. The probability of accepting a new generated solution is given by

$$p(E) = \begin{cases} 1 \, , & \text{if } E \leq 0 \\ e^{-\frac{E}{t}} \, , & \text{if } E > 0 \end{cases} .$$

That means, a better solution is always accepted, and a solution with higher value of the function to be minimized is accepted with probability depending on the current temperature and on how much worse the solution is. Another important parameter of a simulated annealing algorithm is the cooling schedule, i.e., how the temperature is decreased. The authors of [90] claim to achieve good results with Lundy and Mees schedule:

$$t_i = \frac{t_{i-1}}{1 + \frac{t_{\max} - t_{\min}}{K t_{\max} t_{\min}}} \, ,$$

where K is the maximum number of iterations, $t_{\max}$ is the initial temperature, and $t_{\min}$ is the minimal temperature.

The main feature of this method is that it tries to avoid getting stuck in a local minimum.

Balanced SPT/TSP Tree. The solution provided by this algorithm consists of a shortest path tree structure within a certain radius around the sink node and a collection of traveling salesman paths starting from each of the leaves of the SPT. The idea behind this combination is that the leaf nodes contribute most to the total energy cost, since size of a raw data packet is larger than that of an encoded, $s > s_e$. Therefore, in order to minimize the energy consumption, the load coming from leaves of the tree should travel short paths to the sink, which favors the shortest tree paths. In the same time, the raw data from the leaf nodes should travel through as many nodes as possible to enable data encoding at those nodes and reduce the number of leaves, which yields the usage of TSP. This view is consistent with the observation made earlier: If the data correlation is small, the shortest path tree is a (nearly) optimal data gathering topology, or in this case the radius of SPT would be large. When the correlation is large, the costs of transporting encoded packets is negligible, so it is more important to have as many internal nodes as possible, which means the radius of SPT should be small.

In the first step of the algorithm a shortest path tree T rooted at the sink node t is built on a graph induced by the sink node and the nodes within $q(\rho)$ distance from the sink, where ρ is the data correlation coefficient and $q(\rho)$ is inverse proportional to ρ. Then the remaining nodes are attached incrementally in a following way. An edge between a non-tree node $v_0 \in V \setminus V_T$ and a leaf node u_0 of the tree T is added to the tree if (u_0, v_0) is a pair with minimum energy cost. That means if $(u_0, v_0) = \arg\min_{\{u \in L_T, v \in V \setminus V_T\}} w(v, u) + d_T(u, t)$, where L_T denote a set of leaf nodes of the tree T, $w(v, u)$ is the weight of the edge (v, u) and $d_T(u, t)$ is the cost of a path in the tree T from the node u to the sink node t.

Leaves Deletion Approximation. This algorithm is a local optimization strategy. It first constructs an SPT T of the graph G. Then, as long as there is a decrease in total cost, the following operation is performed.

For each leaf node v find a leaf node u adjacent to v that maximizes $s(d_T(v,t)+d_T(u,t))-(s(w(v,u)+d_T(u,t))+s_e d_T(u,t))-I(v)$, where $d_T(v_i,t)$ denotes the weight of the tree path from the node v_i to the sink node t, $w(v,u)$ is the weight of the edge (v,u) and $I(v)$ is a term indicating the cost lost by eventually transforming the parent of node v into a leaf node. The expression to be maximized is exactly the decrease in energy consumption if we would assign a new parent node u to the leaf node v. If the maximizing quantity is positive, the node u becomes a new parent of the node v and the corresponding distances to the sink on the tree are then updated.

The advantage of this algorithm is that it can be implemented in a distributed way.

13.3.3 Simultaneous Optimization for Concave Costs

In [162] Goel and Estrin address the problem of finding a data gathering tree which is a good approximation simultaneously to the optimum trees for a whole class of aggregation functions. It is argued, that usually the data correlation and thus the aggregation function is not known a priori. Therefore, it is important to have a data gathering structure that performs well for different applications.

Problem Definition. The sensor network is again modeled as an undirected graph $G = (V, E)$ with n vertices and m edges, a sink node $t \in V$ and a set $S \subset V$ of $k \leq n$ source nodes. The energy consumption of sending a unit of information between the nodes u and v is represented by the non-negative edge weight $w(u, v)$.

The data aggregation is modeled by stipulating that if information from j sources is routed over a single link, the total information that needs to be transmitted is $f(j)$. A function f defined on non-negative real numbers is called *canonical aggregation function* if f is concave and non-decreasing and $f(0) = 0$. The class $\mathcal{F}$ of aggregation functions considered is the class of canonical aggregation functions.

An aggregation tree is a tree which contains the sink as the root, all source nodes, and may contain additional nodes of the graph. Let $e = (u, v)$ be an edge in an aggregation tree, such that u is a parent of v in the tree. Then the demand routed through e is the number of sources in the subtree rooted at v. If $d(e)$ is the demand routed through an edge e, then the cost of using this edge is defined to be $w(e) \cdot f(d(e))$. It can be easily shown that for canonical aggregation functions, there exists an optimum set of paths that form a tree, and hence it is legible to focus our attention on finding an optimum tree.

Let D be a deterministic algorithm for constructing an aggregation tree, and $C_D(f)$ denote the cost of the resulting tree for the aggregation function f. Let $C^*(f)$ denote the cost of the optimum aggregation tree for the function f. The approximation ratio $\mathcal{A}_D$ of the algorithm D is defined as $\mathcal{A}_D = \max_{f \in \mathcal{F}} C_D(f)/C^*(f)$. The deterministic problem of finding an algorithm D that guarantees a small value for $\mathcal{A}_D$ is referred to as **Det**.

Given a randomized algorithm R constructing an aggregation tree, let $C_R(f)$ be a random variable, which denotes the cost of a tree constructed by R for a function f. There are two ways of extending the deterministic problem:

R1: Find a randomized algorithm R that guarantees a small value for

$$\max_{f \in \mathcal{F}} \mathbb{E} \left(\frac{C_R(f)}{C^*(f)} \right) .$$

R2: Find a randomized algorithm R that guarantees a small value for

$$\mathbb{E} \left(\max_{f \in \mathcal{F}} \frac{C_R(f)}{C^*(f)} \right) .$$

$\mathbb{E}$ denotes the expectation value of a random variable. Problem **R2** subsumes **R1** since $\mathbb{E} \left(\max_{f \in \mathcal{F}} C_R(f)/C^*(f) \right) \geq \max_{f \in \mathcal{F}} \mathbb{E} \left(C_R(f)/C^*(f) \right)$ for any randomized algorithm R.

The Hierarchical Matching Algorithm. Assume, that the graph G is complete and the edge weights satisfy the triangle inequality. If G is not complete, missing edges can be added using the shortest path metric. The assumption of the triangle inequality is fairly strong for the energy consumption weighted graphs. Additionally, assume for now that the number of source nodes k is a power of two, this assumption can be easily removed.

A matching on a graph $G = (V, E)$ is a set of edges $M \subset E$ such that no two of them share a vertex in common. A set of edges M is a perfect matching if every node appears in exactly one edge of M.

The algorithm starts with an empty tree T. Upon the termination T is the set of edges in the aggregation tree. The hierarchical matching algorithm runs in $\log k$ phases. In each phase the following two steps are performed:

1. *Matching Step:* Find a min-cost perfect matching M in the subgraph G_S induced by the source nodes S. Let (u_i, v_i) represent the i-th matched pair, where $1 \leq i \leq |S|/2$.
2. *Random Selection Step:* For all matched pairs (u_i, v_i), choose one out of u_i and v_i and remove it from S.

Due to the assumption that the number of source nodes is a power of two, and G is a complete graph, in every phase there exists a perfect matching in the induced subgraph G_S. In each phase the size of S gets halved. After $\log k$ phases $|S| = 1$. The resulting tree T contains the union of each of the $\log k$ matchings and an edge connecting the last remaining source node to the sink t.

The authors of [162] show first, that for any concave function f the expectation value of the approximation ratio for the tree produced by the algorithm can be bounded by $\mathbb{E} \left(C(f)/C^*(f) \right) \leq 1 + \log k$. That means, that the hierarchical matching algorithm gives a guaranteed value of $1 + \log k$ for the **R1** problem.

Input: complete graph $G = (V, E)$, sink node $t \in V$, source nodes $S \subset V$,
       edge weights $w : E \to \mathbb{R}^+$.
Output: Aggregation tree T
1 $T = \emptyset$
2 **while** $|S| > 1$ **do**
3   find min-cost perfect matching M in G_S
4      $T = T \cup M$
5   **forall** $(v_1, v_2) \in M$ **do**
6   choose randomly $v_j, j \in \{1, 2\}$
7          $S = S \setminus \{v_j\}$
8   **end**
9 **end**

Algorithm 36: Hierarchical Matching Algorithm

Then, the "atomic" canonical aggregation functions are introduced as $A_i(x) = \min\{x, 2^i\}$ for $0 \le i \le k$, and it is shown that a good approximation for just the atomic functions results in a good approximation for all canonical aggregation functions. After the proof of an upper bound on the expected value of the approximation ratio for the atomic functions follows that the hierarchical matching algorithm gives a guaranteed value of $1 + \log k$ also for the problem **R2**.

The time complexity of the algorithm is obviously dominated by the matching step. A minimum weight perfect matching in a complete graph with n nodes can be found in $O(n^3)$ time, [141]. The algorithm performs $\log k$ phases and in each phase the size of S gets halved, therefore the running time can be expressed as $\sum_{i=1}^{\log k} (2^i)^3 = \frac{8}{7}k^3 - \frac{8}{7}$, that is the time complexity is $O(k^3)$.

Derandomization. The deterministic algorithm is constructed from the randomized algorithm in a following way: The basic idea is to use the method of conditional expectations, combined with pessimistic estimators. The random decisions are only made in the selection step, and there are always only two alternatives to select from. Therefore, for the derandomization we could compute both alternatives and deterministically select the one which gives a better estimate for the performance for atomic functions.

The deterministic algorithm is of theoretical interest only, since it is more technical than the randomized algorithm and has worse performance guarantee in terms of constant factors. For the detailed analysis of the algorithms and proofs interested readers should refer to [162].

13.4 Maximum Lifetime Data Gathering

In this section we discuss a slightly different approach to the data gathering problem. Instead of minimizing energy consumption within one round of the data gathering process the goal is now to maximize the lifetime of the network. In this setting the nodes are assumed to have limited non-renewable battery power. The lifetime of the network is defined as the number of rounds the data gathering can be performed until the first node is drained out of energy. Alternatively, the target can be to maximize the number of rounds until a certain portion of nodes dies, or to maximize the number of rounds the data gathering is possible in the network, that is, the nodes with non empty battery can reach the sink node, or to have a long lifetime of the first node together with maximizing the lifetime of the last node.

Having a goal of maximizing the lifetime of a network we not only need to minimize the energy consumption in each data gathering round, but also balance the load between nodes. The most of the algorithms presented in the previous section result in a tree data gathering topology where the nodes close to the sink relay more data packets as the nodes closer to the leafs, and therefore would exhaust their battery sooner than the others.

In this section we describe two simple heuristics LEACH and PEGASIS for the maximum lifetime problem which are both designed for a setting where all nodes are interconnected and the sink node is far away from the source nodes. The third algorithm, MLDA, also requires a complete graph, but makes no assumption on the position of the sink. MLDA can, however, be modified to work on incomplete graph.

13.4.1 LEACH: Low-Energy Adaptive Clustering Hierarchy

LEACH algorithm was introduced in [181]. The network model is slightly different from the model described in Section 13.2. The network graph $G = (V, E)$ is assumed to be complete, that is, there is an edge between any two nodes, and a sink node is positioned far away from the other nodes, which means that direct communication with the sink node is expensive in terms of energy consumption. Data correlation has no influence on the routing scheme.

The idea of the algorithm is that in each round clusters are formed dynamically and cluster-heads gather data locally and possibly perform the data fusion and then transmit the aggregated data to the sink node. Cluster-heads are chosen randomly in each round. Thus, in each round only a few nodes have to perform energy-expensive communication with the sink node and the load is balanced among the nodes due to the random choice of cluster-heads.

The system is supposed to be able to determine, a priori, the optimal number of clusters to have. This number depends on several parameters, such as network topology and the relative costs of computation and communication. If there are fewer than optimal cluster-heads in the system, some nodes have to transmit their data very far to reach a cluster-head, causing the total energy consumption to be large. On the other hand, if there are too many cluster-heads, the distance from a node to the nearest cluster-head does not reduce substantially, yet there are more nodes that have to transmit data along highly weighted edges to the sink.

Every data gathering round consists of a setup phase, where the clusters are organized, and a data transmission phase. First, every node v decides independently with certain probability (v) whether to become a cluster-head for the current round r. The probability depends on the suggested percentage of cluster-heads P and the number of times the node has been a cluster-head so far, and is set to

$$(v) = \begin{cases} \frac{}{1- \cdot(r \mod \lfloor \frac{1}{P} \rfloor)}, & \text{if } v \text{ was not a cluster-head in last } \lfloor \frac{1}{} \rfloor \text{ rounds} \\ 0, & \text{otherwise} \end{cases}.$$

After deciding to be a cluster-head a node broadcasts an advertisement message. The non-cluster-head nodes decide based on received signal strength to which cluster they belong. After selecting the cluster a node informs the corresponding cluster-head, that it will be a member of the cluster.

The cluster-head node creates then a transmission schedule and informs the nodes in its cluster. After collecting data from all nodes in the cluster a cluster head performs data fusion and transmits aggregated data direct to the sink node, see Figure 13.7 for an illustration.

LEACH algorithm is fully distributed.

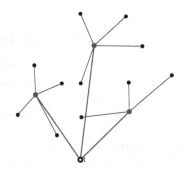

Fig. 13.7. LEACH: Cluster-heads aggregate data from cluster-nodes and communicate with the sink node.

13.4.2 PEGASIS: Power-Efficient GAthering in Sensor Information Systems

PEGASIS algorithm presented in [258] is based on the same network model as LEACH algorithm [181], and aims to improve the performance of LEACH by avoiding the overhead to form the clusters and letting in every data gathering round only one node transmit fused data to the sink node.

The main idea in PEGASIS is for each node to receive from and to transmit to one close neighbor and to take turns transmitting aggregated data to the sink node. Nodes are assumed to be numbered and then placed randomly in the area, therefore the nodes with consecutive indices do not have in general nearby location.

At the setup phase of the algorithm all nodes except for the sink node are organized to form a chain. The chain should be a light weight path connecting all nodes. Since minimizing the cost of such path is equivalent to solving a traveling salesman path problem and therefore an efficient optimal algorithm is not likely to exist, the greedy approach is applied. For simplicity, it is assumed, that every node has a global knowledge of the network. If it is not the case, the nodes may spend some energy to discover their nearest neighbors. The chain construction is started with a node farthest from the sink node to make sure that the nodes farther away from the sink have close neighbors. Then, a nearest neighbor of the last node, that is not in chain yet, gets added to the chain. When a node exhausts its battery power, the chain is reconstructed in the same way to bypass the dead node.

In a data gathering round number r communication with the sink node is performed by the node $i = r \mod n$, where n is the number of source nodes, see Figure 13.8. Remember that the nodes were placed randomly, so the leader of each round is at a random position in the chain. The idea behind it is to make the network robust against failures. The data is transmitted from the ends of the chain to the leader node of the round. Every intermediate node performs data aggregation. After the leader node receives data from both ends of the chain, it transmits the joint information package to the sink node.

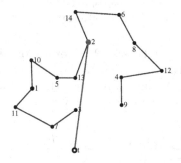

Fig. 13.8. PEGASIS: In every round one node collects the data from all nodes and sends it to the sink, aggregation is performed in every intermediate node.

13.4.3 MLDA: Maximum Lifetime Data Aggregation

The MLDA algorithm, proposed in [205], solves the maximum lifetime problem using an integer program for finding an optimal flow network and then reconstructing a data gathering schedule from the flow network.

System Model. Consider a network of n sensor nodes $1, 2, \ldots, n$ and a sink node t labeled $n + 1$. Similar to the previous two algorithms, it is assumed that all nodes are interconnected, but no assumption is made on the position of the sink node. Every node generates a data packet of size s bits in every data gathering round. Furthermore, each sensor i has a battery with finite energy ε_i, the sink node has unlimited amount of energy available to it.

The data fusion can be performed at any intermediate node. The authors of [205] make a very simplistic assumption that multiple incoming data packets can be aggregated into a single packet of the same size, such data aggregation functions could be average, minimum, maximum or sum of the incoming data.

Since in a radio network the sending node as well as the receiving node consume their energy for data transmission and we are now concerned with the energy each node spends in the data gathering process, we split the costs of a single link into transmission and receiving costs. Let R_{X_i} denote the energy consumed by the node i in receiving s-bit data packet, and $T_{X_{ij}}$ the energy consumed by the node i in transmitting a packet of size s to the node j.

The lifetime T of the system is defined to be the number of rounds until the first sensor exhausts its battery. A *data gathering schedule* specifies for each round how the data packets from all sensors are collected and transmitted to the base station. The schedule can be thought of as a collection of T directed spanning trees, one tree for each round. The lifetime of a schedule is the lifetime of the system under that schedule.

The *maximum lifetime data aggregation problem* is then, given a sensor network as described above, find a data gathering schedule with maximum lifetime.

Reduction to a Flow Network Problem. Consider a schedule S with the lifetime of T rounds. Let f_{ij} denote the total number of packets a node i sends to the node j in S. Since any valid schedule must respect the energy constraints at each sensor, it follows for any node $i = 1, \ldots, n$,

$$\sum_{j=1}^{n+1} f_{ij} \cdot T_{X_{ij}} + \sum_{j=1}^{n} f_{ji} \cdot R_{X_i} \leq \varepsilon_i \ . \tag{13.4}$$

The schedule S induces a flow network $G = (V, E)$. G is a directed graph where $V = \{1, 2, \ldots, n + 1\}$ and an edge (i, j) has the capacity f_{ij} whenever $f_{ij} > 0$. It is shown in [205] that a necessary condition for a schedule to have lifetime T is that each node in the induced flow network can push flow T to the sink node t. According to the Max-flow Min-cut Theorem [83], it is equivalent to the condition that each sensor i must have a minimum i-t-cut of capacity at least T.

Now the problem of finding a schedule with maximum lifetime is equivalent to the problem of finding a flow network with maximum T, such that each sensor node can push flow T to the sink node, while respecting the energy constraints (13.4) for all sensors.

An optimal flow network can be found using an integer program with linear constraints. In addition to the variables for the lifetime T and the edge capacities f_{ij}, let $\pi_{ij}^{(k)}$ be a flow variable indicating for each sensor node $k = 1, 2, \ldots, n$ the flow that the node k sends to the sink over the edge (i, j). The linear integer program maximizes T under the following constraints: the energy constraints for each node i given by Equation (13.4); the flow conservation constraints (13.5) and (13.6) for each sensor $k = 1, \ldots, n$ and each sensor $i = 1, \ldots, n$ with $i \neq k$; the edge capacity constraints (13.7); and constraints (13.8) ensuring that T flow from each sensor k reaches the sink node t.

$$\sum_{j=1}^{n} \pi_{ji}^{(k)} = \sum_{j=1}^{n+1} \pi_{ij}^{(k)} \qquad \forall i, k = 1, \ldots, n, \ i \neq k \tag{13.5}$$

$$T + \sum_{j=1}^{n} \pi_{jk}^{(k)} = \sum_{j=1}^{n+1} \pi_{kj}^{(k)} \qquad \forall k = 1, \ldots, n \tag{13.6}$$

$$0 \leq \pi_{ij}^{(k)} \leq f_{ij} \qquad \forall i, k = 1, \ldots, n \text{ and } \forall j = 1, \ldots, n+1 \tag{13.7}$$

$$\sum_{i=1}^{n} \pi_{i\,n+1}^{(k)} = T \qquad \forall k = 1, \ldots, n \tag{13.8}$$

Finally, all variables are required to take non negative integer values.

The linear relaxation of the integer program, that is the variables are allowed to take fractional values, can be solved in polynomial time. A good approximation for the optimal flow network can be found by first fixing the edge capacities to the floor of their values obtained from the linear relaxation, so that the energy constraints are satisfied, and then solving a linear program maximizing T subject to the constraints (13.5)–(13.8).

Constructing a Schedule from a Flow Network. Recall that an aggregation schedule is a collection of directed spanning trees with one tree for each round. If an aggregation tree is used for f rounds, f is called the lifetime of the aggregation tree and is associated with each of the edges of the tree.

Let A be a directed spanning tree pointing to the sink node t with lifetime f, then the (A, f)-*reduction* of the flow network G is a flow network, which results from reducing the capacities of all edges in G that are also edges in A by value f. An (A, f)-reduction G' of G is called *feasible* if the maximum flow from any sensor node i to the sink node t in G' is at least $T - f$.

The aggregation schedule is reconstructed in a simple iterative procedure, where in each step a feasible aggregation tree is found and the network G is reduced by the lifetime of the tree. Construction of a feasible tree $A =$

(V_A, E_A) is done in a following way. First the tree contains only a sink node t, $V_A = \{t\}$. In every iteration step an edge (i, j) with $i \in V \setminus V_A$ and $j \in V_A$ is added to the tree A if the reduction by $(A', 1)$ of G is feasible, where A' denotes A with the edge (i, j) added. After a spanning tree is constructed, the minimum capacity f of its edges is defined to be the lifetime of that tree, and (A, f)-reduction of G is performed.

The worst case running time of MLDA algorithm is $\mathcal{O}(n^{15} \log n)$, where n is the number of sensor nodes. The running time is dominated by the solution of the linear program with $\mathcal{O}(n^3)$ variables and constraints.

In order to reduce the running time clustering heuristics are proposed. The idea is to partition the nodes in m clusters each having at most c sensors for some constant c, and the sink node t being a single node in its cluster, a cluster is called a super-node. The energy available at a super-node is assigned to be the sum of the battery capacities of the node in that cluster. The transmission cost between two clusters ϕ_i, ϕ_j is defined to the maximum transmission cost between any two nodes u, v, such that $u \in \phi_i$ and $v \in \phi_j$. The linear program is then solved for the m super-nodes. The aggregation trees for the cluster nodes are extended to comprise individual sensor nodes in a greedy way. With appropriate choice of m and c the running time of the greedy clustering algorithm can be reduced to $\mathcal{O}(n^3)$. For details of the algorithm and for an incremental clustering algorithm interested readers can refer to [205].

13.5 Chapter Notes

In all algorithms described above the cost of data aggregation was assumed to be negligible compared to the communication cost. However, some networks may require complex operations for data fusion, e.g. acoustic signal fusion or image fusion, which consume significant amount of energy. The authors of [270] propose a network and aggregation model for the minimum energy data gathering problem, which takes into account the data fusion costs. They also present a factor $\mathcal{O}(\log n)$ approximation algorithm, which optimizes both transmission and fusion costs.

Other problem statements related to data gathering include minimizing energy consumption subject to some latency constraints as in [401] and optimizing placement density under latency and energy constraints as in [91]. The work in [162] deals with a universal solution to the minimum energy data gathering problem. The authors propose randomized and deterministic algorithms constructing a routing tree, which is a good approximation simultaneously to the optimum trees for all concave, non-decreasing aggregation functions.

For the Slepian-Wolf coding it is shown in [89] that as a consequence of the possibility to exploit data correlation without explicit communication, the problem of minimum energy data gathering can be separated into finding first the optimal routing structure and then optimal coding scheme, where the routing structure depends only on the link weights. Further it is shown that for the single source model the optimal routing structure is given by shortest path tree, and the optimal coding strategy can be found in polynomial time using linear programming. Those results hold for linear separable cost functions as we have considered in this chapter, that is the cost of sending s bits along a link with weight w is equal to $s \cdot w$. For non-linear separable cost functions the coding structure can be found using Lagrange multipliers. For the multiple sink setting the problem of finding an optimal routing scheme is $\mathcal{NP}$-complete. Though, if the routing structure is approximated, finding the optimal coding scheme is simple.

It has been shown that the problem of minimum energy data gathering with explicit communication is $\mathcal{NP}$-hard under self-coding strategy even in a simplified setting. On the other hand, a very similar problem setting with foreign coding strategy can be solved optimally in polynomial time. It would be interesting whether a joint problem, where both coding mechanisms are allowed, can be solved efficiently.

In case of single-input coding, the foreign coding strategy seem to be more efficient than the self-coding strategy. An interesting question is: given a routing and coding scheme under self-coding strategy for a network graph is there a routing and coding scheme using foreign coding with not larger total energy consumption for the same graph.

Another research field could be to examine combination possibilities of explicit communication and information-theoretical approach of joint entropy coding schemes, e.g. Slepian-Wolf coding, and the complexity of such combined problem statements.

Most of the algorithms proposed for the maximum lifetime problem either do not have theoretically provable solution quality guarantees or use a very simplistic correlation model and have high degree polynomial running time. A challenging task could be to develop an appropriate network and correlation model together with provable quality algorithms with good running time.

14

Location Services

Benjamin Fabian, Matthias Fischmann, and Seda F. Gürses

14.1 Introduction

A *location service* for an ad hoc or sensor network is an algorithm that on the input of a node's *network identifier* returns as its output this node's physical *location*. A location service sits between the similar sounding concepts of *localization* or *positioning*, which is treated in Chapter 15, i.e., how a node determines its own position (e.g., by using GPS or triangulation relative to other nodes), and *location-based services*, i.e., services that need to be aware of the client's location (e.g., restaurant or gas station recommender systems). A similar, but different problem is the selection of the most conveniently located node.

The advantage and even the necessity of a location service in certain applications are mainly due to the following observation on the topology of wireless ad hoc networks:

> Nodes that are physically close are likely to be close in the network topology; that is, they will be connected by a small number of hops [253].

Accordingly, schemes for geographic routing have been developed which route packets based on information about the physical location of a destination node, which are handled in Chapter 9. To make this work, a source—or at least some intermediate node—needs to know the location of the destination.

There are two trivial solutions to determine the location of the nodes in mobile ad hoc networks. The first one is for each node to flood the complete network with its position updates and to keep track of the position of all other nodes. Such a solution does not scale to large and dense networks because the communication cost for updates (measured in the total number of hops all messages take) is quadratic in the number of nodes.

Another solution is to have one or a few central nodes, which keep track of the location of all other nodes, thus functioning as central location servers.

D. Wagner and R. Wattenhofer (Eds.): Algorithms for Sensor and Ad Hoc Networks, LNCS 4621, pp. 265–281, 2007.
© Springer-Verlag Berlin Heidelberg 2007

This approach results in high costs for location queries, especially for nodes closer to the location servers that need to route a lot of location queries from distant nodes. Also, the costs of communicating with the servers are out of proportion if neighboring nodes that are far away from the location server wish to locate each other. Last, such a model is not robust because it has a single or a few points of failure.

The quest for location service solutions is further complicated by the mobile character of networks. Depending on the mobility of the nodes, we cannot expect a location service to guarantee the correctness of an answer, nor can we even guarantee the availability of corresponding location server nodes themselves. Both are problems a location service solution has to deal with while trying to deliver best-effort results.

All known location service solutions assume that each node knows its own physical location. Motivated by the highly distributed and self-organized nature of ad hoc networks, the following requirements are formulated for location service solutions:

- **Spread tasks among nodes:** The workload should be spread evenly among the nodes in order to avoid bottlenecks and early exhaustion of overworked nodes.
- **Guarantee fault tolerance:** Failure of one node should not affect the reachability of other nodes. The network should be robust enough to deal with a plausible fraction of failing nodes.
- **Minimize traffic:** Queries for the locations of nearby nodes should be satisfied by correspondingly local communication. A central location server is an example that does not fulfill this requirement, because a node distant to the server needs to go a long way to find a node that may actually be very close. Local solutions have the additional advantage of allowing the network to operate in the presence of partitions.
- **Warrant scalability:** The per node storage and communication cost should grow as a small function of the total number of nodes.

Depending on the properties of the network (e.g., density, spread, capacity of the nodes, mobility etc.) different location service solutions may yield the best results. The two main location service solutions discussed in this chapter are GLS (Grid Location Service) and LLS (Locality-aware Location Service). In the discussion of LLS, we also sketch a more recent geographic location service called MLS (Mobile Location Service).

14.2 Grid Location Service (GLS)

14.2.1 Introduction of the Algorithm

In their article *A Scalable Location Service for Geographic Ad Hoc Routing*, Li et. al [253] suggest a solution to the problem of routing in large ad hoc networks of mobile hosts. In order to achieve this, all the nodes keep track of the position of some of the other nodes.

Further, the algorithm is based on what the authors call a grid – a quadtree structure which divides the 2-dimensional plane recursively into four quadrants. This structure is used to index the territory in order to take advantage of the similarity between physical and network proximity among the nodes. The algorithm, called the Grid Location Service (GLS), proves to be more scalable and robust than flooding the network or single-server location service solutions.

In the following, we will introduce the GLS algorithm. We first shortly introduce geographic forwarding, a geographic routing solution, which is the topic of Chapter 9. Then, we will describe the GLS algorithm and conclude by looking at some of the problems of GLS.

14.2.2 The GLS Algorithm

In their study, the authors combine GLS with a geographic routing method called geographic forwarding. In order to understand GLS it is sufficient to know that GLS makes use of geographic routing, which is guaranteed to reach a destination node given its geographic position (provided that the network contains a path from the source to the destination). The location service delivers the position of the destination to the geographic forwarding protocol.

The location service algorithm GLS distributes the *location server* role to multiple nodes in the network. The location servers are not designated prior to deployment, but are distributed throughout the network based on a grid system. This grid is defined as follows:

- The area in which the sensors are located is divided into squares. The squares with smallest granularity (with edge length 1) are called squares of order 1.
- All nodes agree on the lower left corner $(0,0)$ and the upper right corner (M, M) of the grid, where M is equal to the number of squares of order 1 in each edge.
- The combination of four squares of order $n - 1$ make up a square of order n.
- For the algorithm, the squares of order n are limited to those that have corners of the form $(a2^{n-1}, b2^{n-1})$ for $a, b \in \mathbb{N}$. This guarantees that squares of order i do not overlap, that they cover the complete network area, and that each square of order $i - 1$ is fully contained in a single square of order i, called its *home square*.

Figure 14.1 shows a grid with squares of order 1 through 4. Notice that the grey square, which is made up of four squares of order 1, is not a square of order 2, since its starting point is $(1,1)$, and there is no $a, b \in \mathbb{N}$ such that $(1, 1) = (a2^{2-1}, b2^{2-1})$.

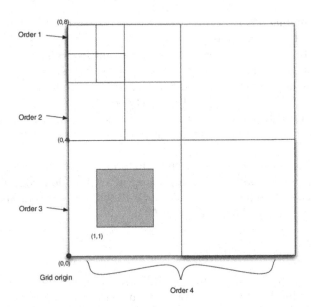

Fig. 14.1. A grid with squares of order 1 through 4

The hierarchy of the squares makes it possible for a node to select more location servers nearby and fewer location servers far away to store its location. How the location servers are selected is described in the following section.

Selecting Location Servers. In order to locate the nodes in a network, nodes address each other through randomly distributed node names. These unique names are generated by hashing from some unique information about a node, such as its host name, IP, or MAC address. With A we (informally) refer to a node whose name hashes to A. Whenever a node A wants to talk to a node B, it already knows the name of B, but not its location. GLS serves A to find the latter.

Initially, all nodes determine their position as well as the globally fixed point of origin $(0,0)$, and compute all layers of the grid. The name space (the set of all possible hash values) is circular, thus every node can find at least one node whose hash value is greater than its own if it can find any node at all. We call the node with the least hash value greater than B also the closest node to B (i.e., Node 3 is closer to node 25 then node 12).

GLS makes location lookups possible by replicating the knowledge of a node's location in a small subset of the network according to an identified rule set. If node A wants to know the location of B it needs to find one of the nodes that know this location. These nodes are called the location servers of B. In GLS the rule set through which B recruits location servers is similar to the way A searches for location servers. Namely, both the selection of and the search for location servers make use of the strategy of finding the nodes with names closest to B. To be more specific, node B selects its location servers as follows. It selects three location servers from neighboring squares (and not its home square) for each level of its grid hierarchy. Hence, for each square γ of order k in which a location server is to be selected,

- B sends a packet containing its location to γ using geographic forwarding (e.g., by sending it to γ's center);
- the first node in γ that receives the packet passes it on to the node closest to B (with respect to hash value order) in γ. This node becomes the location server for B in this square.

The last step of this algorithm, namely, passing the packet on to the node closest to B in γ is not trivial. It requires that all the squares of order less than k have already distributed their locations and hence location servers. If we were to imagine the square hierarchy as a tree, this would mean that all the sibling nodes have to distribute their locations among themselves in order for the distribution of locations at the respective parent nodes to be possible. Here we also assume that nodes in the same order 1 square know the locations of all other nodes in that square (the size of the order 1 squares being selected to be sufficiently small).

Let us give an example. Node 17 is located in the grid as in Figure 14.2. First, 17 sends a packet with its location information to the neighboring squares of order 1 in its home square of order 2. In each of these squares, the nodes closest to 17 are selected as location servers. Next, 17 repeats the same procedure for neighboring squares of order 2, then neighboring squares of order 3 and so on, until the square with highest order is reached. In Figure 14.2 the neighboring squares of each order are emphasized with dashed outlines and the nodes selected as location servers are highlighted.

Querying Locations of Other Nodes. In order to find a node B, the searching node A tries to find one of node B's location servers. This is how it works:

1. Node A sends a query to a node C. C is chosen from all nodes in A's order 1 square such that it is closest to B. (Note that A may actually pick itself, yielding $A = C$.)
2. If C doesn't know B's location, it forwards the query to node C' closest to B in its home order 2 square. If C' doesn't know B either, this step is repeated in home squares of increasing orders, until B is found at the latest in the square with highest order.

Consider Figure 14.3 for an example. Node 76 starts a query for node 17. Since 76 is the only node in its square of order 1, it does not need to take Step 1. It is therefore also the location server for all the nodes in its neighboring squares (within its home square of order 2). This means that 76 knows the locations of the nodes 6, 16, 21 and 83. The node closest to 17 in the home square of order 2 is 21. Since 21 is not a location server of 17, it forwards the query.

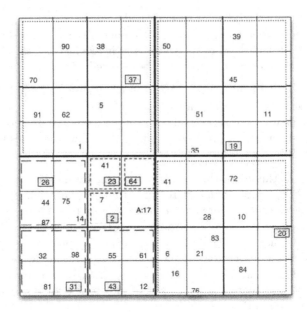

Fig. 14.2. The location servers of the node A with hash value 17 in home squares of all orders.

Note for the following observation that nodes 17 and 21 are in neighboring squares of order 3. The fact that 21 is the closest node to 17 in its square of order 2 and not the location server of 17 has two consequences. First, it means that somewhere in the neighboring squares of order 2 there is a location server of B whose ID is closer to 17. Second, for any such node in one of the neighboring squares of order 2, 21 is the closest in its own square of order 2. Node 21 is hence the location server for all nodes close to 17 and less than 21 in its neighboring squares of order 2. This means that 21 knows all nodes which are closer to 17 than itself in its square of order 3. The only such node is 20, which turns out to be a location server of node 17.

GLS gives several performance guarantees:

- Since the query moves one step at a time up the grid hierarchy, some location server will be found, at the latest when the smallest square containing both nodes has been reached.
- The number of query steps is bounded by the highest order of the squares, which is $\log M$.

In their paper, the authors also claim that GLS fulfills the requirement to minimize traffic. Namely: Going up the grid hierarchy guarantees that the distance the query will travel is bounded by (a function of) the distance between the searching and the searched node. But, especially when two nodes are closely located but on different sides of the border between squares of

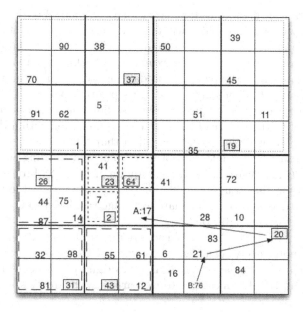

Fig. 14.3. A grid with squares of order 1 through order 4. Node 76 looks up the location of node 17. Node 76 has up-to date location information for $(16, 6, 21, 28, 41, 72)$ of which 21 is closest to 17. 21 has location updates for $(6, 16, 76, 20)$ of which 20 is closest to 17. 20 is a location server of 17 and forwards to it the query of node 76.

higher order, there is no guaranteed bound. The algorithm LLS introduced in Section 14.3 attends to this problem.

Mobility and Location Updates. A location service for a mobile network must provide a location update mechanism. In GLS the rate at which a node sends updates to one of its location servers depends on the distance d the node has travelled since the last update and the order i of the square the server is located in.

Nodes in the same order 1 square are assumed to already know the actual position of a node through the geographical forwarding layer. Actual location servers in an order i square ($i \geq 2$ by the server selection procedure) are updated after movement over a distance of $2^{i-2}d$, so distant servers are updated less frequently.

Location update packets include a time stamp and timeout value, so the freshness of update information can be verified. For this purpose time synchronization between nodes (e.g., by GPS) is a requirement.

Forwarded location information is also cached at every intermediate node, and data packets for regular communication are accompanied by a sender's actual position.

Query Procedure and Possible Failures. Assume a node A wants to communicate with a destination node B. A first checks its local caches and the location table it maintains. If no entry for B is found, A initiates a location query for B by using GLS, which will forward the query to one of B's location servers. B responds directly to A including current location data. In case A gets no response, it will retry using exponential back-off intervals.

If a location server (following an update or lookup procedure) needs the location of a node that has not updated its position before the timeout has expired, a query may fail. A query might also fail when a location server forwards a packet to the square of the next closest node well within the timeout, but the next node is no longer positioned in that square.

To compensate for the latter problem, GLS uses forwarding pointers that a moving node B may leave at its order 1 square of origin by broadcasting it to all other nodes in that square. These pointers are piggybacked to the HELLO messages that nodes in the old square exchange regularly, so new nodes may learn of their existence later on.

A packet arriving for B in its old square due to out-of-date location information at remote servers can then be forwarded to the new square. Clearly, if there are no other nodes at the square of origin, or all nodes leave before they could send HELLO messages to new arrivals, the query still fails.

14.2.3 Discussion

The authors suggest in their paper that the GLS algorithm lives up to the requirements that we listed at the beginning of this section. Their simulations with 100-600 nodes show that the rate of successful queries (fulfilling failure tolerance and scalability) at first try are quite high, the packets to be processed per node per second are bounded (fulfilling spread of tasks among nodes). That the distance travelled by a query between two nodes is proportional to the actual distance between those nodes, as suggested by the authors of the algorithm, does not hold (failing to always fulfill the minimization of traffic). Later studies [96, 171] also show that GLS has its weaknesses for less than 100 and more than 600 nodes.

Some of the reasons for these problems are mentioned by the authors but left open for future research. These are: When nodes are not uniformly distributed, a small number of nodes in a high order square may end up responsible for a large number of nodes in neighboring squares, the update costs may increase when nodes move a short distance but cross the borders of two high order squares, and high node density in squares of smaller order may cause congestion and heavy traffic among these nodes.

14.3 Locality-Aware Location Service (LLS)

In the previous section, we have discussed a classical location service algorithm and hinted to some of its performance limitations. In particular, there is no tight correlation between the distance to be covered (resp. between a searching and a searched node, or between the old and new locations) and the cost (resp. of the lookup, or the update operation). In this section, we will describe the locality-aware location service algorithm (LLS) proposed by Abraham et al. [1] that attempts to guarantee such correlations. Further, we will demonstrate that the bounds that are claimed in the paper cannot be reached in general.

14.3.1 A Note on Points vs. Nodes

In our discussion we will use the coordinates of some point and the node located at those coordinates synonymously. Mapping nodes to coordinates in this way is harmless, but the reverse direction needs a brief explanation: What if we send a message to a point that is not occupied by any node? (This is the least surprising outcome if we pick a point independently of the node distribution, which we will do in a moment.)

The problem we are facing here is one of partitioning a plane occupied by a network of nodes in such a way that for each point there is a unique node that is responsible for it. Intuitively, we want to select the node closest to the point (cases of equal distance can be resolved lexicographically). Decentralized plane partitioning algorithms to negotiate responsibilities in such a setting can be constructed using Voronoi diagrams, and algorithms for routing to whichever node is closest to some point can be built by simple extensions to geographic routing algorithms.

14.3.2 The Grid

Similarly to GLS, LLS works on a grid of squares of exponentially growing edge lengths. However, the origin of this grid is not a global constant, but LLS sets it relative to the node that publishes its location, or the node whose location information is requested, respectively. As the location of the node changes and is not known during lookup, the grid origin is determined by the name alone, independently of its current location. This is done by hashing the name of the node to coordinates: If h_1, h_2 are two hash functions mapping node names to coordinates, then the grid for node B of order $i \in \mathbb{N}_0$ is defined as

$$G_i(B) := \{(h_1(B) + a2^i, h_2(B) + b2^i)\} \quad \forall a, b \in \mathbb{Z}.$$

The result looks very similar to the GLS grid; only its origin is determined by the name of the node to be located and thus is different for every node.

As with GLS, the area of the network is finite and ranges from $(0, 0)$ to (M, M). This means that only squares up to order $\log M$ need to be considered.

On this hierarchy of grids $G_0(B), \ldots, G_{\log M}(B)$, corner points of selected squares are responsible for locating B. This explains why the origin of the grid must not be globally fixed: If all nodes would use the same grid, nodes that happen to be sitting on large squares would have a much higher workload than those sitting "in the middle of nowhere", leading to bottlenecks and early resource exhaustion in the busier parts of the network.

On each layer i, the final algorithm requires two sets of nodes to store a mapping from node names to information on node locations: $W_i(B, x) \subset G_i(B)$ is the order i square surrounding x, and $Z_i(B, x) \subset G_i(B)$ is the superset of $W_i(B, x)$ containing a belt of eight surrounding order i squares, see Figure 14.4. Then, each node in $W_i(B, x)$ stores $B :: x$ (the location of B), and each node in $Z_i(B, x)$ stores $B :: W_{i-1}(B, x)$ (the locations of nodes that know x).

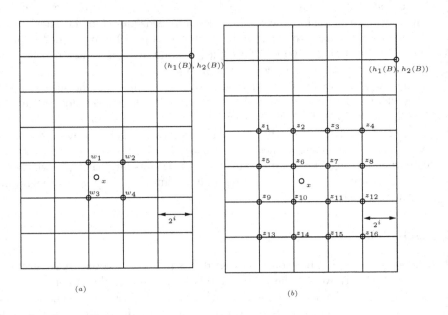

Fig. 14.4. The set $W_i := \{w_1, ...\}$ of nodes that have to store the current location $B :: x$ (a), and the set $Z_i := \{z_1, ...\}$ of nodes that have to store the locations $B :: W_{i-1}(B, x)$ of nodes that know x (b).

14.3.3 Spirals

A simplified variant of the LLS algorithm uses exclusively the above W-sets and works as follows: In order to publish or update its new location x, a node B iterates i over $\{0, \ldots, \log x\}$ and in each step:

1. removes $B :: x$ from all nodes in $Z_i(B, x)$;
2. uploads $B :: x$ to all nodes in $W_i(B, x)$.

Note that this doesn't require location service lookups, as the nodes are anonymous and only identified by their grid coordinates.

As the square of order i surrounding y shares one corner with the corresponding square of order $i - 1$, B only needs to contact three nodes in each iteration. The sequence of nodes can be thought of as spanning a spiral starting at B, and going outwards at an exponentially growing pace. See Figure 14.5. With Spiral(B, y) we denote the sequence of nodes lying on the spiral in the grid hierarchy originated in $(h_1(B), h_2(B))$ and starting from y. In a slight abuse of terminology, we say that this spiral has origin B and center y.

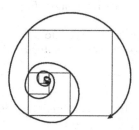

Fig. 14.5. Spiral(B, y) (Source: [1]).

If node A wants to locate B, it spans a corresponding spiral with the same origin B (which is known to A as A knows B's name), but with center z, where z is the current location of A. Given the Euclidean distance $d(A, B)$ between A and B, we know that the lookup request finds a location pointer to B at the latest on the level whose edge-length exceeds $d(A, B)$. Thus, the effort of locating B correlates nicely with the distance between A and B.

The bad news is that publishing new locations always takes $\log M$ steps (the entire spiral needs to be rebuilt). However, short distances should be associated with low-cost location service updates. The spiral should only be updated in the center, to an extent proportional to the distance moved.

14.3.4 Full LLS

The spirals hosting the location servers of B in the full version of LLS need to have the following properties:

1. **Small publish stretch:** If B moves only a small distance, it should be required to publish its new location to few location servers on the spiral only.
2. **Efficient forward:** If A hits the spiral at some point that hasn't been updated for a while, the spiral should still be able to locate B efficiently.

These two goals can be achieved by using redirects: If the location servers of order i point to the location servers of order $i - 1$, and so on, then A can still find B in a number of redirects logarithmic in the distance between A and B.

Input: Name B of the node changing its location (or entering the network), old position x ($\bot$ if B is new), new position y.

Output: Nothing.

```
 1 begin
 2     if all nodes W₀(B,y) contain a location pointer for B then
 3         RETURN
 4     end
 5     if x = ⊥ then
 6         for each node Z in Z₀(B,x) do
 7             store the direct location pointer B :: y in Z
 8         end
 9         for each node Z in Z_log M(B,y) do
10             store the indirect location pointer B :: W_log M−1(B,y) in Z
11         end
12     end
13     i := 1
14     while not all nodes in Wᵢ(B,y) contain a location pointer for B do
15         for each node Z in Zᵢ(B,x) do
16             remove the indirect location pointer B :: W_{i−1}(B,x) on Z
17         end
18         for each node Z in Zᵢ(B,y) do
19             store the indirect location pointer B :: W_{i−1}(B,y) on Z
20         end
21         i := i + 1
22     end
23     for each node Z in Zᵢ(B,y) do
24         update the existing indirect location pointer in Z to B :: W_{i−1}(B,y)
25     end
26     RETURN.
27 end
```

Algorithm 37: The LLS publish algorithm.

If, on the other hand, B moves into another order $i - 1$ square that lies in the same order i square, it only needs to inform the squares up to order i of the change, but not above.

In the following, we look at the full LLS algorithms in more detail, which are shown in Algorithm 37 and 38. The two important changes with respect to the naive approach described above are: (a) The publish spiral is now spanned using Z_i and not W_i; and (b) the location servers on the spiral do not contain direct information on x, but only pointers to other location servers closer to the center.

If B is new ($x = \bot$), then two things happen prior to the main **while**-loop: The order 0 location servers record a direct pointer to B's current position, and the outermost location servers (on $G_{\log M}$) record an indirect pointer to the location servers one layer below (on $G_{\log M - 1}$). (The latter makes sure

---

Input: Name A of the searching node, location z of searching node, name B of wanted node.

Output: Location y of B.

1 **begin**
2   **for** *each node S in the sequence Spiral(B,z), until S knows B* **do**
3 | $y :=$ ask S for B
4   **end**
5   **while** y *is an indirect location pointer to B* **do**
6 | $y :=$ query y for B
7   **end**
8   RETURN y
9 **end**

Algorithm 38: The LLS lookup algorithm (without interleaving broadcast steps).

that the main loop of the publish algorithm terminates, while the former guarantees termination of lookup.) Then, the algorithm iterates over the grid hierarchy, starting from order 1 and terminating once an order i square has been found of which all corners have information on B. In each step, the records in the old order i location servers are removed, and those in the new order i location servers are created. Finally, on the level, say, i on which all four nodes of $W_i(B,y)$ contain location information for B, the update process is stopped, and all location pointers of $Z_i(B,y)$ are updated to point to the square of level $i-1$ in which B is located.

If B has just moved ($x \neq \bot$), the smallest square containing both x and y is the one terminating the main loop and thus providing an upper bound to the cost of the update operation.

Lookup is similar to the naive approach. Only the searching node A may hit an indirect reference y to B rather than a direct one. In this case, it merely needs to dive down the publication spiral until it hits B, starting with coordinate pair y. As soon as a direct reference is found, it is returned. This takes logarithmically many steps in the distance between first indirect and last (direct) reference y.

14.3.5 Complexity

The model underlying LLS is a unit disk graph in which node mobility is limited in two ways:

– If a node publishes its new location, all nodes on the publish spiral need to be stationary, until the spiral has been updated completely. (In between publish operations, a node holding location pointers for other nodes may leave a point if it trades the location information with the node that becomes responsible for this point.)

– Nodes may not update their position information while lookup requests are being executed. This presumes some kind of global synchronization, where all nodes are either in a mobility phase or in a lookup phase.

Publish. The cost of publishing is bounded under the assumption that routing has linear average overhead. Let $X = (x_0, x_1, \ldots, x_j)$ be a sequence of locations published by B, and let $d = \sum d(x_i, x_{i+1})$ be the sum of all distances between neighboring locations on this sequence. The expected cost of publishing all these locations is $O(d \log d)$. The proof amortizes over the costs for individual updates and takes into account that movements within the 16 squares surrounding B do not necessitate any updates at all.

Lookup. The cost for LLS lookup is $O(d(A, B))$ on the average, with a worst-case of $O(d(A, B)^2)$, which is achieved by interleaving spiral traversal with flooding phases. Spiralling phase and subsequent flooding phase have the same cost. The complexity of neither spiral search nor flooding is affected, but flooding may catch situations in which spiral search fails. This provides us with a quadratic upper bound, which is the cost of flooding.

The authors give one reason where this happens: If the routing cost from the source to the destination is low, the routing cost from the source to the first location server pointing to the destination may still be arbitrarily high. In other words, routing may not have constant stretch. Therefore, the lookup spiral traversal is interleaved with broadcasts of limited depth such that the overhead imposed on the lookup by broadcasting is constant.

However, there is another reason why flooding is necessary, even if routing has constant stretch. We will conclude the discussion of LLS by outlining a situation that can yield arbitrarily high overhead for pure spiral search without flooding, no matter what the routing algorithm's performance is.

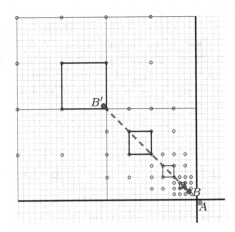

Fig. 14.6. There is no upper bound for the distance overhead of pure LLS spiral search.

Consider Figure 14.6.[15] Node B has moved to its current position from position B', leaving a spiral of location servers behind that is bent towards north-west on the outside. Now, A, that is at a distance of slightly more than the radius of the unit disk graph, runs a lookup for B. It immediately hits the location server at the location server at the corner separating A and B, but this server has only indirect location information. The next location server on the way down the publish spiral to B has not been updated for a while, so it is far up in the north-west of the map. This distance between the first location server s_1 (the one between A and B) and the second one s_2 (the one lagging behind because of lazy publishing) cannot be bounded: As the number of indirections required for A to finally meet B (and therefore, the $d(s_1, s_2)$) is arbitrarily high depending on how the situation is configured, but the distance between A and B remains the same in all configurations, the distance overhead for LLS lookup is arbitrarily high, even with optimal routing.

14.4 Mobility-Aware Location Service (MLS)

To conclude this chapter, we will at least briefly mention MLS [131], an algorithm that has been proposed during the later stages of writing this book, and that improves on GLS and LLS in various ways:

1. As the first published algorithm for mobile location services, MLS not only takes some limited form of node mobility into account, but provides an upper bound on the node speed at which nodes may freely move independently of the other nodes, while the algorithm still works.
2. The unit disk graph allows for areas that are not covered by any nodes (so-called lakes).
3. LLS selects 16 nodes per level to store location pointers. MLS only chooses one node per level. This allows for faster updating the pointers and therefore permits higher mobility.
4. MLS has $O(d)$ lookup stretch in the worst case, where LLS has $O(d^2)$.

As in GLS, the map of MLS is structured in squares of exponentially decreasing edge length, starting from some square L_M with edge length 2^M that covers the entire world.[16]

Like in LLS, hash functions are used to ensure a uniform distribution of location service load. But where LLS used the hash functions to perturb the origins of otherwise fixed spiral routes, MLS spans a fixed route of squares of decreasing size for any target node B (namely, all squares that contain B).

---

[15] Figure and idea have been provided by Roland Flury.

[16] For the sake of simplicity, we omit an additional factor ρ that accounts for the radio range.

It then uses hash functions $h_1(B), h_2(B)$ to select one specific location inside each square on the route that must perform the location service.

To be more specific, order i square L_i is split into four order $i-1$ squares, one of which, say L_{i-1}, contains B. The location server in L_i points to L_{i-1}, leaving the job of locating the location server in L_{i-1} to the algorithm. Without lakes, the location of the location server in L_i can be found by moving a fraction $h_1(B)$ left and a fraction $h_2(B)$ down the square.[17]

If A wants to locate B, it spans a spiral of squares of increasing order around its own position and consults the location server responsible for B in each square until one is found that (indirectly) knows B. From there, A simply follows the indirections.

To avoid constant high publish cost for arbitrary small oscillations between two large squares, a technique for lazy publishing is developed using forwarding location servers. A forwarding location server in L_i does not point to the L_{i-1} it contains that in turn contains B, but to a neighboring square L_i' into which B has moved.

MLS achieves a strict bound of $O(d(A, B))$ for lookup messages; simulations show that the hidden constant is roughly 6. Publish has amortized costs in $O(d \log d)$, with a hidden constant of roughly 4.3. Also, nodes may freely move around with a maximum speed of $1/15$ of the routing speed (without lakes).

14.5 Outlook

The problem of distributed location services has inspired a number of algorithms that show good peformance characteristics. Although the models are quite abstract and lack many characteristics of a realistic setting, this is a necessary first step towards their future embedding in applications. Besides prototyping such applications, at least two fields of research jump to mind.

Foremost, complexity: There are various examples for algorithms that touch the theoretical complexity bound for the problem they solve, but that nobody has ever implemented because the constants in their cost functions are tremendously high. LLS is likely not one of them, but it is equally likely that there are simpler algorithms with worse complexity characteristics, that nevertheless perform better on certain realistic network sizes. This will be especially relevant in the deployment of hybrid systems, in which the number of nodes is in the order of thousands, but only a small subset of those nodes with more expensive hardware are functioning as location servers.

Also, a deficiency of all location service algorithms we know of that will hurt in many applications is that they have no threat model (see Chapter 16): If a fraction of the nodes in the network is malicious, or if there is a powerful

---

[17] Assume that both hash functions have domain $[0, 1]$. If lakes come into play, it suffices to consider only the land parts when calculating fractions.

adversary launching attacks from outside the network by other means, nothing is known about their performance and robustness. Defense measures taylored specifically for attacks against location services, or at least the impact of various classes of attacks, will need to be given more thought.

14.6 Chapter Notes

This chapter has presented three major approaches to location services for ad hoc and sensor networks, i.e., distributed algorithms that map network identifiers to node positions and offer corresponding lookup and update services.

The Grid Location Service (GLS) proposed by Li et. al [253] is one of the pioneering algorithmic solutions for distributed location services. Weaknesses of GLS have been pointed out in [96, 171].

The Locality-aware Location Service (LLS) by Abraham et al. [1] attempts to guarantee better bounds on its operations in correlation to distance.

Mobility-Aware Location Service (MLS) by [131] in turn improves on LLS in many aspects, especially in better and proven bounds.

15

Positioning

Daniel Fleischer and Christian Pich

15.1 Introduction

Many applications of sensor and ad hoc networks profit from the knowledge of geographic positions of all or just some of the sensor nodes. E.g., geometric routing can use positional information to pass packets efficiently to target nodes. Tasks like measuring temperature or humidity levels of a certain area can only be fulfilled, if each sensor node knows its geographic position or at least an approximation of it. Attaching a GPS (global positioning system) to each sensor node is an undesirable solution, since GPS is quite costly and clumsy, compared to the small sensor nodes. Furthermore, the perception of GPS signals might be disturbed when there is no direct connection to the satellites due to some obstacles (buildings, trees,...).

Thus, there is a need to obtain the positions of sensor nodes, just from the network structure, without each sensor node being equipped with GPS. There exist quite a lot of approaches. Some of them demand the existence of a few sensor nodes that know their position (*landmark nodes*). Others do not need any absolute positions at all and just compute *virtual coordinates* instead, i.e. coordinates that do not try to exactly mirror the original coordinates, but rather aim to yet improve algorithms that rely on positional information. E.g., for geometric routing this has been shown to be sufficient [328]. The approaches can be further subdivided into methods where

- each sensor node only knows its neighbors ([290], Section 15.3.1),
- each sensor node furthermore knows the distance to all of its neighbors ([20], Section 15.3.2),
- each sensor node knows the angle between each two of its neighbors ([52], Section 15.3.6).

In general, the underlying graph model for sensor networks is given by *unit disk graphs* (in the following we will also consider the more realistic model of quasi unit disk graphs). Each sensor node is supposed to have a finite scan radius r, within which neighbors can be detected. Hence, by scaling, there

D. Wagner and R. Wattenhofer (Eds.): Algorithms for Sensor and Ad Hoc Networks, LNCS 4621, pp. 283–304, 2007.
© Springer-Verlag Berlin Heidelberg 2007

is an edge between two nodes, if and only if the Euclidean distance between them is less or equal to 1. The problem UNIT DISK GRAPH RECOGNITION is to decide, whether a given graph is a unit disk graph. This problem was first shown to be $\mathcal{NP}$-hard in [50]. This implies that there cannot exist a polynomial time algorithm that computes the positions of sensor nodes, given only their adjacencies (unless $\mathcal{P} = \mathcal{NP}$), because otherwise this algorithm computes valid coordinates for a given graph if and only if it is a unit disk graph. Despite this hardness result there exist a lot of heuristics to determine the positions of sensor nodes that work quite well.

Outline of this chapter. In Section 15.2 proof sketches of the $\mathcal{NP}$-hardness of UNIT DISK GRAPH RECOGNITION and some related problems are shown. Section 15.3 introduces some heuristics for positioning and obtaining virtual coordinates. Finally, Section 15.4 concludes with some chapter notes.

15.2 Hardness Results

In this section we consider the complexity of the problem of computing the positions of nodes in a sensor network within various scenarios. In Section 15.2.1 only the network $G = (V, E)$ is known, Section 15.2.2 deals with approximations, Sections 15.2.3 and 15.2.4 additionally consider distance and angle information, respectively. We will use the following definitions.

Definition 15.2.1 (Embedding). *An* embedding *of a graph $G = (V, E)$ in the Euclidean plane is a mapping $p : V \longrightarrow \mathbb{R}^2$.*

Definition 15.2.2 (Realization). *A* realization *of a unit disk graph $G = (V, E)$ in the Euclidean plane is an embedding p of G, such that*

$$\{u, v\} \in E \Longrightarrow \text{dist}(p(u), p(v)) \leq 1 , \tag{15.1}$$
$$\{u, v\} \notin E \Longrightarrow \text{dist}(p(u), p(v)) > 1 . \tag{15.2}$$

15.2.1 Unit Disk Graph Recognition

We derive that the problem of computing the positions of nodes in a sensor network, given only its adjacencies, is $\mathcal{NP}$-hard. The basic decision problem with respect to positioning is UNIT DISK GRAPH RECOGNITION.

Problem 15.2.3 (UNIT DISK GRAPH RECOGNITION). Given a graph $G = (V, E)$. Is G a unit disk graph?

Theorem 15.2.4 ([50]). *UNIT DISK GRAPH RECOGNITION is $\mathcal{NP}$-hard.*

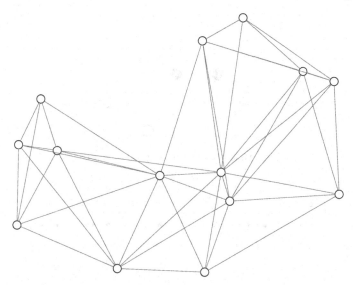

Fig. 15.1. Realization of a unit disk graph. Finding such an embedding is $\mathcal{NP}$-hard

Proof. We present a proof sketch of the reduction from 3SAT with the additional condition that each variable is contained in at most 3 clauses, leaving the problem $\mathcal{NP}$-complete. The basic building block for the proof is called a *cage*, see Fig. 15.2. A cage is a cycle with some additional independent nodes called *beads*, i.e. a short path that is attached to two nodes of the cycle. Note that only a bounded number of nodes can be embedded inside the cycle because otherwise some not adjacent nodes would be too close. If several cages (and some extra nodes to ensure the beads can only flip in two directions) are connected to a chain, see Fig. 15.3, embedding the leftmost beads inside the cycle leads to an embedding of the rightmost beads outside the cycle. This effect is used to propagate the information between variables and clauses of an instance I of 3SAT. Apart from the cages used for the chains, there is a different type of cage for each clause, see Fig. 15.4. Not all the beads of a clause cage can be embedded inside the cycle. The beads that have to be embedded outside the cycle take care that this clause is true. For each variable, there is a structure of several cages, see Fig. 15.5. The 5 nodes are embedded either in the left or right cycle, forcing the other beads of the corresponding cycle to be embedded outside. This corresponds to a truth setting of the variable. Together with a structure of cages for crossings of chains, now all basic gadgets are available to create a graph G that models an instance I of 3SAT; G is a UDG if and only if I is satisfiable. □

Corollary 15.2.5. *There is no polynomial time algorithm that computes a realization of a given unit disk graph (unless $\mathcal{P} = \mathcal{NP}$).*

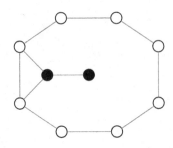

Fig. 15.2. Basic building block is a cage

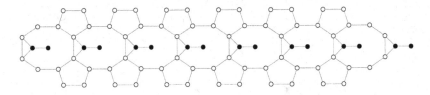

Fig. 15.3. Information propagation through chains of cages

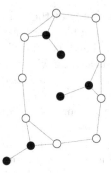

Fig. 15.4. Modeling a clause. Not all beads can be embedded inside the cycle

Proof. Assume there is such a polynomial time algorithm $\mathcal{A}$, and run $\mathcal{A}$ on an arbitrary graph G. If $\mathcal{A}$ does not terminate within its given upper time bound with respect to the input size, or the embedding does not fulfill conditions (15.1) and (15.2) (which can be checked in polynomial time), G is not a UDG. Otherwise, if the embedding fulfills both conditions, G is a UDG. Hence, a UDG can be recognized, a contradiction to Theorem 15.2.4.   □

Note that Corollary 15.2.5 implies that there cannot exist a polynomial time algorithm that computes the positions of nodes in a sensor network, given just its adjacencies, since these positions form a realization. Computing the actual coordinates of the sensor nodes is thus at least as hard as finding a realization of a UDG, which proved to be $\mathcal{NP}$-hard.

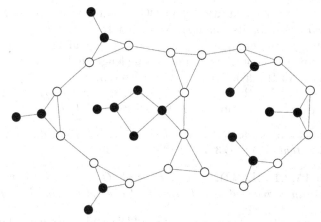

Fig. 15.5. Modeling a variable. Embedding the inner 5 black nodes on the left or right-hand side, corresponds to a truth setting

15.2.2 Unit Disk Graph Approximation

The last section showed that it is $\mathcal{NP}$-hard to decide whether a given graph is a unit disk graph. In this section we introduce d-quasi unit disk graphs, where d is a constant in $[0, 1]$, and show that their recognition is also $\mathcal{NP}$-hard (for $d \geq 1/\sqrt{2}$, Theorem 15.2.10). Theorem 15.2.9 states that for a unit disk graph no approximate embedding (in the sense of Definition 15.2.8) with quality $q \leq \sqrt{3/2} - \varepsilon$ can be found in polynomial time.

Definition 15.2.6 (d-Quasi Unit Disk Graph). *A graph $G = (V, E)$ is called a d-quasi unit disk graph (d-QUDG, $0 \leq d \leq 1$), if there is an embedding p, such that*

$$\{u, v\} \in E \implies \text{dist}(p(u), p(v)) \leq 1 \;, \tag{15.3}$$

$$\{u, v\} \notin E \implies \text{dist}(p(u), p(v)) > d \;. \tag{15.4}$$

Note that this definition does not define whether there is an edge between two nodes at a distance in $(d, 1]$. This yields a more realistic model of sensor networks, since nodes, whose distance is within that interval, still may or already may no more detect each other.

Definition 15.2.7 (Realization of a d-QUDG). *An embedding p of a d-QUDG $G = (V, E)$ is called a realization of G, if it fulfills both (15.3) and (15.4).*

Definition 15.2.8 (Quality). *The quality $q(p)$ of an embedding p of a graph $G = (V, E)$ is defined by*

$$q(p) = \frac{\max_{\{u,v\} \in E} \text{dist}(u, v)}{\min_{\{u,v\} \notin E} \text{dist}(u, v)} \;.$$

Thus, the quality of a realization of a UDG is smaller than 1 (a *good* embedding has a *small* value for its quality). Note that the two concepts — d-QUDG and quality — are very similar. Let   be an embedding of a graph G with quality , where w.l.o.g. the maximum edge length is 1 (scale   by a factor of $1/\max_{\{\ \}\in}\ \mathrm{dist}(u,v))$. Then,   is a realization of G as a $(1/\ -\varepsilon)$-QUDG for every $\varepsilon > 0$. On the other hand, if   is an embedding of a graph G as d-QUDG, then   is an embedding of quality ( ) $1/d$.

Theorem 15.2.9 ([235]). *Given a UDG G, it is $\mathcal{NP}$-hard to find an embedding , such that ( ) $\leq \sqrt{3/2} - \varepsilon(n)$, where $\lim_{n\to\infty} \varepsilon(n) = 0$.*

Theorem 15.2.10 ([235]). *Given a graph G, it is $\mathcal{NP}$-hard to decide, whether G can be realized as a d-quasi unit disk graph with $d \geq 1/\sqrt{2}$.*

Proofs of these theorems can be found in [235] and are omitted here.

Note that Theorem 15.2.10 is more interesting from a practical point of view than Theorem 15.2.9. If sensors at a distance in $(d, 1]$ may or may not be connected, the sensor network is no more a UDG as assumed for Theorem 15.2.9 but only a d-QUDG.

Corollary 15.2.11. *Given a graph G, it is $\mathcal{NP}$-hard to find an embedding , such that ( ) $\sqrt{2}$.*

15.2.3 Distance Information

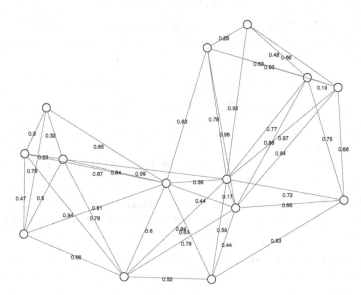

Fig. 15.6. Scenario: each node knows the distances to its neighbors

Assume for this section that distance information is available, i.e. each node knows the distances to all of its neighbors. Even in this scenario, it is $\mathcal{NP}$-hard to find a realization.

Theorem 15.2.12 ([20]). *Given a graph G and distances between all pairs of adjacent nodes, it is $\mathcal{NP}$-hard to decide, whether G can be realized as a unit disk graph.*

Proof. We very briefly sketch the proof of [20], a reduction from CIRCUIT SATISFIABILITY. The basic building block is called a *binary hinge*, see Fig. 15.7, i.e. a structure that has exactly two embeddings (modulo translation, rotation and reflection). Note that each triangle has exactly one embedding when distances are given. Binary hinges with some more rigid structures are then used to form NOT and implication gates, and junctions to connect the gates. These gates already form a universal set, hence for an instance I of CIRCUIT SATISFIABILITY one can construct a graph G modeling the circuit that has a realization if and only if I is satisfiable.   □

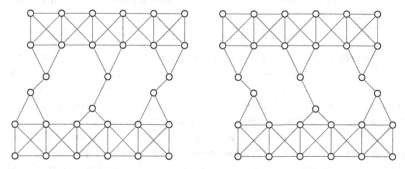

Fig. 15.7. A binary hinge has exactly two embeddings. Upper and lower bars consisting of several triangles are rigid, i.e. there is only one embedding

15.2.4 Local Angle Information

Assume for this section that local angle information is available, i.e. each node v with two neighbors u and w knows the angle (u, v, w). But also in this scenario it remains $\mathcal{NP}$-hard to find a realization. Note that the knowledge of both distances and angles immediately delivers the actual coordinates modulo translation, rotation and reflection (if G is connected).

Theorem 15.2.13 ([52]). *Given a graph G and local angles for each node, it is $\mathcal{NP}$-hard to decide whether G can be realized as a unit disk graph.*

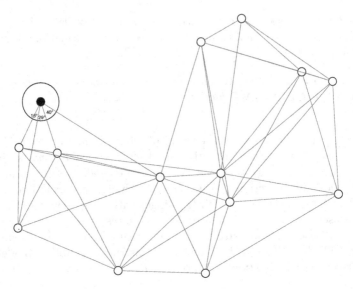

Fig. 15.8. Scenario: each node knows local angle between neighbors (only displayed for black node)

Proof. We very roughly present a proof sketch of a reduction from 3SAT found in [52]. Three basic structures are used to build up a graph G with given local angle constraints, corresponding to an instance   of 3SAT that is realizable if and only if   is satisfiable. The basic building blocks are *springs, amplifiers* and *propagators*, see Fig. 15.9. A spring is not fixed in its length, e.g. the spring of Fig. 15.9 can have an arbitrary length from the interval]1,4]. An amplifier fixes the ratio between two edge lengths and a propagator is used to propagate edge lengths horizontally and vertically. Using these 3 basic building blocks one can construct 0/1 blocks that correspond to a variable, and clause components that are connected in such a way that there is a realization if and only if a given instance of 3SAT is satisfiable.   □

15.3 Algorithms

The last section showed that computing the positions of sensor nodes is $\mathcal{NP}$-hard in general. Even if we consider approximations or additional information is available, the problem remains $\mathcal{NP}$-hard. Anyhow, this section presents several heuristics that deliver quite good results in practice.

15.3.1 Virtual Coordinates

This section roughly describes the algorithms by Moscibroda et al. [290] and Rao et al. [328]. The algorithm by Moscibroda et al. was the first algorithm to

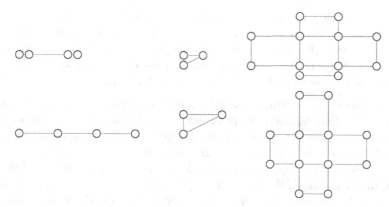

Fig. 15.9. Basic building blocks with different embeddings from left to right: spring, amplifier, propagator

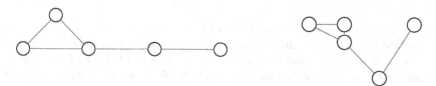

Fig. 15.10. Two realizations of the same unit disk graph

compute a realization of a given UDG with provable approximation ratio and polynomial running time. Apart from the adjacencies this algorithm needs no further input, in particular no distance or local angle information need to be available. Since no absolute positions are available, this algorithm can only compute *virtual coordinates*, i.e. nodes are assigned to positions in an arbitrary coordinate system, where close nodes in this system are supposed to be close in real coordinates. Often, this is sufficient to improve algorithms that depend on geographical positions, e.g. geometric routing [328].

The algorithm can be divided into four steps.

Step 1: solve an LP to obtain a metric ϱ on the nodes,
Step 2: compute a volume respecting embedding into $\mathbb{R}^n$,
Step 3: random projection into $\mathbb{R}^2$,
Step 4: local refinement of the random projection.

We briefly go through these steps.

Step 1. A metric ϱ on the input graph is computed, satisfying certain additional constraints, e.g. $\varrho(u,v) \leq 1$ for all $\{u,v\} \in E$ and $\varrho(u,v) \leq \sqrt{n}$ for all pairs of nodes. These constraints together with equations such that ϱ satisfies positiveness, triangle inequality and some *spreading constraints* given by

$$\sum_{v \in S} \varrho(u,v) \geq \kappa |S|^{3/2}, \text{ for all independent sets } S \subseteq V, \forall u \in V, \kappa \text{ const.} \quad (15.5)$$

form the LP that has to be solved. The spreading constraints (15.5) ensure that the sum of pairwise distances of nodes of independent sets is sufficiently great, such that they will be well spread over the plane in the following steps. Note that there are possibly exponentially many equations and it is not trivial how to reduce their number. For details we refer to [290].

Step 2. Volume respecting embeddings into high-dimensional spaces were introduced in [124]. When nodes from this high-dimensional space are projected into a random lower dimensional space, arbitrary volumes spanned by tuples of nodes are well preserved, i.e. roughly speaking large volumes are supposed to be projected to large volumes, small volumes to small volumes. This step finds an embedding of the nodes of the given graph into $\mathbb{R}^n$ that has the property that the Euclidean distances of the nodes in $\mathbb{R}^n$ closely match ϱ and that volumes (and distances, respectively) will be well preserved when randomly projecting to a lower dimensional subspace. The next step consists of such a random projection into $\mathbb{R}^2$.

Step 3. A vector in $\mathbb{R}^n$ is expected to be scaled by a factor of $\sqrt{2/n}$ by a projection to a random subspace of dimension 2 and the probability is concentrated around this expectation. Hence, the distances between the nodes in $\mathbb{R}^n$ and between their projections to a random subspace will be well preserved up to a scaling factor of $\sqrt{2/n}$.

Step 4. The final step consists of a local refinement process. First, the projected nodes of step 3 are enclosed in a grid of cell width $1/\sqrt{n}$. For each cell C a maximal independent set is computed. Let M be the maximal cardinality over all these independent sets. The grid is further refined to cells with width $1/\sqrt{nM}$. The nodes of each maximal independent set S of a cell C are then arbitrarily assigned to different centers of cells of the refined grid. Nodes in a cell C that are not in the maximal independent set S, are arbitrarily assigned to one of its neighbors v in S and positioned on a circle of radius $1/(3\sqrt{nM})$ around v. Non-neighbored nodes are equally distributed on the circle while neighbored nodes may be placed on the same position. Finally, the plane is scaled by a factor of $\sqrt{nM}$, such that the cell width of the refined grid is 1.

Theorem 15.3.1 ([290]). *Steps 1 to 4 yield a polynomial time algorithm, whose quality is in $\mathcal{O}(\log^{2.5} n \sqrt{\log \log n})$ with high probability.*

Another heuristical algorithm can be found in [328]. There do not exist any proven bounds on the quality of this algorithm, but experimental studies show quite good results. The algorithm can be seen as the last member of a sequence of three algorithms with decreasing information available. While the first algorithm needs all *positions of perimeter nodes*, the second assumes that *perimeter nodes are known* (to be perimeter nodes), the last needs *no additional information* apart from the adjacencies of the given graph.

Positions of perimeter nodes are known. If the positions of perimeter nodes (i.e. the nodes that span the convex hull of the network) are known, the remaining nodes can be set to positions according to the following *spring-embedder* model. Assume that each edge of the given graph acts like a spring, whose attracting force is proportional to the distance between its end nodes. In this model a node v will be stable in its position (x_v, y_v), when all forces effecting it sum up to zero, i.e. when v's position is the barycenter of its neighbors

$$x_v = \sum_{u:\{u,v\}\in E} x_u/d(v) \ ,$$

$$y_v = \sum_{u:\{u,v\}\in E} y_u/d(v) \ . \tag{15.6}$$

To set all remaining nodes to stable positions, their positions are all initialized by an arbitrary point inside the perimeter nodes, and then equations (15.6) are iteratively applied for these nodes, i.e. an iteration round consists of computing updated positions for all nodes based on their old positions. The iteration will converge to stable positions, and this forms the final embedding of all nodes. This procedure will not put any nodes outside of the convex hull. Alternatively, the positions can also be computed by solving those lines of the two linear equation systems

$$Lx = 0, Ly = 0,$$

where L is the Laplacian matrix of the given graph corresponding to nodes that are not perimeter nodes. If the given graph is connected, this yields a unique solution (see, e.g. [367]).

Perimeter nodes are known. If it is just known which nodes are perimeter nodes, but no positions of them are available, virtual coordinates for these perimeter nodes are first computed, and then the previous algorithm is applied. So, how can virtual coordinates for the perimeter nodes be obtained? First each perimeter node v determines the hop-distances, i.e. the number of edges of a shortest path, to all other perimeter nodes by simply flooding the network with a message like $(v,1)$, where the index is increased by one when forwarding this message and where only the message with least index from a specific node v is forwarded/accepted. Depending on the number of iterations the following step should be done by all perimeter nodes parallel, or all the distance information is passed to one designated node that executes the following step. Again we use a spring-embedder model, but this time, the springs are relaxed when having a length of the hop-distance of the previous step. This can again be solved iteratively and after a sufficient number of iterations the positions are used as virtual coordinates for the perimeter nodes.

No additional information. If perimeter nodes are not known, it is first tried to identify them, and afterwards the previous procedure is applied. To identify perimeter nodes, one designated bootstrap node v floods the entire network, such that each node knows its distance to v. Then each node that is farthest away among all its distance-2-neighbors, decides to be a perimeter node.

15.3.2 Distance Information

For the next sections it is assumed that distances between adjacent nodes are available, i.e. sensor nodes are able to measure the distance to their neighbors, e.g. via signal strength. A straight-forward method is to obtain positions by *triangulation*, e.g. see [341], i.e. a node v can compute its position when having 3 neighbors that know their positions (that are not collinear) because there is only one position satisfying the 3 distances to its neighbors. If initially there are 3 nodes knowing their position, this method can eventually be applied incrementally to deliver positions for all nodes of the network. If n nodes are placed independently, uniformly at random on a square of maximum length $\sqrt{n/8 \log n}$, this is possible with high probability [115]. Anyhow, the disadvantage of this incremental method is that it is extremely sensitive with respect to errors in the distance measurements because of error propagation.

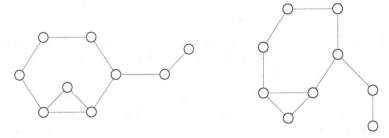

Fig. 15.11. Two realizations of a UDG with the same distances between adjacent nodes

15.3.3 Force Based Optimization

In 2003, Priyantha et al. were the first to propose an algorithm that assigns coordinates to the nodes in a sensor network by using only the measured distances d_{ij} between every pair of adjacent nodes n_i, n_j with $\{n_i, n_j\} \in E$ and that works fully distributed and without landmark nodes (sometimes also called *anchor nodes*), hence called AFL (Anchor Free Location). The algorithm basically uses a distributed mass-spring optimization method that tries to adjust the location of any node such that the global energy of the system (penalizing differences between measured and calculated distances) is minimized.

Through simulation, the authors observe that the crucial point in the optimization process is the initial arrangement of the system. When there are multiple solutions to the system of distance constraints, they mainly differ in *foldovers*, meaning that there are larger parts of the graph that fold over other parts without violating any of the distance constraints. To prevent the optimization from converging into a folded local minimum, a *fold-free* initial arrangement is needed.

Hence, the algorithm consists of two phases: In the first phase, a heuristic method constructs a fold-free placement that is already "close" to the real one. Then, in the second phase, this placement is refined by the optimization process.

Phase 1: Initial fold-free placement. A heuristic selects five reference nodes: n_1, n_2, n_3, n_4 should be on the periphery of the sensor network while n_5 should be roughly in the center. First, let $h_{ij} = h_{ji}$ denote the *hop count*, i.e. the graph-theoretical distance, between n_i and n_j, and assume that the maximum radio range of every node equals 1. For tie-breaks, nodes can be assigned unique identifiers.

- Select a node n_0 randomly, then choose n_1 to maximize h_{01}.
- Select a node n_2 maximizing h_{12}.
- From all n_3 minimizing $|h_{13} - h_{23}|$, choose the one maximizing $|h_{13} + h_{23}|$.
- From all n_4 minimizing $|h_{14} - h_{24}|$, choose the one maximizing h_{34}.
- From all n_5 minimizing $|h_{15} - h_{25}|$, choose the one minimizing $|h_{35} - h_{45}|$.

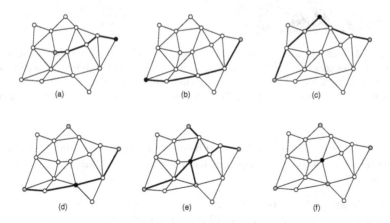

Fig. 15.12. Finding reference nodes n_1 to n_5. Subfigures (a) through (e) correspond to the five steps, and (f) shows the resulting reference nodes. The relevant shortest paths are drawn with bold line; grey nodes have been found in previous step, black nodes are found in the current step.

Thus, the roughly antipodal pairs (n_1, n_2) and (n_3, n_4) should be perpendicular, and n_5 should be in the center. Fig. 15.12 depicts an example of finding the reference nodes; the graph is due to [322]). The hop counts from nodes n_1, n_2, n_3, n_4 to all other nodes n_i are then used to approximate the radial coordinates (ρ_i, θ_i), thereby arranging the n_i in concentric circles around n_5:

$$\rho_i = h_{5i}, \qquad \theta_i = \arctan\left(\frac{h_{1i} - h_{2i}}{h_{3i} - h_{4i}}\right)$$

This method it is not guaranteed to produce embeddings without foldovers, but the authors state that it can indeed reduce the probability of convergence towards local minima. Fig. 15.13 shows an initial embedding generated by AFL. Interestingly enough, it could be used for computing rough virtual coordinates.

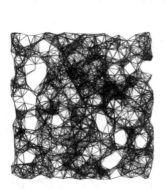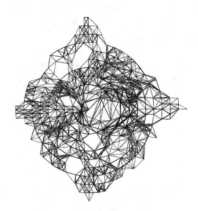

Fig. 15.13. Left: Randomly generated sensor network, initial positions. Right: Initial embedding constructed by AFL. It can be seen how the nodes are placed around in a concentric fashion, many of them having identical positions due to equal distances to the five reference nodes. The images are taken from the original article.

Phase 2: Mass-spring optimization. Once the initial embedding is calculated, the distributed mass-spring optimization is started. At any time, each node n_i knows

- an estimate $\hat{p}_i$ of its own position,
- the *measured* distances d_{ij} to all its neighbors n_j,
- the estimates $\hat{p}_j$ of positions of all neighbors n_j,
- and hence the *estimated* distances $\hat{d}_{ij} = \|\hat{p}_j - \hat{p}_i\|$ to all its neighbors n_j.

and periodically sends its own estimated position $\hat{p}_i$ to all of its neighbors n_j. Then, expressed by means of the spring-model, the spring between n_i and n_j exerts a force

$$F_{ij} = \left(\hat{d}_{ij} - d_{ij} \right) \cdot \frac{\hat{p}_j - \hat{p}_i}{\| \hat{p}_j - \hat{p}_i \|}$$

on node n_i towards n_j. Summing up the force vectors for all springs attached to n_i results in the *net force vector*

$$F_i = \sum_j F_{ij}.$$

The differences between estimated and measured distances are quadratically penalized by the *energy*

$$E_i = \sum_j \left(\hat{d}_{ij} - d_{ij} \right)^2$$

of node n_i, so the total energy E of the system is

$$E = \sum_i E_i.$$

The energy E_i of node n_i, and hence the global energy E, can be reduced by moving the node into the direction of its net force vector F_i. This approach is known as *gradient descent*. To prevent the optimization process from running into a local minumum, the amount by which the node is moved must not be too large, but in turn, a too small value will result in slow convergence. The authors propose to normalize the amount of movement by $\deg(n_i)$, the number of neighbors of n_i:

$$\hat{p}_i \leftarrow \hat{p}_i + \frac{F_i}{2 \deg(n_i)}$$

The distributed optimization is iterated until the change in position p_i stays below a previously defined threshold value for every node n_i. The resulting embedding still has degrees of freedom in rotation and reflection, but can be pinned down when at least three nodes know their absolute positions.

15.3.4 Spectral Methods

Another algorithm using distance information was introduced by Gotsman and Koren in 2004 [166] and is based the distributed computation of certain eigenvectors of a matrix associated with the sensor network.

It is very similar to the AFL method described above in that it first computes an initial embedding which is then refined by some distributed optimization technique. However, it tries to overcome some of AFL's disadvantages by using more sophisticated techniques in both phases. For example, it uses only direct communication between adjacent nodes instead of multi-hop messages. Moreover, it seems to be more stable against errors in distance measurements.

Phase 1: Initialization. For the ease of presentation, we keep to the original article and start presenting the one-dimensional layout, which is then extended to higher-dimensional cases. Since the algorithm is heavily based on methods from linear algebra, it makes sense to write the x-coordinates as vectors $x \in \mathbb{R}^n$, with x_i being the x-coordinate assigned to node n_i.

Again, the key to success in constructing a good initial configuration that is free of foldovers is to use an energy function whose minimization keeps adjacent nodes close together and non-adjacent nodes apart, and which can be optimized in a distributed fashion. Given the measured distances d_{ij} between every adjacent pair $\{ij\} \in E$, the *energy function* for a given placement x is defined as

$$E(x) = \frac{\sum_{\{ij\} \in E} w_{ij} \|x_i - x_j\|^2}{\sum_{i<j} \|x_i - x_j\|^2}$$

where the w_{ij} denotes the weight of node pair i, j the enumerator. This weight can be interpreted as the similarity of two adjacent nodes i, j, derived from the measured distance d_{ij}; the default setting is $w_{ij} = \exp(-d_{ij})$. The denominator is to be maximized, thereby trying to achieve a good scattering of all nodes over the whole placement by taking into account all pairs of nodes, regardless of there being a connecting edge.

The problem of finding a placement x minimizing $E(x)$ can be directly solved using results from standard linear algebra: Let the square matrix W have entries w_{ij}, and let D denote the diagonal matrix with the ith diagonal entry being the sum of the entries in W: $D_{ii} = \sum_j w_{ij}$. Then, the global minimum of $E(x)$ (which is known to exist and to be zero since the Euclidean distances are derived from a measurement) is the eigenvector of the related Laplacian matrix $L = D - W$ associated with the *smallest* positive eigenvalue. However, the closely related *transition matrix* $D^{-1}W$ is used in practice. This is mainly because, unlike the Laplacian, the highest eigenvector of the transition matrix is 1, which is convenient for numerical computations.

The desired eigenvector is computed by iteratively adding to a node value x_i the weighted average value of all its adjacent nodes j by

$$x_i \leftarrow x_i + \frac{1}{2}\left(\frac{\sum_j w_{ij} x_j}{\sum_j w_{ij}}\right)$$

where the coefficient $\frac{1}{2}$ damps the growth of $\|x\|$. This method of iterating averaging is called *power iteration* and in general it will converge to the eigenvector associated with the eigenvalue with the *highest* absolute value. Reversing the order of eigenvectors by using $I - D^{-1}W$ (I being the unity matrix) in the power iteration yields the desired *lowest* eigenvectors of the original matrix $D^{-1}W$.

The top eigenvector of $D^{-1}W$ is $v_1 = \mathbf{1}_n = (1, 1, \ldots, 1)$ and is associated with the eigenvalue 1, so we are rather interested in the second eigenvector v_2, which is the actual solution. Thus, we first compute v_2 by starting the power iteration with a vector u that is D-*orthogonal* to v_1 (i.e., $u^T D v_1 = 0$; see below). Then, the next highest eigenvector v_3 can be computed by starting the process with a vector D-orthogonal to v_2. In the end, we will be able to use v_2 as x-coordinates and v_3 as y-coordinates for the initial embedding.

Since computations can only be done locally, there is the need for a distributed way of orthogonalization against x when computing y. This is done by initializing the power iteration with $y = Au$, where u is a random vector, and A is a "local" matrix (with non-zero entries only for adjacent nodes) defined as

$$A_{ij} = \begin{cases} \frac{-x_j}{D_{ii}} & \text{if } \{i,j\} \in E, \\ 0 & \text{if } \{i,j\} \notin E, i \neq j, \\ -\sum_k A_{i,k} & \text{if } i = j. \end{cases}$$

which ensures that $A^T D x = 0$ and thus $u^T(A^T D x) = (Au)^T D x = y^T D x = 0$, i.e. y is D-orthogonal to x. This vector is D-orthogonal to $x = v_2$ but not to $v_1 = \mathbf{1}_n$. The v_3 component must hence be prevented from vanishing against the v_1 component in the power iteration. This can be achieved by carefully scaling and translating y in the process. See [166] for details.

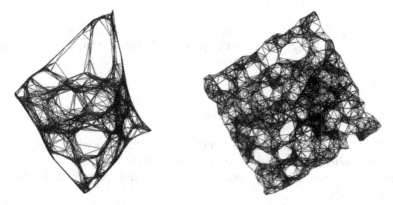

Fig. 15.14. Left: An initial spectral embedding of the graph in Fig. 15.13 found by the eigenprojection. Right: The embedding after the optimization procedure, being very close to the actual configuration. Image from the original article.

Phase 2: Energy minimization.. In contrast to the AFL approach which uses gradient descent to minimize the local energy in each step, a distinct approach called *majorization* is used. This is mainly because it does not depend on the scale, and it does not require any heuristically fixed damping factor for the optimization step size, as opposed to the mass-spring optimization.

The objective function to be minimized when reconstructing the true node coordinates is the *localized stress*

$$\text{stress}(p) = \sum_{j:\{i,j\}\in E} (\|p_i - p_j\| - d_{ij})^2,$$

at node n_i. Once a good initial configuration is found, there is good chance for the optimization to converge to the global minimum, in which $d_{ij} = l_{ij}$ for all i, j, i.e. stress$(p) = 0$, and that is guaranteed to exist because the measured distances have been derived from an existing (yet unknown) placement.

Majorization is very common in multidimensional optimization and based on the fact that it is possible to find a sequence of layouts $p(0), p(1), p(2), \ldots$ with non-increasing stress, converging to the desired solution layout. The main idea is to use a *majorizing function* $M(p, q)$ of stress(p), satisfying stress$(p) \le M(p, q)$ for all layouts q, and stress$(p) = M(p, p)$. Starting at $p(0)$, the function $M_0(p, p(0))$ is minimized as a function of p. The resulting minimum $p(1)$ is then used to minimize the new majorizing function $M_1(p, p(1))$, and so on. This is repeated until convergence; practically, the iteration is terminated when the absolute difference between successive layouts falls below a given threshold.

Improving a layout $p(t)$ to $p(t + 1)$ is done using a distributed version of Jacobi iteration at each node i, namely

$$p_i \leftarrow \frac{1}{\deg(i)} \sum_{j: \{i,j\} \in E} (p_j + d_{ij} \cdot (p_i(t) + p_j(t))) \cdot \text{inv}\left(\|p_j - p_i\|\right),$$

where inv$(x) := x^{-1}$ for $x \ne 0$, and inv$(0) := 0$. Eventually, this yields $p(t + 1) = p_i$. The sequence of the $p(t), t \in \mathbb{N}$ (each computed by such an iterative process), indeed converges to the minimum, as shown, e.g., in [166] or [146].

15.3.5 Multidimensional Scaling

In 2004, Shang and Ruml [345] presented another distributed method (without landmark nodes) of localizing based on multidimensional scaling, being an extension of a centralized version called MDS-MAP presented by the same authors earlier.

The method consists of three phases: First, each node constructs a local map of its two-hop neighborhood. Second, all local maps are successively merged into a global map. Third, the computed map is further refined. Thus, the first and second step can be viewed as an initialization similar to the ones in the algorithms described above, and the final refinement step corresponds to the optimization processes.

When only connectivity information is available, i.e., if each node knows which other nodes are within reach, the method can provide *virtual* coordinates that can be used, e.g., for routing. When the measured distances are also known, they can be used for refining both the local and the resulting global maps to get an estimate of *absolute* coordinates. In addition, [345] presents an approach to dealing with only partial measurements (when distances between nodes are only known in parts of the ad hoc network).

Constructing the local maps. In the first phase of the algorithm, each node constructs the subgraph induced by the set of nodes at most k steps away. In the original article, $k = 2$ proved a good compromise between accuracy and computation costs.

To construct such subgraphs, it suffices for each node to collect the directly available information related to all of its neighbors (who they are, and, if available, how far they are away) and then pass this collection to all of its neighbors. Thus, each node can construct the distance matrix related to its subgraph, and, by multidimensional scaling using the first two eigenvectors, generate a "local embedding".

Once this embedding is computed, it gets refined using the measured distances. The energy to be minimized in this step is a weighted version of the energy used in the AFL algorithm

$$E = \sum_{i<j} w_{ij} \left(d_{ij} - \hat{d}_{ij} \right)^2,$$

where $\hat{d}_{ij}$ represents the "estimated" distance resulting from the positions assigned to nodes i and j, and d_{ij} is the measured distance, or, if it is not available or j is more than one hop away from j, the shortest-path distance, i.e. $d_{ij} = h_{ij}$.

The weights w_{ij} can be used to lower the importance of nodes that are not direct neighbors, for example:

$$w_{ij} = \begin{cases} 1 & \text{if } j \text{ is direct neighbor of } i, \\ \frac{1}{4} & \text{if } j \text{ is two hops away from } i, \\ 0 & \text{otherwise.} \end{cases}$$

Since the MDS and the refinement are computed locally and no further node-to-node communication is needed, any standard optimization method can be used, such as gradient descent, simulated annealing, etc.

Merging the local maps into a global map. We first describe how two partial maps are merged into one larger map. Then, we briefly discuss different strategies of constructing the entire global map from the local patches.

Consider two partial maps p, q, which can be read as position vectors over $\mathbb{R}^2$ of varying length. Let I denote the set of nodes present in both p and q. Then our goal is finding a *linear transformation* $T : \mathbb{R}^2 \to \mathbb{R}^2$ that transforms the nodes in I from their position assigned by q as closely as possible to positions assigned by p, by minimizing the sum of squared errors

$$\sum_{i<j} (p_i - T(q_i))^2.$$

This linear transformation T translates, reflects, rotates, and scales the one map to fit best (in the sense given above) over the other, as can be seen in

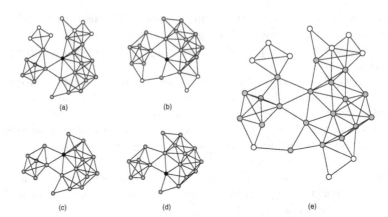

Fig. 15.15. Merging the two local maps (a) and (b). Nodes that are present in both maps are depicted by grey color; the source nodes are drawn in black. Removing the white nodes results in maps (c) and (d), which are merged and augmented by the white nodes, yielding map (e).

Fig. 15.15. The authors of the original article report that using only non-scaling transformations is slightly better than scaling. The resulting position r_i in the resulting new map r is then given as

$$r_i = \begin{cases} (\ _i + T(\ _i))\,/2 & \text{if } i \in \ , \\ \ _i & \text{if } i \text{ was present in} \quad \text{but not in} \quad , \\ T(\ _i) & \text{if } i \text{ was present in} \quad \text{but not in} \quad . \end{cases}$$

The straightforward way of composing the local patches to a global map is a greedy approach: A core map is successively extended by "sewing on" new patches until the local patches of all nodes have been integrated. A new patch can be selected randomly, or by choosing the candidate patch with a maximum number of nodes common with the core map.

However, since merging operations in an ad hoc network can be carried out in parallel, another more balanced and efficient way of constructing the global map is distributed map merging. Here, merge operations form a binary merge tree in which maps of the same size are merged. This strategy appears less error-prone, usually works faster, and requires less communication between nodes. Once the global map is composed by integrating the local patches, it can be further refined by using one of the techniques presented in this section.

15.3.6 Linear Programming

For the scenario that each sensor with two or more neighbors knows the local angles between them (i.e. each sensor node is able to measure the difference angle of signals from two of its neighbors) Bruck et al. [52] presented a heuristic

Fig. 15.16. Two realizations of a UDG with the same local angles

based on linear programming. Section 15.2.4 showed that obtaining an exact solution in general is $\mathcal{NP}$-hard.

Observe that local angles can easily be transformed to absolute angles θ_{e_i}, i.e. angles between adjacent nodes and some x-axis that is defined, e.g. by one designated edge e_1, if the given graph is connected.

Lemma 15.3.2. *If local angles between adjacent nodes of a unit disk graph are known, all pairs of crossing edges can be determined.*

Proof. We check whether two edges $e_1 = \{u_1, v_1\}$ and $e_2 = \{u_2, v_2\}$ cross. By simple geometrics, two edges can only cross, when there is a node, let's say u_1, that is connected to all three other nodes. Now, if e_1 does not cut the triangle formed by e_2 and u_1 (which can be checked by the angles in u_1), e_1 and e_2 do not cross. Otherwise, it can be checked by the angles in u_2 or v_2, whether v_1 is inside the triangle (e_1 and e_2 do not cross) or not (e_1 and e_2 cross). □

The variables of the following linear program are the edge lengths $\ell(e_i)$. If they are also known, all positions can straight-forward be computed (modulo translation and rotation).

– *Edge length constraint*: for all edges

$$0 \leq \ell(e_i) \leq 1 \ .$$

– *Cycle constraint*: for any cycle $e_1, e_2, \ldots, e_p$

$$\sum_{i=1}^{p} \ell(e_i) \cos \theta_{e_i} = 0 \text{ and } \sum_{i=1}^{p} \ell(e_i) \sin \theta_{e_i} = 0 \ .$$

– *Non-adjacent node pair constraint*: for any path e_1, e_2, whose end nodes are not adjacent

$$\ell(e_1) + \ell(e_2) > 1 \ .$$

– *Crossing edge constraint*: for any two crossing edges $e_1 = \{u_1, v_1\}$ and $e_2 = \{u_2, v_2\}$, where u_1 is a node connected to all three other nodes

$$\ell(e_1) \geq \ell(e_2) \sin(\alpha)/\sin(\alpha + \beta),$$

where α denotes the angle between u_1, u_2, v_2 and β denotes the angle between u_2, u_1, v_1.

These four types of constraints are used as basic constraints. Their number can be reduced by the following observation. Assume three edges e_1, e_2, e_3 form a triangle. Hence, their lengths are not independent from each other. Keeping just $\ell(e_1)$ and fixing $\ell(e_2) = c_2\ell(e_1)$ and $\ell(e_3) = c_3\ell(e_1)$ for some constants c_2, c_3 reduces the number of variables. This can further be applied to larger *rigid subgraphs*, e.g. a sequence of triangles, where succeeding triangles share an edge. Another advantage of this method is to deliver tighter bounds. For any two non-adjacent nodes of a rigid subgraph their distance can be expressed as $c\ell(e_1)$, where $\ell(e_1)$ is the independent edge length of this subgraph. Since this distance has to be greater than 1, we can give a tighter edge length constraint $1/c < \ell(e_1) \leq 1$. For dense graphs these methods drastically reduce the number of variables (and thus constraints). Note that the number of cycle constraints is bounded by the size of a cycle base, since it is sufficient to ensure these constraints on pairwise independent cycles, instead of all possibly exponentially many cycles.

15.4 Chapter Notes

This chapter gave an overview over the complexity of the problem to compute positions of nodes in a sensor network within various scenarios (only adjacencies given [50], d-quasi unit disk graphs [235], given angles [52], given distances [20], ...) and presented several heuristics [290, 328, 166, 345, 52]. Although the problem is $\mathcal{NP}$-hard in each scenario, the heuristics deliver quite good results. In general one can say that they work well when the network is sufficiently dense. Density means that there is more information available that can be used to compute positions. Additional knowledge like angles or distances drastically improves the results of the heuristics. Anyhow, sparse networks remain a challenge.

16

Security   .

Erik-Oliver Blaß, Benjamin Fabian, Matthias Fischmann, and Seda F. Gürses

16.1 Introduction

Security is one of the fundamental problems in wireless sensor and ad hoc networks. Properties specific to such networks make it hard to apply traditional security solutions and require the design and analysis of new security mechanisms.

Many attacks are very powerful, but either not specific to wireless sensor and ad hoc networks or of little algorithmic interest, for example eavesdropping or injection of faulty traffic, radio jamming, or tampering with the hardware. Although of independent interest, we will not consider these issues in the following, and instead delve into the problem of distributing cryptographic keys securely.

The problem of distributing symmetric or asymmetric keys has always been a nuisance in applied cryptography: Before two parties can establish a secure communication channel C, they first need another secure communication channel C_k over which they can negotiate the key required for establishing C.[18] In wireless sensor and ad hoc networks, this problem turns out to have solutions that look quite different from those in more traditional settings.

We start with giving a brief overview of security issues and to the cryptographic basics that are required to understand why distribution of cryptographic keys is a necessary condition for secure communication. Then, we delve into the *key distribution* issue. Section 16.2 deals with the symmetric case, and Section 16.3 with the asymmetric case.

---

[18] If asymmetric cryptography is an option, no secret needs to be communicated for key agreement. But this requires considerably more expensive hardware, and man-in-the-middle attacks still threaten the authenticity of the exchanged key, and thus confidentiality and integrity of the communication.

D. Wagner and R. Wattenhofer (Eds.): Algorithms for Sensor and Ad Hoc Networks, LNCS 4621, pp. 305–323, 2007.

16.1.1 Obstacles to Sensor and Ad Hoc Network Security

Many of the security problems of traditional distributed networks (e.g., authentication and confidentiality issues, fault tolerance, or clock synchronization) also occur to wireless sensor and ad hoc networks. Yet, old solutions do not work well due to the differences in the latter. Some of the differences are:

Limited CPU and Memory/Storage: Typical sensor network platforms are based on 8 Bit microcontrollers, like the MICA Motes common at the time of writing this book,[19] built from 8 Bit Atmel ATMEGA128L CPUs with only 4 KByte RAM and 128 KByte ROM. Clocked at 7 MHz, they reach about 7 MIPS. Such hardware limitations constrain the size and complexity of software that can run on a sensor node. This means that the resources for security-related code are extremely limited.

Power Limitation: Power is an important constraint. Once deployed, it may not be possible to replace or recharge sensor batteries. In order to extend battery life, code often needs to be highly optimized for power saving.

Unreliable Transfer: Tiny wireless radio interfaces are prone to static noise and their radio range is small. Resulting unreliable transfer may cause packet loss, corrupt packets and non-reachable regions of the network at a far larger scale than in wired networks. Collision of messages may cause packet loss (on a network layer where cryptography can provide very little protection). This yields new variants of denial of service attacks. Hence, e.g., data integrity needs to be reconsidered.

Latency: Through multi-hop routing, network congestion, and node processing latency may occur. This may cause divergence in node clocks and prevent reliable synchronization on which many security protocols depend. The issue of latency becomes crucial with real time safety critical systems.

Dynamically Changing Topology: In realistic application scenarios, thousands of sensor nodes may need to be deployed at the same time in a network with rapidly changing topologies. Nodes may enter or leave the network or may change positions. Even in static networks, nodes may fail or may have to be replaced due to changes in system requirements. In chapters 9, 10, 11, we have seen that routing protocols need to take these dynamics into account.

Exposure to Physical Attacks: Node capture or failure is one of the central problems. Many wireless sensor and ad hoc networks are deployed far away from the reach of its maintainers (e.g., at the bottom of an ocean, embedded in complex machinery, or in military applications next to the adversary). Consequently, sensor networks are sometimes easier to tamper with than desktop computer hardware.[20]

Managed Remotely: As sensor networks are managed remotely, it is difficult to identify tampered nodes. Further, it is not possible to decide whether a node

---

[19] http://www.tinyos.net/scoop/special/hardware

[20] The opposite may be true as well: The bottom of an ocean may be harder to reach for an adversary than a desktop located in a low-security office. In this case, cryptographic protection may be more effective and require less effort.

has been captured or whether it is simply failing temporarily (e.g., because of poor radio connectivity. It is also difficult to recognize a remotely organized malicious use of the sensor network to collect data.

No Central Management Point: Many researchers propose keeping the responsibilities of the nodes distributed. Single points of failure are generally considered bad practice, and more so if any candidate for such a point is relatively unreliable. This is a challenge for all aspects of security, as it requires an autonomous setup and operation of security protocols. This is a problem especially for key distribution – a single trusted network entity would be a very convenient point to receive keys securely from, and if there is no such entity, good alternatives are hard to find.

Advanced Attacker Model: Classic security protocols assume the so-called Dolev-Yao attacker model, where an attacker can intercept all messages sent through the network or inject arbitrary ones herself. She cannot decrypt encrypted messages without having the encryption key, and she may not compromise nodes already in the network. In other words, nodes are assumed to be *tamper-proof*. This model might hold for PC/Internet-networks, because physical access to nodes in a network is difficult. However, in sensor networks access to sensor nodes is relatively easy, because they are often publicly accessible, i.e., they can be stolen, and an attacker can read out all its contents. Because of simple cost and feasibility reasons, tamper-proof hardware is not adequate in sensor networks. Therefore, security protocols, especially protocols for a secure key exchange, have to cope with an attacker being able to compromise and control a fraction of the network. The size of this network fraction, like the size of the keys used to encrypt communication or assumptions on the computational capabilities of the adversary, needs to be considered as an additional security parameter and set in accordance with a specific application at hand and the expected nature of an adversary.

16.1.2 Security Requirements for Wireless Sensor and Ad Hoc Networks

Before we understand *how* we can achieve protection, we need to understand *what* we want to protect, or what the security requirements are. The following incomplete list is of particular interest in the context of sensor and ad hoc networks, and should be kept in mind for the remainder of this chapter:

- **Confidentiality:** Guarantees that communication remains secret. Any outside observers of the network (or ideally also nodes in the network other than those talking to each other) will learn nothing about the transmitted data (or ideally even about the existence thereof).
- **Integrity or Authenticity:** A communication channel is integrity-preserving if the data transmitted is not modified by unauthorized parties. It is authenticated if the receiver of a message always can securely identify the sender. Neither of these requirements is logically possible without the

other: If a third party can modify the messages after they have been digitally signed by the sender, the receiver has no benefit from identifying the original sender, and if the receiver has no way of knowing who sent the message, it may be a perfectly intact message from an adversary. Hence, integrity and authenticity should be treated as equivalent.

- **Freshness:** Ensures that the messages the nodes send or receive are recent instead of old messages replayed by an adversary.

- **Availability:** Guarantees that the nodes and the data they collect are available when needed.[21]

To meet these requirements, communication protocols running over radio links susceptible to eavesdropping, deletion, and injection of data need to be hardened using cryptographic building blocks.

16.1.3 Building Blocks

A cryptographic building block is a distributed algorithm for a set of parties that enforces some security requirement (such as confidentiality or integrity) in the presence of an adversarial party. For example, confidentiality is usually enforced by a pair of keyed encryption and decryption functions E_k, D_k: The sender encrypts her message m and sends $E_k(m)$ to the receiver. The receiver then decrypts $E_k(m)$ obtaining $D_k(E_k(m)) = m$. The adversary, who is not in possession of D_k, cannot decrypt the secret message, even if she manages to intercept it.

In fact, encryption and decryption functions are usually public knowledge and subject to years of academic research and analysis. The secret that sender and recipient share and that protects them from the adversary is represented as a parameter k, which is called the *cryptographic key*. A key is a string of a small fixed number of bits and is generated by a random number generator. The problem of key negotiation between two parties that trust each other, but that do not trust their communication link, is the subject of the bigger part of this chapter.

Here is a brief, incomplete, and only loosely formalized list of the most common cryptographic building blocks in cryptographic applications.[22] This overview is given to motivate the need for secure key distribution, but will not be developed any further in the rest of this chapter because it is not a subject specific to wireless sensor and ad hoc networks.

---

[21] Availability (that a communication link is not interrupted) is harder to achieve than confidentiality or integrity (that the communication is not understood or changed by the adversary), and cryptography can only help to some extent. Usually, an integrity-preserving link is established and packets are labelled with serial numbers so that if a packet is lost, it can be rerouted.

[22] The curious reader is referred to the standard literature on cryptography that is not specific to our context, but to data networks in general; see Chapter Notes.

- **Symmetric Encryption:** A pair E_k, D_k of an encryption and corresponding decryption function such that $D_k(E_k(m)) = m$ and $D_k(\cdot)$ is hard to compute without knowledge of k. Thus, an adversary who can intercept the communication cannot learn anything about its content. The most prominent example for symmetric encryption protocols is the advanced encryption standard AES.
- **Asymmetric Encryption:** A pair E_{k^*}, D_k of encryption and decryption functions such that $D_k(E_{k^*}(m)) = m$ and $D_k(\cdot)$ is hard to compute without knowledge of k. In contrast to symmetric encryption, the *public key* k^* can be freely distributed, even to the adversary. This makes key negotiation easier, but because of their complex mathematical structure E, D are harder to compute than their symmetric counterparts (see Section 16.1.4). Important asymmetric encryption schemes are RSA, Diffie-Hellman, XTR and NTRU.
- **Message Authentication:** A pair A_k, C_k of authentication and verification functions such that $C_k(A_k(m)) = m$, but it is hard to compute $A_k(\cdot)$ without knowing k. Thus, an adversary cannot inject or delete traffic from a link that she has complete control over (e.g., because she has captured a node on the link's route). (Message authentication can be achieved by using an algorithm for symmetric encryption in a specific way.)
- **Digital Signatures:** The analogous counterpart of message authentication for the asymmetric case. As before, the key needed for verification can be freely published, because it does not help an adversary trying to authenticate her own messages. Digital signatures are more practical than message authentication codes if many parties need to be able to verify the integrity of a message, but the same performance problems as in asymmetric encryption are involved in its use.

These and other building blocks are used in different combinations in widely deployed classical security protocols as IPsec, HTTPS, S/MIME, or PGP. But as was discussed in the previous section, we need to consider carefully which mechanisms could be implemented given the limited resources of ad hoc and sensor networks.

16.1.4 Performance

The use of symmetric cryptography for sensor and ad hoc networks has been researched extensively. As this is usually based on small circuits operating on a few hundred bits, its performance and the additional cost involved is often acceptable (even if the circuit is usually a piece of software for a general-purpose CPU). Intelligent choices for key distribution algorithms are essential for this to work in as many applications as possible.

On the other hand, asymmetric algorithms are based on complex mathematic structures such as residue class fields or elliptic curve groups. The size of elements of these structures is up to a few thousand bits, and the operations performed on those elements are far more costly to implement. For the older generation of well-established algorithms like RSA and Diffie-Hellman, these elements are 1024 bits long and need to be taken to the power of a large d for decryption and signature generation.[23]

Newer algorithms such as NTRU or XTR promise better performance using smaller operands while still providing similar levels of security. Current research approaches focusing on these new technologies achieve encryption and decryption as well as signature generation and verification in times between below one second up to a few seconds on relevant hardware. As an example, Table 16.1.4 compares different execution times for the most critical operations of different public-key algorithms. All algorithm parameters are claimed to provide equal security, i.e., comparable to RSA using 1024 bit operands.[24]

Algorithm	Parameter	Operation	Time
RSA	$\lvert n\rvert = p^2 q = 1024$	Signature	18
ECC	$\lvert p\rvert = 160$	Point mult.	5.55
HECC	$\lvert g\rvert = 2, \lvert q\rvert = 80$	Divisor mult.	9.1
XTR	$\lvert p\rvert = 170, \lvert q\rvert = 160$	$S_n(c)$	0.35
NTRU	$\lvert N\rvert = 167, \lvert p\rvert = 3, \lvert q\rvert = 128$	Polynomial mult.	0.40

Table 16.6. Time in seconds for public-key operations. (Source: Blass, Junker, Zitterbart 2005)

Asymmetric cryptography is particularly interesting if a relatively long-term relationship between two nodes is to be established. In this case, a hybrid approach can be used in which a symmetric key is negotiated over an asymmetrically secured channel. Link establishment is much more expensive than in the symmetric case, but because public keys can be freely distributed, the reduced costs of key management may outweigh communication costs.

16.2 Symmetric Key Distribution

Consider two parties in a newly deployed network that want to establish a secure communication link. Due to limited hardware, they have no option but to use symmetric algorithms. Hence, they face the problem of negotiating a key in the presence of an adversary that can intercept and modify anything

---

[23] Currently, RSA-1024 bit keys are under scrutiny and may be declared unsafe in the near future, which makes longer keys recommendable. Many applications are already switching to a more conservative key length of 2048 bits.

[24] The track record of vulnerabilities in the design of NTRU gives reason for pessimism (see Chapter Notes). All published attacks have been repaired, but the fact that there were so many in the first place is usually a good hint that the system is unstable and one should expect more problems in the future.

they say to each other. This leaves two alternatives: *Key pre-loading* in a secure environment such as the factory and, somewhat surprisingly, *plaintext key broadcast*. The next two sections are dedicated to these two alternatives, respectively. Then, to demonstrate that problems are sometimes better solved in a not too abstract fashion, we will discuss tree-structured sensor networks and outline a solution for the key distribution problem in this special case.

16.2.1 Key Pre-loading

The problem of symmetric key distribution can be addressed in a few different ways. The two most straight-forward approaches are one single *Mission Key* for the whole network or loading the keys of all nodes into the memory of every node before deployment. Both solutions have a number of severe drawbacks.

Security: In the case of a single mission key, the whole network gets compromised as soon as a single node is compromised. Typically, sensors' hardware is not tamper-proof, so it is quite likely that an attacker might capture a node and read-out its memory state. Therefore, a single mission key might easily fall into the hands of an attacker. Naïve key pre-loading is only more secure if we have a priori knowledge on which keys to skip (because every key owner will only talk to a small number of others). If every node knows every key, compromise of a single node reveals all key material to the adversary as well.

Performance: Pre-loading all keys to all nodes prior to deployment scales poorly for large networks. Because every node is likely to need direct links to a fraction of the network only, it has a lot of overhead. Finally, if requirements change and new nodes need to be added to the network that the old nodes do not know about, old nodes have no way of distinguishing legitimate additions from an attack.

One possible improvement is the construction of intermediaries that can introduce nodes to each other if they do not share a common symmetric key. One of the first and among the most cited work was published by Eschenauer and Gligor in [117]. The basic idea is to randomly distribute subsets of larger key pools to different nodes. With a given probability, two nodes share at least one common key within their key pools. Whenever two nodes do not share a common key, they can establish a *key path* using intermediary nodes.

This idea has been refined by Chan and Perrig in the PIKE protocol. In the PIKE protocol (Peer Intermediaries for Key Establishment), each of the n nodes in the network is only loaded with $O(\sqrt{n})$ keys before deployment, and not n. Two nodes A, B that want to establish a secure channel, but do not share a key, can always find a node C that shares one key with A, and one with B. C can create a key on request that is then shared by A and B, at least temporarily.

Figure 16.1 illustrates how the keys are assigned. In order to make sure that an intermediary C always exists, all nodes are arranged in a square, and each row and each column is assigned a key. A node at coordinates (x, y) is

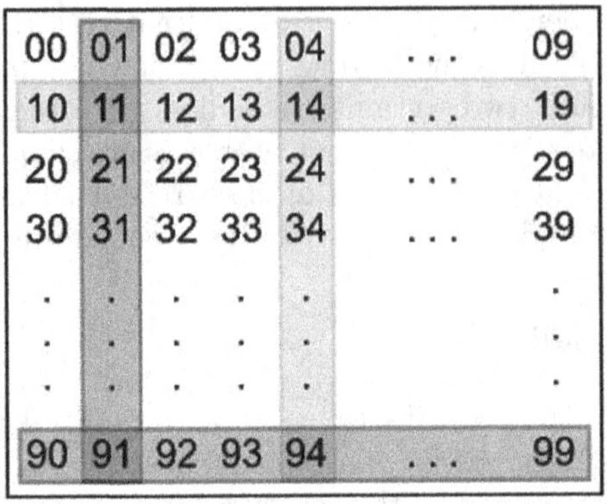

00	01	02	03	04	. . .	09
10	11	12	13	14	. . .	19
20	21	22	23	24	. . .	29
30	31	32	33	34	. . .	39
.	.	.	.	.		.
.	.	.	.	.		.
.	.	.	.	.		.
90	91	92	93	94	. . .	99

Fig. 16.1. Key assignment in PIKE. (Source: Chan, Perrig 2005)

assigned all keys from row x and from column y.[25] After deployment, if node $A = (x_1, y_1)$ wants to talk to $B = (x_2, y_2)$, it sends a fresh key to $C = (x_1, y_2)$, and C forwards it to B. Both links can be encrypted, because C is in the same row as A, and in the same column as B.

In this setup, communication between any pair of nodes becomes much cheaper at the cost of modest traffic overhead, and the effect of a single node compromise is reduced. If the adversary attempts to eavesdrop on a connection, it needs one of two keys. Which one is chosen is not determined by PIKE, so the adversary needs to be lucky. Thus, if the adversary compromises k out of n nodes, the probability it can intercept a uniformly chosen communication link is $\frac{k}{2\sqrt{n}}$.

A simple extension to PIKE addresses the issue of unplanned additions to the network: A new square of nodes neighboring the original one can be initialized such that at least one intermediary node is still guaranteed to exist.

An earlier idea is random key predistribution [67, 117]: Before deployment, each node picks a set of keys from a pool at random, and the size of the pool is adjusted to the size of the network such that any two nodes are likely to share a common key. Simulations show that PIKE performs better than random key predistribution, so we will not go into the details here.

Both approaches share the deficiency that one still needs to load a large keyring to every node before switching the network on. This is the case even if it is clear from the application that most of these keys will never be used in the life-time of the network. One solution is to impose more structure on the network and use this additional structure to reduce the amount of keys that need to be exchanged and stored. We will say more about this approach

---

[25] The coordinates are somewhat arbitrary and should rather be thought of as a name than as a location in a 2-dimensional grid, once all keys are loaded to the nodes.

in Section 16.2.3, but if there is no such structure, it is of limited use. What is often needed is something much more light-weight.

16.2.2 Key Infection

Why not, instead of pre-loading keys to nodes before deployment, simply have each node broadcast its symmetric key when the network is deployed? Of course, the problem with this idea is that the adversary, whose exclusion from the communication channels is the main point of secure key distribution, can eavesdrop on these broadcasts, and then not only intercept all communication, but also impersonate any legitimate node. However, it turns out that this naïve approach is amazingly effective against *realistic* adversaries [16].

The key infection method is very simple. Each node broadcasts its name or address; if B receives a messages from A, it computes a corresponding key, sends it to B in the clear, and B returns an acknowledgement. In short (let H be any previously agreed-on hash function[26], and N_1 a nonce that makes sure nothing is ever hashed twice[27]):

1. A broadcasts its name, B receives the signal.
2. B : $k_{A,B} = H(\{A, B, N_1\}) \rightarrow A$
3. A : $\{\text{ACK}\}_{k_{A,B}} \rightarrow B$

Now, A, B can communicate using $k_{A,B}$, and all attackers that enter the stage after this point have no way of recording or altering the plaintext. The protocol is iterated until no new nodes are found.

The key observation is that the only way for an adversary to record a large fraction of the keys negotiated in this way is to deploy a network of black (i.e., under control of the adversary) nodes similar in size to the network under attack. Any small number of attacking nodes will only intercept a small number of key broadcasts. Key pre-loading, although intuitively safer, is often just as easy to break. If the attacker has the resources to deploy a black network, she could just as well capture a few white nodes after deployment of the white network, and extract the keys by tampering with the hardware.

Fortunately, in many applications, the adversary only appears on the scene after the deployment phase, or has limited knowledge about the location and time of deployment of her target, or, if she does, she has no convenient access to the location. If the black network isn't in place and operational during the few seconds the key infection requires, the attack fails.

---

[26] Simply put, a hash function is a function with a small fixed-size image that cannot be inverted. In other words, given a hash value the adversary cannot compute any input to the hash function that produces this value.

[27] Nonces are common in cryptography to prevent replay attacks. Assume that a server executes commands that are sent to it over an encrypted channel. Then the adversary can intercept an encrypted command and send it to the server, thus tricking it into executing the command a second time without having the authorization to do so.

There are a few improvements that give yet better results at the cost of a moderate communication overhead. The simulations in [16] have shown that if the fraction of monitored traffic $\alpha \leq 0.03$, the fraction of intercepted keys is roughly the same as α.

- **Key Whispering:** Nodes do not shout their names and keys, but broadcast repeatedly on increasing energy levels, starting from a very low whisper. As soon as a node receives a response from a new neighbor, it registers the key together with the name of that node, sends an acknowledgement, generates a new key, and starts again. This way, an adversary-node E can intercept the key negotiated between A and B if it is sitting closer to A than B.

- **Secrecy Amplification:** If nodes A and B have secured a link for them, they contact a third (hopefully honest) node D to harden their link further. If a black node C can hear A and B, but not D, then A and B end up with a secured link despite C.
  1. A, B, D establish keys $k_{A,B}, k_{A,D}, k_{B,D}$ as described above.
  2. $A:$ $\{A, B, N_2\}_{k_{A,D}} \rightarrow D$
  3. $D:$ $\{A, B, N_2\}_{k_{D,B}} \rightarrow B$
  4. $B:$ $k'_{A,B} = H(\{k_{A,B}, N_2\}); \{N_2, N_3\}_{k'_{A,B}} \rightarrow A$
  5. $A:$ $\{N_3\}_{k'_{A,B}} \rightarrow B$

- **Multihop Keys:** If two nodes need to communicate, but have no direct radio link, they use other nodes in the network to route their packets. Negotiating a key between themselves and establishing an end-to-end secure link, thus protecting their communication from the routing intermediary nodes, has two advantages. (1) The routers need to do fewer cryptographic operations and can save scarce computation and communication resources. (2) A compromise of the direct radio links does not affect security if it happens *after* key negotiation. In the case of sensor networks with a central instance that gathers all information, direct links to this central data sink are particularly useful to increase security.

16.2.3 Using Aggregation for Efficient Key Distribution

Most cited symmetric key distributions in sensor networks follow the general idea to establish pairwise keys between all n nodes within the network. This might hold for classic networks, where, ideally, every node needs to communicate securely with every other node. In practice, however, sensor networks often follow a different communication paradigm called *aggregation*. Sensor send their measured data, e.g., the actual room temperature, to a so called *aggregation node*. This node collects measurements from a bunch of sensors, processes it, and sends a resulting *aggregate* further towards the sink. Aggregation can be cascaded: there can be multiple levels of aggregation. For example, one aggregation node might compute the average room temperature, for the next level of abstraction, another aggregation node might use different average room temperatures to compute an average floor temperature and so on. Typically, aggregation forms a hierarchy, i.e., a tree – the so called *aggregation tree* with the sink as root node.

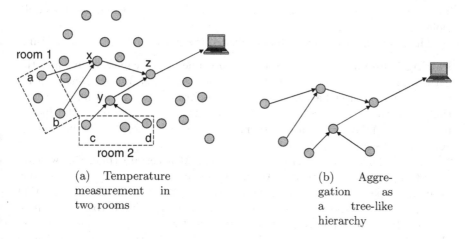

(a) Temperature measurement in two rooms

(b) Aggregation as a tree-like hierarchy

Fig. 16.2. Example data aggregation in a sensor network. (Source: Blass, Zitterbart 2006)

An example sensor network is shown in Figure 2(a). A data sink is interested in the mean temperature of a complete building. Sensor nodes and b measure the temperature in room 1 at different positions, e.g., at the ceiling and the floor, and nodes and d measure the temperature in room 2, respectively. Aggregation node collects temperature measurements from nodes and b and computes their mean value. The resulting aggregate is forwarded to another aggregation node, i.e., z. Aggregation node does the same for nodes and d. Finally, node z computes the mean temperature for the whole building and reports it to the sink. As shown in Figure 2(b), this communication scheme forms a tree.

This fact can be used for a energy and space efficient key establishment. From a security perspective, sensor nodes do not need to arbitrarily communicate which each other, but only with some nodes on their aggregation tree. For example, different temperature sensors do not need to securely exchange information with each other, neither do a temperature sensor and a humidity sensor. However, a measuring node needs to securely communicate with its aggregation nodes, as well as, e.g., for data-origin verification, the aggregation node's aggregation node and so on; to all nodes lying on the node's aggregation path towards the sink.

This results in logarithmic memory and time behavior: It is fair to assume that the aggregation tree is not *degraded*, i.e., the nodes within this tree have more than one successor – otherwise aggregation would not make sense. If the mean degree of each node is supposed to be d, the height of the tree can be estimated with $h = \lceil \log_d n \rceil$. Therefore, only h instead of the typical n keys are required to be stored in expensive sensor memory.

In general, an *extended* Dolev-Yao attacker, cf. Section 16.1.1, is assumed. This is a stronger assumption compared to related work which assumes only Dolev-Yao attackers.

The following gives a rough overview of the key establishment algorithm.

Pairing Using a Master Device. Before a new node can securely enter the network, it has to be prepared. The user or a so-called *Master Device* (MD) representing the user has to deploy initial information, e.g., an initial secret, that helps the new node to introduce itself securely to the network. Such a pairing is mandatory, because otherwise, an attacker would also be able to add malicious nodes to the network.

The whole scheme is defined inductively. At a certain time, a new node i wants to enter the network. At this point, all nodes in the network have established pairwise secret keys with their predecessors in the aggregation tree. The MD hands out two initial tickets, similar to Kerberos, to i. These tickets allow i to securely introduce itself to two different nodes, d and e, already in the network. These nodes will assist i to further establish pairwise keys with the nodes lying on i's aggregation path.

Splitting Keys. If now i has to establish a new pairwise key with another node, say f, lying on its aggregation path, it will ask d and e to assist. Node i generates a new key K and *splits it perfectly* into two halves $K = K_1 \oplus K_2$. This is done by choosing K_1 randomly and computing $K_2 = K \oplus K_1$. Now, K_1 and K_2 are called *key shares* of K and can be distributed to two different nodes. Knowing only one share K_1 or K_2, no node can gain any information about K. K can only be restored by knowing K_1 and K_2.

Node i sends K_1 to node d and K_2 to node e. Both nodes will again perfectly split their received key shares into two key shares. Node e splits $K_1 = K_1^1 \oplus K_1^2$ and node d $K_2 = K_2^1 \oplus K_2^2$. Both nodes securely forward these shares tree-upwards up to nodes which know f and their respective parents. This basic setup is shown in Figure 3(a). Because node b knows, and therefore shares a key with, node f, node e sends K_1^1 to b and K_1^2 to b's father, i.e., a. Similar, node d sends K_2^1 to a and K_2^2 to a's father (not shown in Figure 3(a)).

Finally, the nodes knowing f and who have received the key shares, securely send these to node f. This is shown in Figure 3(b). As a result, node f can now restore K by computing

$$K = K_1^1 \oplus K_1^2 \oplus K_2^1 \oplus K_2^2.$$

It can be shown, that, even in the presence of $k = 1$ malicious node in the network, there are two distinct paths within the network, so that the malicious node will never learn the complete key $K_{i,f}$, i.e., all key shares, but only a

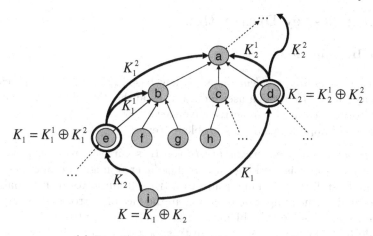

(a) Distribution of key shares tree upwards

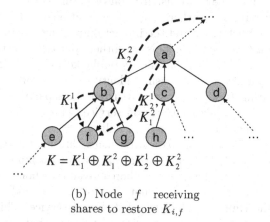

(b) Node f receiving
shares to restore $K_{i,f}$

Fig. 16.3. Key split, distribution and receipt. (Source: Blass, Zitterbart 2006)

proper subset of all shares. The scheme can also easily be extended to cope with $k > 1$ malicious nodes.

As a result, every node in the network only stores the $= \lceil \log n \rceil$ keys it would need because of its mission, anyway. This scales better compared to the $O(n)$ many keys necessary to related work. Also, the number of transmissions necessary to establish a new key scales with $O(\log n)$.

16.3 Public-Key Distribution

16.3.1 Introduction

The *key distribution problem* of public-key based security protocols can be stated as follows: How can a user u obtain the authentic public key of another user v in the presence of an active adversary?

The main approaches to this problem are:

- *Cryptographically generated Identities*: To secure the routing process by security protocols working on a higher layer it might become necessary to first establish a genuine route to a trusted authority or non-malicious peer node, which might be infeasible in the face of a strong adversary. The main idea to solve this dilemma is to derive the identity or address of a node from its public key, so identity spoofing will not work without the real node's private key. No distribution of public keys or certificates will be necessary, but on the other hand the generation of network addresses or identities is not arbitrary and the impact on routing has to be considered.

- *Identity-based Cryptography*: This is another approach to establish a strong binding between node identity and its public key. Here we derive the public key from an already established identity. To prevent adversaries from simply imitating this procedure, a trusted third party is necessary that operates as a private-key generation service. This service itself could be distributed to avoid a single point of failure.

- *Distributed Certificate Authorities*: By using threshold cryptography [283, p. 525] it is possible to split a private key among a number of nodes, so that a subset of them that is larger than a given threshold has to cooperate to use it. If an adversary does not control enough nodes to reach this threshold, she cannot control the distributed service. This idea is used to establish a distributed CA to sign certificates that bind identities to public-keys.

- *Web of Trust*: This self-organizing approach dispenses with any kind of infrastructure and relies on the network participants instead to sign and store certificates. By trusting certain other nodes to correctly issue certificates one node can follow the resulting chain of trust to verify a communication partner's public key. The existence of such a chain, however, is not guaranteed, but depending on the application it may be sufficiently probable.

In the remainder of the section we focus on the last, the web of trust approach.

16.3.2 Trust Graphs and the Shortcut Hunter Algorithm

Hubaux et al. propose a self-organized public-key infrastructure similar to the web of trust approach of Pretty Good Privacy (PGP), but without central certificate servers; the certificates are stored and distributed by the users instead. An important requirement is that users are assumed to be *cooperative* and *honest*. User u issues a certificate for v if and only if u believes in the correspondence of the given public key and the user v.

The users and certificates are modelled by a directed graph $G = (V_G, E_G)$ called the *trust graph*, where the set V_G of vertices models the users and the set E_G of edges represents the issued certificates. There is a directed edge $(u, v) \in E_G$ if and only if user u issues a public-key certificate to user v. A *certificate chain* from user u to user v is represented by a directed path from u to v in the underlying trust graph G, denoted by $u \rightsquigarrow_G v$.

Since mobile devices may suffer from memory and communication constraints, it is not reasonable to assume that a user u knows the whole graph G, so every user u just stores a particular subgraph G_u of G. If u wants to communicate with v securely, u needs to verify the public key of v. To accomplish this, u and v merge their local certificate repositories and u tries to find a certificate chain from u to v in the joint repository. In the trust graph model, u tries to find a directed path $u \rightsquigarrow_H v$ in the merged subgraph $H := G_u \cup G_v \subseteq G$.

The central question is how each node should construct its local subgraph G_u. An important assumption is that a trust graph – seen as a representation of social relationships – is a so-called *small world* graph, which can informally be described as a graph with many clusters but small average diameter. We need the following definitions:

Definition 16.3.1. *The neighbourhood Γ_v of a vertex v is the subgraph of $G = (E_G, V_G)$ that consists of the vertices adjacent to v, not including v itself.*

Definition 16.3.2. *The clustering coefficient γ_v of a vertex v characterizes the extent to which vertices adjacent to v are adjacent to each other:*

$\gamma_v = \frac{|E(\Gamma_v)|}{\binom{k_v}{2}}$, *where $|E(\Gamma_v)|$ is the number of edges in the neighbourhood Γ_v of v, k_v the number of vertices in Γ_v, and $\binom{k_v}{2}$ the total number of possible edges in Γ_v.*

Definition 16.3.3. *The clustering coefficient γ of a graph G is the average of γ_v for all $v \in V_G$.*

Definition 16.3.4. *The characteristic path length L of a graph $G = (E_G, V_G)$ is the median of the means of the shortest path lengths connecting each vertex $v \in V_G$ to all other vertices.*

Definition 16.3.5. *A small world graph G has the following two properties:*

1. *High clustering. Local groups of friends are well interconnected. Formally: G has a high clustering coefficient.*

2. *Logarithmic length scaling.* The average length of shortest paths is very small compared to the graph size. Formally: The characteristic path length L of G is $O(log(|V_G|))$.

The small diameter of such small world graphs results from the existence of *shortcuts*. Shortcuts are edges (w_i, w_j) whose removal makes the length of the shortest path between vertices w_i and w_j strictly larger than two. The preference of vertices with shortcuts improves the probability that a certificate chain between *arbitrary* nodes can be reconstructed by just using the merged repositories (outbound path from u intersects inbound path to v, and vice versa – assuming a small world).

The *Shortcut Hunter* algorithm (Algorithm 39) tries to construct an outbound and inbound path from u of fixed length s to or from some arbitrary other nodes. It prefers intermediate vertices with the highest number of shortcuts. Only the construction of an outbound path is shown below – together with a similarly constructed inbound path it will form the subgraph G_u of G representing the local certificate repository at u.

Additional requirements for the algorithm are that every user w knows its own outgoing and incoming edges (i.e., certificates about w) and the outgoing and incoming edges of every w_i adjacent to w – this last requirement is to determine the number of shortcuts w has.

Discussion. The Shortcut Hunter has been tested on real PGP graphs. In the literature, the main evaluation criterion for a repository construction algorithm $\mathcal{A}$ is the fraction of directed paths $u \leadsto_G v$ that can be reconstructed as paths $u \leadsto_H v$ using only the union H of the subgraphs of u and v that have been built by $\mathcal{A}$. This criterion has been called "performance" of $\mathcal{A}$ in the literature, which should not be confused with the actual performance of its implementation on the devices.

Definition 16.3.6. *The "performance" $p_\mathcal{A}(G)$ of the subgraph selection algorithm $\mathcal{A}$ on the trust graph G is defined as follows (# denotes set cardinality):*

$$p_\mathcal{A}(G) = \frac{\#\{(u,v) \in V \times V | u \leadsto_H v\}}{\#\{(u,v) \in V \times V | u \leadsto_G v\}}.$$

A "good performance" $p_\mathcal{A}(G)$ is close to 1. The basic Shortcut Hunter performs well on smaller trust graphs G (up to around 3000 vertices) even if users store only a small number s (e.g., $s = 10$ each) of outbound and inbound certificates locally.

There is an extension of the Shortcut Hunter algorithm for larger trust graphs G: the *Star Shortcut Hunter*. The Star Shortcut Hunter algorithm constructs more than just one outbound and inbound path at each node u resulting – loosely spoken – in two "stars" with center u, e.g., the directed outbound tree T_{out} where only root u has more than one child. It also prefers vertices with the highest number of shortcuts. A third version, the *Maximum Degree* algorithm, also constructs many paths at each node u like the *Star*

Input: A graph $G = (V_G, E_G)$, a length parameter s.
Output: An outbound path $S = (V_S, E_S)$ from u in G of maximum length s.

```
 1  begin
 2  |   V_S ⟵ {u}
 3  |   E_S ⟵ ∅
 4  |   N ⟵ ∅
 5  |   w ⟵ u
 6  |   i ⟵ 0
 7  |   while i < s do
 8  |   |   T ⟵ {(w, z) ∈ E_G | z ∉ V_S and z ∉ N}
 9  |   end
10  |   if T = ∅ then
11  |   |   if w = u then
12  |   |   |   Output the path S = (V_S, E_S) and stop.
13  |   |   else
14  |   |   |   N ⟵ N ∪ {w}
15  |   |   |   Take v ∈ V_S such that (v, w) ∈ E_S
16  |   |   |   E_S ⟵ E_S \ {(v, w)}
17  |   |   |   V_S ⟵ V_S \ {w}
18  |   |   |   w ⟵ v
19  |   |   |   i ⟵ i - 1
20  |   |   end
21  |   else
22  |   |   Choose (w, z) ∈ T such that z has the highest number c of shortcuts.
        |   |   (If there are several, choose randomly from them.)
23  |   |   if c = 0 then
24  |   |   |   Choose (w, z) ∈ T such that z has the highest number of outgoing
        |   |   |   edges. (If there are several, choose randomly from them.)
25  |   |   end
26  |   |   E_S ⟵ E_S ∪ {(w, z)}
27  |   |   V_S ⟵ V_S ∪ {z}
28  |   |   w ⟵ z
29  |   |   i ⟵ i + 1
30  |   end
31  |   Output the path S = (V_S, E_S) and stop.
32  end
```

Algorithm 39: The Shortcut Hunter Algorithm

Shortcut Hunter, but selects vertices with the highest (inbound respectively outbound) degree, which makes node selection more straighforward. Both extensions perform well on larger trust graphs, but need a larger local repository size s than the basic version to reach the same depth into G.

To take dishonest users into account, so-called *authentication metrics* have been in the discussion to weight edges of the trust graph according to the fraction of ideal complete trust they represent. However, no definite results have been achieved following this idea, and more research needs to be done.

The assumption of small world properties on node trust relationships is surely a strong one and may not be applicable to many application scenarios. Similarly, no evaluation of the practical performance of the algorithm and the corresponding communication schemes on resource-constrained devices like sensor nodes is given.

Finally, there is the issue of *privacy*. Because the main application paradigm stated is ad hoc networking of personal devices, building a trust graph depends heavily on the ability of every participant to track social networks of others, which might constitute a conflict of security versus privacy interests.

16.4 Open Questions

Privacy is an important and neglected issue for sensor and ad hoc networks. The sensitivity of the data that are planned on being collected by WSNs and their remote availability and vulnerability make privacy incommensurable. Respecting privacy requires minimal collection and delicate handling of data, which can be articulated in terms of confidentiality or integrity requirements (e.g., anonymity, unobservability, accountability, linkability). Further, knowledge on the identity and location of the sensor nodes may make a WSN more vulnerable. Using such knowledge may allow attackers to do elaborate traffic analysis in order to initiate precise attacks (e.g., by locating the sink or aggregation nodes close to the sink). It is for these reasons that privacy may be a valid and important security requirement. So far decentralizing sensitive data, changing data traffic and node mobility have been suggested for guaranteeing privacy in WSNs. Nevertheless, privacy in sensor and ad hoc networks is still immature and will remain a hot topic in upcoming research.

16.5 Chapter Notes

There are several popular text books on security in general. [220, 164, 165] provide solid theoretical foundations, [344] is a more direct approach for the practitioner with a more light-weight mathematical background, and [283] is a reference book for cryptographic algorithms and problems.

New security challenges and differences to security research in traditional networks are discussed in [381]. The Dolev-Yao attacker model mentioned in the introduction has been proposed in [102].

Typical building blocks for security protocols are symmetric ciphers such as AES [93], and asymmetric cryptography such as ECC [220]. NTRU [307] and XTR [250] are newer and more fragile asymmetric candidates: NTRU has suffered a series of successful attacks, e.g. [159, 187], and comparatively little research has been done on XTR.

The feasability of asymmetric cryptography is arguable and has been addressed in numerous performance studies [37, 275, 175].

Ideas around random key pre-distribution have been proposed in [117, 67]. PIKE has been published in [66]. The idea of key infection, i.e., plaintext key broadcast while the adversary is not listening, has been proposed in [16]. An optimized key distribution scheme for aggregation trees has been published in [38]. Some inspiration for these results has been drawn from Kerberos [288], and they make use of a well-established technique for key splitting that is due to [348].

The key distribution problem has been stated in, e.g., [189], and the categorization of answers in this chapter is due to [208, pp. 103-114].

Cryptographically generated Identities have to our knowledge first been proposed in [41]. Identity-based Cryptography was presented in, e.g., [213] and [283, p. 561]. Distributed Certificate Authorities are discussed in [406]. Finally, the Web of Trust idea is treated in, e.g., [189] and [61]. The famous PGP crypto system that has inspired the idea of a web of trust has a user manual that also introduces the term [407].

Small worlds have been researched in [393]. The shortcut hunter algorithm was invented in [189]. Research on authentication metrics has been published in [331].

Effects of wide-spread use of sensor and ad hoc networks on privacy has seen little research so far, but a preliminary discussion can be found in [320].

Figure 16.1 is borrowed from Chan and Perrig's paper on PIKE [66]; table 16.1.4 from [37]; and Figures 16.2, 16.3 from [38].

Trust Mechanisms and Reputation Systems

Erik Buchmann

17.1 Introduction

Most of the approaches for ad hoc networks assume that the nodes readily and honestly follow the protocol. Unfortunately, this assumption does not always hold in reality. Misbehaving in ad hoc networks can take place at different levels of the system architecture:

1. On the **communication level**, nodes can refuse to forward messages of others in order to save bandwidth and energy. Many misbehaving nodes would lead to a *low reliability* of the system. But such nodes can be detected, because every node can eavesdrop the network traffic of adjacent nodes ([279], 'Watchdog'). Most of the research on reputation and trust models in ad hoc networks focus on that issue.
2. In contrast, misbehavior on the **application level** covers spoof query results and valueless services from nodes that would be able to provide high-quality ones. It is challenging to automatically detect this kind of adverse[28] behavior. Typically it is the end-user who recognizes that a service provided by a certain node does not meet its expectations. But it is very important for open ad hoc networks to detect and expel such nodes. Otherwise, the system can *become suspicious*, and the users can loose their motivation to participate in the system [317].

In the following, we leave aside communication level misbehavior and focus on the application level. At first we ask: Why should a node participate in the system, but deny to follow the protocol properly? We see four reasons for that kind of behavior:

Costs. In most application areas, the economic dominant strategy in a decentralized system with many participants using changeable pseudonyms is to limit the quality of the offered services [139].

---

[28] We use the terms *adverse* and *malicious* interchangeably.

D. Wagner and R. Wattenhofer (Eds.): Algorithms for Sensor and Ad Hoc Networks, LNCS 4621, pp. 325–336, 2007.

Attacks. Nodes in ad hoc networks can be subject to attacks. A node might try to expel peers it competes with, block particular services or data, or tamper with transactions of other nodes. This can be done with denial-of-service attacks, by generating Byzantine faults [63] or by Sybil attacks [302].

Social Issues. Collaboration in ad hoc networks has many social aspects [278]. For instance, nodes could take revenge for failed transactions or try to gain attraction by disturbing the network.

Technical Problems. Finally, misbehavior can be the result of technical problems. It is hard to distinguish between nodes with technical difficulties and nodes which intentionally do not follow the protocol. However, from the perspective of other nodes it is not important whether a node is faulty by intention or not.

17.1.1 Dealing with Misbehavior

It is challenging to design protocols that rule out misbehaving nodes. Adverse nodes usually try to hide their intention. Unlike faulty nodes (in most cases), they will not come up with "connection refused" or similar statements when unable to process messages properly. Ad hoc networks are designed for spontaneous communication of users which are generally unknown for other participants in the system. Thus, adverse nodes can change their identity frequently at little cost by leaving and re-joining in the system [139]. Furthermore, a malicious participant can operate multiple nodes, the same node could introduce itself under multiple identities [302], or malicious nodes can collude. Finally, the nodes can switch between cooperative and adverse behavior at any time. While it is not desired to give advances to adverse nodes, a node that had ceased from misbehavior should be able to participate in the ad hoc network again.

Current approaches deal with these issues by introducing a central authority, by pre-configuration, with quorum-based techniques or by using trust and reputation. A *central trust authority* can provide central authentication, i.e., it provides immutable identities. In addition, a central instance can observe the behavior of the nodes and expel misbehaving ones. The downside of centralized systems is that they are single point of failure, i.e., they can either fail by an accident or due to attacks. Therefore, approaches based on centralized infrastructures are not applicable in ad hoc networks.

Another way to cope with the problem of malicious nodes is to use *pre-configuration.* Before organizing to an ad hoc network, the nodes are equipped with a set of identifiers or public keys of trustworthy nodes. But pre-configuration requires ex-ante knowledge about future communication partners, which conflicts with the core intentions of ad hoc networks.

In scenarios where multiple nodes offer comparable services, e.g., in settings with replication, *quorum-based* approaches can be applied. Here, the participants issue queries to multiple nodes, and accept the query answer that was returned by the majority of the nodes. Unfortunately, quorum-based methods require a significant amount of additional costs regarding network traffic and data processing, they give statistical guarantees only and they are not applicable in scenarios without replication.

A promising approach to cope with the problem of misbehaving nodes is to adapt the concepts of *trust and reputation* to ad hoc networks. Trust enables the nodes to estimate a *trust value*, i.e., the probability that a transaction with a certain other node yields a satisfactory result.

17.1.2 Trust and Reputation

We define *trust* as the subjective belief of one node in the willingness and ability of another node to follow the protocol of the ad hoc network in order to process a certain transaction. The definition has the following implications:

- Trust is *unidirectional.* A node that trusts another one might not be trusted from the other side in turn.
- Trust is *subjective.* Each node computes the trust values from its own criteria and thresholds. Even when all nodes have the same information, one node could trust a particular other one, while others do not.
- Trust is bound to a certain *domain of interest.* For instance, a node that is trusted to give excellent recommendations for restaurants could provide poor information on movies.
- Trust is the *estimated* probability that a transaction succeed. The notation of trust cannot give guarantees and contains the risk of frauds. Thus, security-related applications (banking, medical appliances etc.) must not solely rely on trust management.

When we leave aside simple settings where nodes immutably trust or mistrust any other node, the trust value is usually derived from observations of the behavior of the node in question. Therefore, each node maintains a *local history* containing the outcomes of past transactions. A local history is helpful only if (1) it contains at least one transaction with a node in question or (2) to estimate the global level of trust of the society of all nodes. In ad hoc networks the nodes join and leave the system very frequently. Thus, each node gets in touch with strangers very often and trust systems based on local histories tend to show outdated information.

The concept of *reputation* enables reputation-based trust systems to respond quickly to changes in the ad hoc network by exchanging trust-related information among the nodes. We define reputation as the general opinion of a society of nodes towards a certain node in a specific domain of interest. Reputation comes with the following characteristics:

- Reputation is the aggregated sum of observations made by many nodes.
- Reputation is shared among the nodes. Usually each node has the same information on the standing of any given other node.

– Reputation is expressed in a generalized, normalized manner. Nodes that want to exchange reputation information have to agree to a common set of domains and measures.

The concepts of trust and reputation *reduce the complexity* of systems where many users collaborate. In general, it is quite simple (or inexpensive, in economic terms) to estimate a trust factor from an incomplete set of data, e.g., from the outcome of a few recent transactions with a certain node. In contrast, it would be very expensive to maintain a comprehensive database of all transactions of all nodes, and it is impossible to obtain data concerning future behavior. [278] features an exhaustive analysis of trust and reputation in general.

Trust management systems are an important cornerstone of a broad range of collaboration systems. However, the special needs of ad hoc networks prevent from the immediate adoption of trust systems used in other collaboration environments. For example, trust systems designed for electronic markets, e.g., ebay.com, depend on a central coordinator. Multi-Agent systems often reproduce market situations with a few suppliers that remain in the system for long times and many consumers which change frequently. Peer-to-Peer systems in turn are characterized by a huge number of participants that are connected with high-speed networks. The following sections describe promising approaches for trust management and review its applicability to ad hoc networks.

17.2 General Trust Models

This section outlines trust measures derived from game theory, Bayesian belief networks and collaborative filtering. It focuses on the principles and the algorithmic background. Particular realizations of trust models in ad hoc networks are described in Section 17.3.

17.2.1 The Game-Theoretic Approach

Trust management systems can be seen as an instance of the *iterated prisoners dilemma*, which is a well-known and well-understood problem in game-theory [29]. In the prisoners dilemma two players seek to maximize their advantages in a selfish way without concern for the other. Each player can defect or cooperate, and the players have to decide simultaneously. A player obtains the highest possible payoff if it defects and the other cooperates, and receives the minimal payoff if it cooperates and the other defects. The global optimal payoff is realized if both players cooperate. From the perspective of ad hoc networks, 'cooperation' means that a node honestly handles transactions while 'defection' means that a node successfully cheats the transaction partner. Table 17.7 shows a sample payoff matrix for the prisoners dilemma.

	Player A cooperates	Player A defects
Player B cooperates	A: 3, B: 3	A: 5, B: 0
Player B defects	A: 0, B: 5	A: 1, B: 1

Table 17.7. Example for the payoffs in the prisoners dilemma.

The iterated prisoners dilemma tells us that one of the best deterministic strategies is Tit-for-Tat [29]. A node using this strategy first cooperates, and then always copies the action of its opponent in the previous round.

Trust systems based on game theory strive for a *Nash equilibrium* where cooperation is the optimal strategy for each player. A Nash equilibrium is a set of strategies such that no node can better off by changing its behavior on its own. For the simple (non-iterated) prisoners dilemma the Nash equilibrium is achieved when both players defect. Dominant cooperative strategies require knowledge about the past behavior. Therefore, by adapting game-theory to ad hoc networking, each node has to maintain a local history (a.k.a. transaction log) that contains the outcomes of the past transactions, and the trust system specifies a reactive strategy which is intended to guarantee that 'cooperate with cooperative nodes' forms the *only* Nash equilibrium.

However, there are remarkable differences between the settings in game theory and ad hoc networks in the wild. In particular, optimal strategies in game theory assume an infinite number of games while nodes tend to have one-shot interactions on the application level, i.e., they do not request for the same information twice. In real life it would be oversimplified to divide the outcomes of transactions simply into 'successful' and 'not successful', and the payoff can change from transaction to transaction. In addition, in many ad hoc networks the nodes can choose their transaction partners from a set of nodes offering comparable services (disclosed prisoners dilemma). Thus, pure strategies from game theory cannot be applied directly. But game theory is a well-understood playground to evaluate new strategies, e.g., with genetic algorithms [95].

17.2.2 The Bayesian Approach

An approach for trust management derived from graph theory and machine learning is the Bayesian network-based trust model [387]. The approach is based on Bayesian learning in distributed communities. A Bayesian network is a relationship network using statistic methods to represent probability relationships between the users, i.e., it is a graph where the participants of the ad hoc network are the nodes[29]. These nodes are weighted with the probabilities for successful transactions under various conditions.

---

[29] In order to avoid confusion, in this subsection the terms *node* and *network* always refer to the graph representation of the Bayesian network. Nodes in the ad hoc network are denoted as *participants*.

Every participant of the ad hoc network maintains one simple Bayesian network for each participant it has interacted with. The network has a root node which can have the values $T = 0$ ("unsatisfying") and $T = 1$ ("satisfying"). The probability $p(T = 1)$ in the root node represents the overall amount of trust in the participant, from the point of view of the maintainer of the particular Bayesian network:

$$p(T = 1) = \frac{\text{number of satisfying interactions}}{\text{number of all interactions}}$$
$$= 1 - p(T = 0)$$

(17.1)

The leaf nodes below the root node represent the trustworthiness of the participants in different aspects. For example, consider a scenario where mobile devices located in cars provide traffic reports. Assume it is important to know the accuracy of the traffic density, the delay between the occurrences and the report and the precision of the locations reported by different participants. Figure 17.1 shows a Bayesian network that corresponds with these concerns.

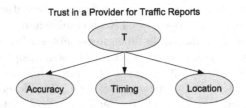

Fig. 17.1. Example for a Bayesian Network.

Each node contains a conditional probability table. The table for the root node provides information about the overall capabilities of the nodes, and the tables at the leaf nodes contain more specific information. Table 17.8 shows an example conditional probability table for the Node "Location".

Location	$T = 0$	$T = 1$
Berlin	$p(L = \text{"Berlin"} \mid T = 0)$	$p(L = \text{"Berlin"} \mid T = 1)$
Hamburg	$p(L = \text{"Hamburg"} \mid T = 0)$	$p(L = \text{"Hamburg"} \mid T = 1)$
Karlsruhe	$p(L = \text{"Karlsruhe"} \mid T = 0)$	$p(L = \text{"Karlsruhe"} \mid T = 1)$
Magdeburg	$p(L = \text{"Magdeburg"} \mid T = 0)$	$p(L = \text{"Magdeburg"} \mid T = 1)$

Table 17.8. The conditional probability table for Node "Location".

Now the participants can determine the trustworthiness of others with regard to various conditions. To be more precise, let $p(h)$ be the 'prior probability' for of the hypothesis h and $p(e)$ be the prior probability of the evidence e, i.e., $p(e)$ is the ratio of successful transactions and $p(h)$ is the percentage of transactions involving a certain feature. Thus, $p(h|e)$ is the probability h given e. The Bayes rule tells us that:

$$p(h|e) = \frac{p(e|h) \cdot p(h)}{p(e)} \qquad (17.2)$$

In our example, $p(L = "Berlin"|T = 1)$ is the conditional probability for a traffic report from Berlin with the condition that the report will satisfy the demands. According to the Bayes rule, the probability can be computed as follows:

$$p(L = "Berlin"|T = 1) = \frac{p(L = "Berlin", T = 1)}{p(T = 1)} \qquad (17.3)$$

The term $p(L = "Berlin", T = 1)$ denotes the probability that the location of the report is "Berlin" and the transaction is successful, i.e.,

$$p(L = "Berlin", T = 1) = \frac{\text{successful interactions involving "Berlin"}}{\text{number of all interactions}} \qquad (17.4)$$

Other examples for probabilities that can be computed from the Bayesian network are the probability that the rated participant is trustworthy when providing timely traffic reports $p(T = 1|D = "fast")$, or the probability that it is trustworthy if it releases reports for accurate locations in Magdeburg $p(T = 1|A = "accurate", L = "Magdeburg")$.

After each transaction, the participants update the weights of their corresponding Bayesian networks. For example, if a participant receives a new traffic report, it could increase the weights for $p(T = 1|D = "slow")$ and $p(T = 1|L = "Berlin")$.

17.2.3 The Collaborative Filtering Approach

A drawback of trust systems based on local histories is that each node has to make its own bad experiences. Thus, a reputation system where the nodes exchange information about the behavior of others is highly desired. However, it is challenging to integrate trust-information coming from other nodes. First, malicious participants can spread spoof information. Second, the participants have to agree to a common set of measures and domains. But this is a complex task: One participant might find a certain service acceptable, while others would complain about it. Thus, reputation management systems intended to drive complex scenarios have to model complex relationships between the nodes as well.

Collaborative filtering is a method to detect common preferences among the opinions of a large number of different users. For example, collaborative filtering is used at www.amazon.com. Based on a set of observations, collaborative filtering makes recommendations to users which have similar preferences. [402] shows an approach for reputation management that has been inspired by collaborative filtering. It assumes that nodes with a high rating in a certain domain submit ratings that can be merged into an substantial trust value for that domain.

Unlike classical collaborative filtering scenarios, ad hoc networks do not provide a centralized data pool containing the (node, object, rating)-tuples generated from the participants. Instead, the ratings are propagated among the nodes, i.e., a node that disseminates a certain rating value introduces its very own opinions and creditability to the value. Therefore, the ad hoc network is regarded as a directed graph ("Web-of-Trust" model, cf. Section 16.3), and the participants are expressed as nodes in the graph. A reputation value is assigned to each participant. Transactions between two participants form new edges between the nodes, and the edges are weighted with the rating of the outcomes of the associated transactions. If a participant rates the same node again, the new edge supersedes the old one. Figure 17.2 shows an example for a directed graph representation of trust relationships between 6 nodes.

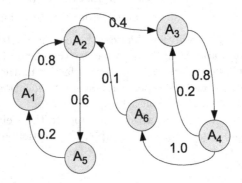

Fig. 17.2. A graph representation of trust relationships.

When user A_3 queries the reputation value for user A_4, the system determines all directed paths connecting A_3 and A_4 with a length less than n by using *Breadth First Search*. The graph in Figure 17.2 contains two paths for $n \geq 3$: $A_4 \rightarrow A_3$ and $A_4 \rightarrow A_6 \rightarrow A_2 \rightarrow A_3$. For each of these paths the weights of the edges are used to determine an auxiliary trust value. The personalized trust value of A_4 is the aggregate of all auxiliary trust values of the paths. The trust system can be optimized by modifying the functions that calculate the auxiliary- and personalized trust values. But the biggest obstacle for any collaborative filtering based trust system is not the calculation of trust values. Instead, it is hard to find the paths between the nodes in loosely connected communities. However, in contrast to Peer-to-Peer Networks or Multi-Agent Systems, ad hoc networks usually do not consist of thousands of nodes. Thus the cost of finding paths between the nodes may be acceptable.

17.3 Reputation-Based Trust Protocols in Ad Hoc Networks

Obviously, the generalized trust models presented in Section 17.2 have accentuated strengths and weaknesses. Therefore, implementations of trust models tend to build upon a mixed model. Every trust system consists of four components:

- a component for observation gathering,
- one that estimates the probabilities for successful transactions with particular nodes,
- a component that adapts the behavior of the node based on these probabilities, and
- a component for the dissemination of observations among nodes that might be interested in.

This section exemplarily presents two selected protocols for reputation-based trust systems designed for ad hoc networks. The approaches are selected according to prominent solutions in one or more components.

17.3.1 Enhanced Reputation Mechanism for Mobile Ad Hoc Networks

The proposal [261] offers a reputation-based trust model that features trust values from -1 (completely untrustworthy) to 1 (completely trustworthy). The trust value evolves over time, i.e., the approach reflects by using a variant of exponential moving averages that nodes can change their behavior at any time: Let r_t be the trust value for a particular node at time t and b be the observed behavior, then $r_t = r_{t-1} \cdot p + b \cdot (1 - p)$. Factor p adjusts the recency of the trust value: the lower value p, the more weight is given to recent observations.

The approach provides a very fine-grained trust model that (1) differentiates between the ability to provide good services and trustworthy recommendations and (2) comes with an ontology tree that provides information about the similarity of domains. Differing between quality of service and quality of recommendations is helpful with credulous nodes that disseminate tampered reputation information, but are still offering good services. The benefit of an ontology tree for service domains is that a node can estimate trust values from adjacent domains. For example, given a node wants to obtain diet tips from another one which is only known to provide excellent cuisine recipes. If the node knows the *distance* between cuisine recipes and diet tips, it would be able to make a reasonable estimation for the probability to receive a satisfactory query result. The reputation on context $c \in T$ in a ontology tree T is calculated as follows:

$$r(c) = \frac{\sum_{c' \in T} r(c') \cdot q^{|c'-c|}}{\sum_{c' \in T} q^{|c'-c|}} \tag{17.5}$$

Parameter q adjusts the impact of adjacent service domains, i.e., $q = 0$ means that the node does not consider trust values from similar contexts while $q = 1$ values all competences equal.

After each interaction, the nodes update their trust values with its transaction partners. The recommendations of others are weighted with the trust value for recommendations mentioned above. Though, the reputation propagation component is not highly developed.

17.3.2 The Effect of Rumor Spreading in Reputation Systems for Mobile Ad Hoc Networks

It is challenging to establish trust networks 'from scratch', i.e., in situations where many users meet the first time. Obviously, a malicious node would lie, thus react to each complaint with a counter-complaint. Once established, a network with dense trust relationships can differentiate between an insider that did not lie before and a possibly untrustworthy newcomer.

[54] proposes a distributed, reputation-based trust system designed to cope with false information. It uses a Bayesian model (cf. Subsection 17.2.2), extended by a method to merge the observations of others into the view of the individual node.

The paper discusses some approaches to join the observations of others. With *independent opinion pools*, the prior probability is the product of the probabilities of other nodes $p(\theta) = k \cdot \left[\prod_{i=1}^{m} p_i(\theta) \right]$. Factor k is an external tuning parameter. In the case of *linear opinion pools*, weights w_i with $1 = \sum_{i=1}^{m} w_i$ are assigned to the information sources, thus $p(\theta) = \sum_{i=1}^{m} w_i \cdot p_i(\theta)$. Alternatively, probabilistic modeling could be applied. A trust component weights the direct observations of a node as compared to second-hand observations.

In addition, a further countermeasure against wrong accusations is to exclude the trust values of nodes that provide exceptional ratings. Let $R_{i,j}$ be the trust value of node j provided by node i. Then the approach excludes a trust value $R_{k,j}$ from the model merging if it deviates from $R_{i,j}$ by more than a threshold u.

The evaluation of [54] tells us that – at least in their artifical scenario – settings that include second-hand and third-hand observations help to adopt to malicious nodes in a very short time and with a remarkably low number of total observations. However, it would be important to extent the evaluation to a more detailed model of adverse behavior, e.g., nodes which selectively disseminate wrong accusations.

17.4 Open Problems

Although many different approaches promise to deal with the problems induced by adverse nodes, there are numerous challenges that need to be solved.

One of the biggest obstacles on the way to a reliable and open ad hoc network is that there are no practical experiences with malicious nodes in that kind of networks. Large-scale Peer-to-Peer Systems (gnutella[30], Kazaa[31]) and huge electronic market places (eBay[32], Amazon[33]) exist for years, and the emergence of various kinds of unwanted behavior among the participants of these systems have been examined for years as well. In contrast, the only wide-spread ad hoc network protocols are *Bluetooth* and *WLAN*. Bluetooth is commonly used to build enclosed networks of a tiny number of small devices owned and operated by the same user, e.g., cellphone, headset, keyboard or laptop, and WLAN-driven networks usually provide centralized Internet-access or connect small networks of workstations. Besides of a few research prototypes, there are no implementations of large, open communities based on ad hoc networking.

Furthermore, it seems that there is no approach that provides each feature desired for a reputation-based trust system for ad hoc networks. A perfect trust system for ad hoc networks should be highly dynamic in order to react quickly to changes in the behavior of the nodes and the infrastructure, it should cope with loosely connected communities and produce a sharp borderline between honest nodes and doubtful ones. In addition, such a system has to consider different domains of applications, it must be tamper-resistant and it should not induce (more than absolutely unavoidable) additional costs for honest nodes.

It is hard to evaluate the characteristics of existing approaches. Thus, an important step on the way to find a suitable trust system would be to define a generalized framework for evaluation and comparison. The framework should include profiles for various kinds of applications and behavior, e.g., characteristics of queries in mobile ad hoc communities or properties of collaboration attacks.

17.5 Chapter Notes

This chapter has shown that misbehavior on application level is a serious issue in ad hoc networks. There are many reasons why nodes could refuse to handle incoming requests properly. As a result of misbehaving nodes the quality and reliability of the services provided by the ad hoc network would be decreased. Reputation-based trust systems are promising approaches to tackle with this kind of challenges. The fundamentals of the concept of trust in electronic societies are shown in [278]. The paper derives a heuristic and a formal notation of trust provides a basic understanding of trust-supported cooperation mechanisms. There are many ways to obtain trust – and an appropriate strategy for

---

[30] http://www.gnutella.com
[31] http://www.kazaa.com
[32] http://www.ebay.com
[33] http://www.amazon.com

further actions – from experiences gained during the past transactions. The chapter discussed proposals based on Game Theory [29], Bayesian Networks [387] and graph-ased trust models [402]. However, a general flaw of all of these proposals is that none of them have proved its applicability in a large-scale real world application until now.

18

Selfish Agents and Economic Aspects

Leonid Scharf

18.1 Introduction

Routing algorithms in wireless networks can be developed using different techniques. In a network consisting of selfish nodes, traditional approaches will fail, since nodes are not interested in forwarding foreign data and might refuse to cooperate. In this chapter we discuss algorithmic mechanism design, which provides a possibility to deal with selfishness of network nodes.

Nodes in a network are called selfish, if they are not providing own resources to other nodes for free, which is usually bandwidth for transit data transfers. Forwarding foreign data in a wireless network means for a battery-powered node spending valuable energy, which can be considered as an expence for a node. We will call such expences *costs*. A *rational* node will provide a service when offered compensation for its costs, wich could be some kind of virtual currency. The *value* of such a transaction to a node is the difference between the amount of compensation and the own costs of providing this service. In this chapter we will not consider malicious nodes, for example, nodes with the intent of jamming or tapping the network.

Selfish entities are called agents and are assumed to be rational and responding to incentives when asked to cooperate. For each network model and given routing problem, the aim is to design a cheating-proof mechanism, which computes outcome (the route) and payoff function in polynomial time and minimizes, if possible, the overall route cost.

18.2 Mechanism Design

In Mechanism Design multiple independent agents in a game-theoretic setting are considered. For each given problem, the aim is to develop a mechanism, which computes an output and a payoff vector depending on the agents' strategies. Properties of the output depend on the system goal. In the case of unicast

D. Wagner and R. Wattenhofer (Eds.): Algorithms for Sensor and Ad Hoc Networks, LNCS 4621, pp. 337–358, 2007.
© Springer-Verlag Berlin Heidelberg 2007

network routing the output could be the path with the least energy cost from source to destination.

Consider a system with n independent selfish agents. Each **agent** $a_i, i \in 1, \dots, n$ is described by its **type** $t_i \in T_i$, which is some private information, known only to the agent himself. This information could be, for example, the cost, the agent has to pay to forward data from another agent in the network. Types of all agents define a **type vector** $t = (t_1, \dots, t_n)$ in the game. In each round, agents are asked to select their strategies from corresponding sets, i.e. agent a_i selects a **strategy** s_i from the **strategy space** S_i, which is given to the agent by the mechanism. A strategy is in words a plan or a decision rule, which defines an action of an agent in a given state of the system. Agents' strategies define a **strategy vector** $s = (s_1, \dots, s_n)$. After the mechanism has collected all strategies from the agents, it computes the **output** $o = o(s), o \in O$ (set of all possible outputs) and the **payoff vector** $p = p(s), p = (p_1, \dots, p_n)$.

Preferences of an agent a_i are represented by its **valuation function** $v_i(t_i, o)$, which computes a monetary value of the output o to the agent. This value can be, for example, the cost of the agent (a negative number) in the case it is selected in the output o and zero otherwise. The **utility** of the agent is a function $u_i(t_i, o, p_i)$, $u_i : T_i \times O \times \mathbb{R} \to \mathbb{R}$ and is often assumed to be the sum of valuation function and payment, i.e, $u_i(t_i, o, p_i) = v_i(t_i, o) + p_i$. In other words, utility of an agent is its payment minus its cost and depends on its type, the outcome and the payment. Agents are assumed to be **rational**, which means that each agent always selects a strategy, which maximizes its expected utility. A strategy s_i is called **dominant** if it maximizes the utility of agent a_i regardless of what other agents do. A great benefit of dominant strategy is that the agent does not need to take the behavior of other agents into account, which simplifies the agent's algorithm.

Definition 18.2.1 (Mechanism). *A mechanism* $M = (o, p)$ *defines a strategy space* S_i *for each agent* a_i, *the output function* $o = o(s)$, $o : S_1 \times S_2 \times \dots \times S_n \to O$, *and a payment vector* $p = (p_1, \dots, p_n)$ *with* p_i *being the payment to the agent* a_i.

A mechanism is called direct revelation mechanism, if strategy spaces consist of all possible types of corresponding agents.

A mechanism is **strategyproof** or **truthful** if the agents' types are part of the strategy space and revealing the true type is always a dominant strategy for each agent [384]. Such a mechanism removes speculations and counter-speculations among agents. The most important class of mechanisms are **Vickrey-Clarke-Groves** (VCG) mechanisms by Vickrey, Clarke and Groves.

Definition 18.2.2 (VCG Mechanism). *A direct revelation mechanism belongs to the VCG family if the computed output* o *maximizes the objective function* $g(o, t) = \sum_i v_i(t_i, o)$ *which is called social welfare. Furthermore the* *payment is calculated as the added value of the agent to the system plus its declared cost. Formally*

$$p_i(t) = \sum_{j \neq i} v_j(t_j, o) + h_i(t^{-i})$$

where $t^{-i} = (t_1, \ldots, t_{i-1}, t_{i+1}, \ldots, t_n)$ and h_i is an arbitrary function of t^{-i}.

The most positive property of VCG mechanisms is their truthfulness, which is proved by Groves in [170]. For many problems, though, it is intractable to compute the exact maximization, and since VCG mechanism can lead to untruthfulness if used with sub-optimal solutions [384], it is necessary to design custom mechanisms for such problems.

18.3 Network Model

We model a sensor network as an undirected graph $G = (V, E)$ with $V = \{v_0, v_1, \ldots, v_{n-1}\}$, the set of nodes corresponding to wireless nodes in the network and E, the set of links (v_i, v_j) such that direct communication between v_i and v_j is possible. We assume G to be node bi-connected, for the purpose of truthfulness, since the bottle-neck node could otherwise define arbitrary high cost for relaying foreign data.

Depending on device features we could assume that the nodes have fixed but not necessairly the same transmission range and no sending power regulation. The resulting model is then a node-weighted graph, since devices incur the same cost when sending a unit sized packet, independent of destination node, as long as it is in the transmission range. Most modern wireless transmitters are, however, capable of adjusting emission power and therefore regulating their transmission range. In such networks the edge-weighted model should be selected. We consider the unicast routing problem for both edge and node weighted networks and concentrate on edge-weighted graphs in the multicast case.

Further, we assume that each transmission can be received by any node in the range and each node is able to send data to other nodes, to forward data for other nodes, and to accept data from other nodes.

18.4 Unicast in Selfish Networks

18.4.1 Ad Hoc-VCG

In this section we present the mechanism by Anderegg and Eidenbenz [15]. This mechanism is designed for ad hoc wireless networks, consisting of nodes, which are able to regulate their emission power. The protocol works in two phases: route discovery and data transmission. During the first phase the cost-efficient route between two nodes is found and the payment to intermediate nodes is computed. In the second phase the actual data is transmitted and relay nodes are paid for forwarding data.

The model is an edge-weighted graph.

Model. We consider a network of selfish wireless nodes. As mentioned before, the nodes are assumed to be rational, i.e. they always try to maximize their utility, which is the payment minus the cost, a node incurs when forwarding foreign data. Each node is able to control its emission energy when sending data. Nodes are also able to determine the strength of received signal. During the route discovery phase nodes are sending the value of the emission power in the header of each packet, providing the receivers with a possibility to compute the minimal energy required to send data between these two nodes. Each node maintains its cost-of-energy parameter, which may vary depending on, e.g., battery level. Actual cost of sending a data packet for a node is then the product of its cost of energy parameter and the emission power for this packet.

Definition 18.4.1. *Let v_i be a selfish wireless node. We denote its emission power as P_i^{emit}, its individual cost of energy (private information or type) as c_i and the payment, the node receives, as p_i.*

Nodes are assumed to have omnidirectional antennas, i.e. they cannot control the direction, in which the data packet sent. Emission power P^{emit} can be selected by nodes, defining the transmission range, where data package can be received.

Definition 18.4.2. *When a signal from node v_i is received by node v_j, signal strength is denoted as $P_{i,j}^{\mathrm{rec}}$ and is*

$$P_{i,j}^{\mathrm{rec}} = \frac{K}{d^{\alpha}} P_i^{\mathrm{emit}}, \tag{18.1}$$

where K is a constant and α the distance-power gradient varying between one and six depending on the environment condition of the network.

Definition 18.4.3. *If node v_i sends a packet to v_j at power P_i^{emit} and includes P_i^{emit} in packet header, v_j is able to compute the minimal emission power required to send data from v_i to v_j:*

$$P_{i,j}^{\mathrm{min}} = \frac{P_i^{\mathrm{emit}}}{P_{i,j}^{\mathrm{rec}}} \cdot P_{\mathrm{min}}^{\mathrm{rec}} \tag{18.2}$$

If this power exceeds the minimum level $P_{\mathrm{min}}^{\mathrm{rec}}$, a node v_j can successfully receive the data. It is assumed, that $P_{\mathrm{min}}^{\mathrm{rec}}$ is constant for all nodes in the network. Nodes are able to transmit packets with power up to $P_{\mathrm{max}}^{\mathrm{emit}}$, which is assumed to be $P_{\mathrm{max}}^{\mathrm{emit}} = \infty$ for the simplicity. Considering other values for

$P_{\max}^{\mathrm{emit}}$ would not compromise the results of the mechanism, except for upper bounds on overpayment. It must, however, be guaranteed, that the network is node-biconnected, since the bottle-neck node could declare arbitrary high cost of energy. It is although still possible for nodes to cheat, if they collude, trying to maximize the common utility. Thus, the mechanism is not group-strategyproof. However, due to selfishness of agents it will be difficult for them to agree on how to split up the gain, since each agent could argue, others would not achive this gain without him, so he should get more, than the others.

Definition 18.4.4. *The real cost of sending a unit data packet from node v_i to node v_j is denoted as $w(v_i, v_j)$ and is*

$$w(v_i, v_j) = c_i \cdot P_{i,j}^{\min}. \tag{18.3}$$

This can be also considered as link weight.

The proposed mechanism deviates from the classical mechanism design model in following: agents do not know their type. More precisely: if a node v_i sends a packet in the route discovery phase, it appends P_i^{emit} and c_i to the header of the packet. Each node v_j, when receiving this broadcast packet, computes the minimal emission power $P_{i,j}^{\min}$, which defines the type of the node v_i on the link (v_i, v_j) to be $w(v_i, v_j)$. Despite this deviation, the mechanism is strategyproof.

The mechanism does not have the a priori knowledge of which or how many nodes belong to the network, nor are the weight function or link set known. We assume that the nodes are embedded in the plane.

Definition 18.4.5 (Utility). *The utility u_i of a node v_i is*

$$u_i = p_i - c_i \cdot P_{ij}^{\min}$$

where p_i is the received payment and v_j is v_i's successor node on the shortest path from source to destination, if this path contains v_i. Otherwise, the utility is zero, since payment and cost are zero as well.

All nodes in the network are acting selfishly: they try to maximize their utility and will forward foreign data, receiving payment in conventional dollars for it. This includes, that nodes would cheat, declaring false cost or altering foreign data, if it would increase the utility.

A natural question is, who pays intermediate nodes for forwarding data? A good candidate is the sending node, since it initializes the transmission. In the following alternative models are proposed:

Source model: if node S wants to send data to node D over several relay nodes, node S pays intermediate nodes a premium in addition to their true cost for forwarding data, if these nodes form a minimum cost path from S to D. Moreover, it is assumed that source node S is acting truthfully when reporting its own cost of energy parameter and emission energy. As authors note, this assumption is quite strong as it essentially requires the source node to act non-selfishly, whenever it is a source; however, there is a tradition in mechanism design to treat the source (or auctioning agent) in such a special way. (Chapter 23.C, pp. 880-881 in [281]).

Central-bank model: source node S pays only the true cost of intermediate nodes, premiums are paid by a special entity, a central bank, which acts as a clearing house. The central bank manages accounts for all nodes in the network and pays premiums to nodes which are participating in data transmission. Paid amount is evenly distributed over all nodes in the network and charged from their accounts. This can be considered as the fee for participating in the network. A detailed analysis of this approach can be found in [167].

In both models, nodes declare their individual cost of energy parameters c_i, therefore the computed cost-efficient path is not necessarily the most energy-efficient one. It is possible to require equal c_i parameters from all nodes; this would lead to energy-efficient routes. Individual cost of energy parameters give nodes the possibility to prolongate their life time by setting c_i higher when the battery goes low, since this would decrease the probability to be on the least-cost-path between some two nodes.

In the route discovery phase nodes are not paid for the packets they send. It is assumed that communication overhead in this phase is negligible compared to the packets transferred in the the second phase, where actual data is transmitted. This scheme leads to following problem: if a node knows from the previous sessions, that in the following session it will not be selected as intermediate node, it might "play dead" and not participate in the route discovery phase, possibly causing higher premiums for other participating nodes. This problem can be solved by selecting an alternative version of the route discovery phase, where nodes get paid even for sending discovery packets, which is described in [15].

Both payment models have their drawbacks: in the second model nodes need to send data to the central bank, which means network overhead and extra costs. One advantage of this model is the fair distribution of the route costs between the source and all other nodes in the network. In the first model it must be assumed, that the source node acts truthfully, which is a strong assumption and clearly a disadvantage, while the lower energy consumption and simplicity are the positive features of the model.

Furthermore, we assume that during the route discovery phase nodes are not moving and not changing their cost of energy parameter.

The Ad Hoc-VCG Protocol. The Ad hoc-VCG protocol by Anderegg and Eidenbenz consists of two phases: route discovery and data transmission. During the first phase the structure of the network is discovered by essentially flooding the network. The process of discovery is initiated by the source node $S := v_0$ and the information about the netwoork structure receives the destination node $D := v_n$. All nodes obtain the partial knowledge of the network structure during the discovery phase. After D builds up the network structure, it computes the most cost-efficient route from S to D as well as the payments to the intermediate nodes and sends this information along the computed route to S. To avoid altering the information by relay nodes, D could cryptographically sign the route reply packet. In the second phase the actual data is transmitted along the computed cost-efficient route from S to D along with payments to relay nodes. The nodes get paid for each unit-sized packet they forward.

Route Discovery. Let node $S = v_0$ be the source and $D = v_n$ the destination nodes. If node S does not already have the information about the route from the previous sessions, it initiates the route discovery by sending a broadcast route discovery packet with the following content:

1. A sequence number $s_{0,n}$
2. The identification 0 of the source node $S = v_0$
3. The identification n of the destination node $D = v_n$
4. The emission power P_0^{emit}
5. The cost-of energy parameter c_0

If a node $v_j, j \neq 0, j \neq n$ receives a route discovery packet from a node v_i it proceeds according to the Algorithm 40.

Input: broadcast packet K_i from v_i received by v_j
Output: broadcast packet K_j sent by v_j
1 **if** *no new edge information contained in K_i* **then**
2 | send nothing
3 **else**
4 | determine power $P_{i,j}^{\text{rec}}$
5 | compute $P_{i,j}^{\text{min}}$ according to Equation 18.2
6 | $K_j := K_i$
7 | replace the pair c_i and P_i^{emit} in K_j with a single value
   | $w(v_i, v_j) = c_i \cdot P_{i,j}^{\text{min}}$
8 | $K_j := K_j + \quad j, P_j^{\text{emit}}, c_j$
9 | broadcast send K_j
10 **end**

Algorithm 40: Route discovery packet processing

After the reception of a route discovery package the node examines whether information about new, in the session with the current sequence number unseen, edges is contained. Each node builds its own view of the network during the discovery phase and notes on which edges it has information so far. If new information is contained in the packet, the node computes the minimal emission power, required to send data from the sender to this node. Minimal emission power is then multiplied with the reported cost of the predecessor node, producing the weight of the link; this value is then written in place of cost value and emission power of the predecessor node. This step is particularly important, since if only the emission power of the predecessor would be rewritten with the minimum emission power, it would make it possible for nodes to cheat (see Lemma 18.4.9). At last, own data is appended to the packet, containing identification of the node, emission power and own cost of energy parameter. This altered and extended packet is then re-broadcasted.

During the discovery phase each node builds up a structure, corresponding to the subset of the whole network. In [103] authors present a structure for efficient caching of such information, which is also applicable in Ad hoc-VCG. In the case of a complete graph, each node has the knowledge of the entire network. Due to possible mobility of the nodes and their ability to change the cost of energy parameter this knowledge is only a snapshot of the entire network, which could change until the next session or even during the current data transmission session. Albeit this knowledge, cheating would not increase the utility of a single node, as we will see later.

In the route discovery phase each node (except for S and D) sends a packet at most $O(m)$ times, since it tracks the information on already seen edges in the current session and only sends packages with new edges. Therefore the maximal number of messages sent in this phase is $O(mn)$ or in the case of complete or near-complete graph $O(n^3)$.

Destination node D collects all incoming packets and builds up the structure, corresponding to the entire network. As soon as all information is collected, D computes the least cost path from S to D using for example Dijkstra's algorithm [83]. This path is the output of the mechanism. Let this path be $SP = S, v_{\sigma(1)}, \ldots, v_{\sigma(k)}$ and $|SP|$ denote the total cost of this path. The next step for D is to compute payments to the nodes $v_{\sigma(i)}$. Payments are computed according to the VCG scheme as:

$$p_{\sigma(i)} := |SP^{-\sigma(i)}| - |SP| + w(v_{\sigma(i)}, v_{\sigma(i+1)}), \qquad (18.4)$$

where $SP^{-\sigma(i)}$ is the least cost path in the network, induced by $V \setminus \{v_{\sigma(i)}\}$ from the original network. Payment to a node v_i not on the least cost path is 0.

The payment to a node on the least cost path is in words its cost for forwarding data on the path plus a premium, which can be described as the added value of the node to the network. The premium is always non-negative, since otherwise SP would not be a least cost path.

Payments can be computed in a straightforward way by computing the first term of the Equation 18.4 for each node using Dijkstra's shortest path algorithm on the graph G without the considered node. In the worst case, it would mean the asymptotic complexity of $n \cdot O(m + n \cdot \log n)$ where m is the number of edges in the network, or in case of complete graph $O(n^3)$.

After the payments are computed, node D composes a route reply packet, which contains a sequence of node identifications $\sigma(1), \ldots, \sigma(k)$, forming the shortest path from S to D, along with minimal emission powers $P^{\min}_{\sigma(i),\sigma(j)}$ for the nodes on this path and payments $p_{\sigma(i)}$ to these nodes. This packet is then sent along the reversed shortest path to the source node S. In order to prevent the relay nodes from altering their payment values to their advantage, destination node D signs the packet with a digital signature.

Data Transmission. As soon as the node S receives the route reply packet, it can start sending data along the least cost path $S, v_{\sigma(1)}, \ldots, v_{\sigma(k)}, D$, which is contained in this packet. Along with the actual data and routing information S sends payments to intermediate nodes, which differ depending on the used model.

In the source model the node S pays a node $v_{\sigma(i)}$ the full amount $p_{\sigma(i)}$. There are different models for such payments, first of which requires a universally accepted financial institution, which issues digital money that can be transferred from node to node, see [405]. The other model is based on some hardware module, which can store money in a tamper-proof way, see [57].

In the central-bank model the node S owes relay nodes only their incurred cost $c_{\sigma(i)} \cdot P^{\min}_{\sigma(i),\sigma(i+1)}$. Premiums are paid by the central bank, which manages accounts for each node. Following scheme is then applied: destination node D records for each intermediate node v_i its payment as well as how much the source node S owes v_i. These records are periodically transmitted to the central bank, which credits and debits accounts accordingly. It can be argued that, since nodes do not report information about own payments or credits directly to the central bank, such a scheme is truthful. The central bank keeps track of how much premiums has been paid (say it is K) and debits periodically all network nodes with the amount K/n, distributing the cost of the premiums evenly across the whole network. This can be considered as the fee for participating in the network.

Route Recovery. An intermediate node is allowed to change its cost of energy parameter during a data session (but not during the route discovery phase), which would alter its payments or even change the least cost path between the current source and destination nodes. In this case, the node sends a "broken link" control packet to the source, which forces the source node to start the route discovery phase anew. A moving node on the least cost path would trigger the same procedure: changing minimal emission power affects the cost for forwarding data by this node. The result would be again a "broken link" packet from the intermediate node followed by a new route discovery phase. The source node could also wish to find a new cost efficient route without an external reason, which would again lead to a new route discovery phase.

Analysis. Luzi Anderegg and Stephan Eidenbenz show in [15] that Ad hoc-VCG protocol is truthful and always finds the most cost-efficient routes. The first observation made is, that truthfulness of the protocol directly implies the cost-efficiency: if all nodes report their costs and emission powers truthfully and do not alter information about the other's costs when forwarding route discovery packets, destination node D computes the least cost path from the source to the destination based on the correct network data. Thus we have:

Theorem 18.4.6. *The Ad hoc-VCG protocol computes the most cost-efficient routes between source and destination nodes, if truthfulness is guaranteed.*

Energy-efficiency can be provided by setting all nodes' cost of energy to the same constant value.

Recall that a direct-revelation mechanism is called truthful, if for each agent revealing is true cost is a dominant strategy. A strategy is called dominant, if playing this strategy maximizes agents' utility regardless of what other agents do. Note, that Ad hoc-VCG guarantees voluntary participation for agents: each agent has a non-negative utility, since not selected agents have zero utility and participating agents have the utility of payment minus the cost, which is the always non-negative premium.

In the case of Ad hoc-VCG, truthfulness is given if and only if it is a dominant strategy for each node v_j to always

1. declare its true cost of energy c_j and its true emission power P_j^{emit},
2. correctly compute and declare the minimal emission power $P_{i,j}^{\text{min}}$ of all its predecessors v_i, and
3. correctly rebroadcast all edge-weight information contained in the route request packets received without alterations or intentional dropping.

We call these items "cheating possibilities", each of which is treated in a separate lemma.

Three types of nodes are distinguished in the following: the source node S, the destination node D, and all other nodes, which are simply called nodes. An adversarial node is assumed to be omniscient, i. e. knowing the structure of the whole network including the edge-weights and the source-destination pair of the next session. If even such a node cannot exploit its knowledge to its advantage, the protocol is truthful.

As mentioned before, in the source payment model the node S is assumed to be truthful, since it could otherwise have incentives to lie about its cost of energy or emission power. The case could be that the energy cost of a single hop from the source to the destination is higher than the cost of the shortest path, but less than the overall payment to the intermediate nodes. An omniscient source node could then underdeclare its own cost to make the single hop route be the shortest path thus ending up paying less. It is a strong, but necessary assumption.

In the central bank model the source node S pays only the total cost of the least cost path, so it is in its interest to find the most cost-efficient path, therefore S will declare its cost of energy and emission power truthfully.

Lemma 18.4.7. *Each node v_j declares its true cost of energy c_j and its true emission power P_j^{emit}.*

Proof. We show that a node v_j cannot increase its utility by lying about its cost of energy or emission power. The node v_j has two possibilities to cheat: to make its edge look cheaper or to make its link look more expensive.

1. To make the link look cheaper the node v_j could either underdeclare its cost of energy or its emission power or both, i.e. claim it to be $\overline{c_j}$ and $\overline{P_j^{emit}}$ respectively with $\overline{c_j} < c_j$ or $\overline{P_j^{emit}} < P_j^{emit}$. We can rewrite the Equation 18.4 as

$$p_j = |SP^{-j}| - \left(|SP| - c_j \cdot P_{j,s(j)}^{min}\right) = |SP^{-j}| + \sum_{k \neq j} v^k(c_k, P_{k,s(k)}^{min}, SP),$$
(18.5)

where $v^k(c_k, P_{k,l}^{min}, SP)$ is the valuation function of v_k, which is 0, if v_k is not on the shortest path SP and $-c_k \cdot P_{k,l}^{min}$ otherwise ($v_{s(k)}$ is the successor of v_k on the SP).

If the node v_j would be on the shortest path without lying, it would stay on it with underdeclared cost or emission power. The payment would stay the same, since both terms in the Equation 18.4 do not depend on the cost of energy and the emission power of v_j. The real cost of v_j would also stay the same, so in this case the node cannot improve its utility by underdeclaring.

If v_j moves itself onto the shortest path by underdeclaring c_j or P_j^{emit}, its utility will be

$$u_j = p_j - c_j \cdot P_{j,s(j)}^{min} = |SP^{-j}| - \sum_{v_k \in \overline{SP}, k \neq j} c_k \cdot P_{k,s(k)}^{min} - c_j \cdot P_{j,s(j)}^{min}$$

$$= |SP^{-j}| - \sum_{v_k \in \overline{SP}} c_k \cdot P_{k,s(k)}^{min} = |SP^{-j}| - |\overline{SP}|,$$
(18.6)

where $|\overline{SP}|$ is the real cost of the path $\overline{SP}$, which would be shortest with underdeclared values. Since $\overline{SP}$ with the real cost and emission powers is not the shortest path, which is SP^{-j}, the value of Equation 18.6 will be negative. Therefore, in this case the node would get negative instead of zero utility if underdeclaring its cost of energy or emission power.

Therefore is is shown, that nodes cannot improve their utility by underdeclaring their parameters.

2. Overdeclaring c_j or P_j^{emit} makes the link seem more expensive. If v_j is not on the shortest path with overdeclared parameters, its utility will be zero, which is not better, than without overdeclaring, since truthful revelation guarantees non-negative utility. The other case is, v_j is still on the shortest path when overdeclaring its values. Again, since the payment does not depend on the declared cost of energy and minimal power (Equation 18.5) and the real cost for v_j is the same, utility of the node v_j does not change even with overdeclared parameters.

Therefore the nodes cannot improve their utility by lying and will declare their cost of energy and emission power truthfully.   □

In the following Lemma it is shown, that a node cannot improve its utility by incorrect declaration of minimum emission power of its predecessor nodes.

Lemma 18.4.8. *Each node v_j correctly computes and declares the minimum emission power $P_{i,j}^{\min}$ of all its predecessors v_i.*

Proof. There are again two cases: v_j could overdeclare or underdeclare $P_{i,j}^{\min}$.

1. If v_j underdeclares the minimum emission power of v_i as $\overline{P_{i,j}^{\min}}$ with $\overline{P_{i,j}^{\min}} < P_{i,j}^{\min}$ it possibly moves itself and v_i onto the shortest path. The node v_i would then use $\overline{P_{i,j}^{\min}}$ as emission power in the data transmission phase. Since this value is smaller than the real minimum emission power, node v_j will not be able to correctly receive the packet from v_i, which also includes the payments. In the case that v_j is not on the shortest path, its utility does not change when underdeclaring $P_{i,j}^{\min}$. Therefore, underdeclaring minimum emission powers of predecessors would not improve the utility of v_j.

2. Overdeclaring the minimum emission power would possibly result in v_j no longer being on the shortest path, in which case the utility of v_j would be zero instead of non-negative value, which is not an improvement. In the other case, when v_j would still be on the shortest path, v_j would receive a smaller payment, since $|\overline{SP}| > |SP|$, thus lowering its utility. If v_j is not on the shortest path when truthfully declaring $P_{i,j}^{\min}$, it will not move onto it, when overdeclaring. Therefore, it would not change its zero utility.

Thus, the nodes cannot improve their utility by manipulating the minimum emission power of their predecessors and will compute and declare it truthfully.

□

The following Lemma shows that nodes cannot cheat by altering parameters of their predecessor nodes in the received packet and rebroadcasting it.

Lemma 18.4.9. *Each node v_j correctly rebroadcasts all edge-weight information contained in route request packets received without alterations or intentional dropping.*

Proof. The node v_j could either overdeclare or underdeclare an edge-weight information $w(v_g, v_h)$ or drop a route discovery packet.

1. Suppose node v_j decreases $w(v_g, v_h)$. Consider following cases:
   - If the edge (v_g, v_h) is on the shortest path before alteration, it is still on it with decreased weight value, but the communication over this edge will be impossible, since the node v_g calculates its emission power $P_{g,h}$ as $w(v_g, v_h)/c_g$ and it will be lower than the real minimum emission power $P_{g,h}^{\min}$.

Note that, if route discovery packets would include c_i and $P^{min}_{i,s(i)}$ values instead of the products $w(v_i, v_{s(i)}) = c_i \cdot P^{min}_{i,s(i)}$, a node v_j on the shortest path could alter the c_i value of node v_i, which is on the shortest path, but is not on the shortest path SP^{-j}, thus increasing its own utility and decreasing the utility of the node v_i, since $|SP^{-j}|$ would then remain the same and $|SP|$ would decrease, which increases the payment to v_j.

- If the edge (v_g, v_h) is not on SP but on SP^{-j} before alteration and node v_j is on SP, decreasing edge weight would either result in smaller payment to v_j, if v_j stays on the shortest path, or SP^{-j} will be the new shortest path, resulting in zero utility for v_j and broken shortest path, since v_g would use too low emission power to send packets to v_h.

- If edge (v_g, v_h) is neither on SP nor on SP^{-j} and v_j is on SP, reducing $w(v_g, v_h)$ will either have no effect, or will cause the selection of a new shortest path, making the utility of v_j zero and as before produce a broken shortest path.

- If node v_j is not on SP and (v_g, v_h) is not on the shortest path, decreasing $w(v_g, v_h)$ would either have no effect or put v_j onto the shortest path, which will be broken, since v_g would again use too low emission power.

Thus, v_j cannot increase its utility by decreasing an edge weight.

2. Now we consider the case, where v_j increases the weight $w(v_g, v_h)$. This could increase its utility only in two cases:

- If node v_j is on SP and (v_g, v_h) is on SP^{-j} before increasing $w(v_g, v_h)$. All nodes on SP^{-j} will forward information truthfully, so the destination node can simply ignore the higher weight information from the cheating node v_j.

- If node v_j is not on SP and (v_g, v_h) is on SP before increase, truthfully reported information from the nodes on SP will reach D and the increased edge weight from the cheating node can be ignored by D.

Thus, v_j cannot increase its utility by increasing an edge weight.

If the node v_j drops packets, information about incoming and outgoing edges of v_j does not reach the destination node D. All other information finds its way to D. The node v_j would then not be on the shortest path, making its utility zero.

Thus, nodes cannot improve their utility by modifying edge weights or dropping packets and will forward all route discovery packets without unallowed modifications. □

Combining cheating possibilities would not help v_j to increase its utility. We consider two cases:

1. v_j is not on the shortest path and will try to move itself onto it. In order to achieve it, it could either try to increase the cost of the true shortest path, which will fail as the nodes on the true SP report their information

truthfully, or v_j could try to decrease the cost of the shortest path containing v_j, which will result in communication failure, if v_j changes edge weights, or in negative utility, if v_j underdeclares own cost or emission power.

2. v_j is on the shortest path and tries to increase its utility. This can be done either by increasing the cost of SP^{-j}, which will fail, since the nodes on SP^{-j} report their information truthfully, or by underdeclaring the edge weights on SP while overdeclaring its own cost, which would again end up in a communication failure, since the first node on the modified edge would not reach its successor.

Thus, it is proved:

Theorem 18.4.10. *Ad hoc-VCG is truthful.*

Overpayment. Ad hoc-VCG is not budget-balanced, i.e. the payments to the nodes are higher, than their actual costs. Anderegg and Eidenbenz showed, that the total overpayment ratio does not exceed $2^{\alpha+1}(c_{\max}/c_{\min})$, where $c_{\max}$ and $c_{\min}$ are the minimum and the maximum cost of energy declared by any network node on the shortest path SP and α is the exponent to the distance with which signal strength is lost.

18.4.2 Truthful Unicast Routing

Another scenario is a wireless network consisting of nodes without a capability to regulate the emission power. In this section we present a truthful mechanism for unicast routing in such networks by Wang and Li [384].

Mechanism Description. This mechanism solves the unicast routing problem in a wireless network consisting of selfish agents, such as PDAs or laptops owned by different users. We assume that wireless nodes have fixed emission power and, consequently, fixed transmission range. That means also, that a node incurs the same cost sending a unit of data to any other node in its transmission range. It is natural in this setting to consider nodes as selfish agents and routing requests as rounds in a game. In each round agents are asked to declare their costs and after computing the least cost path, nodes on this path are paid for each transmitted data unit. We will see that the utility of each individual agent is maximized when it reveals its true cost.

The considered network model is as described in Section 18.3. In addition to this model let $d = (d_0, d_1, \ldots, d_{n-1})$ be the vector of declared costs of all agents and c the type vector (agents' real costs). Since nodes have fixed transmission range, we will consider d as node-bound weights on the graph G.

The output o computed by the mechanism is the least cost path $\Pi(v_i, v_j, d, G)$, connecting the given source and destination nodes v_i and v_j in the node-weighted graph G with node weights d. The payment scheme is VCG-based: the node v_k will be paid nothing, if not on the least cost path; otherwise its payment is

$$p_i^k(d) = w(\Pi(v_i, v_j, d|_{-k}, G \setminus v_k)) - w(\Pi(v_i, v_j, d, G)) + d_k, \qquad (18.7)$$

where $G \setminus v_k$ is induced from G by removing v_k from V and $d|_{-k} = (d_0, d_1, \ldots, d_{k-1}, d_{k+1}, \ldots, d_{n-1})$. $w(\Pi)$ denotes the total cost of the path Π. In other words, payment of the node v_k is its added value to the network plus its declared cost.

This payment scheme falls into the VCG class and since the output also maximizes the social welfare $g(o, d) = \sum_{k \in \Pi(v_i, v_j, d, G)} (-d_k)$, this mechanism belongs to the VCG family and is therefore strategyproof. That means, each participating agent maximizes its utility $p_i^k(d) - c_k$ by revealing its true cost c_k.

It is clear, that the output of the mechanism can be computed with Dijkstra's algorithm in $O(n \log n + m)$ time.

Payment Computing. Centralized payment computation can be done in a straightforward way by calculating each node's payments with Dijkstra's shortest paths algorithm, which would have $O(n^2 \log n + nm)$ time complexity. The authors of this mechanism propose a faster algorithm, which computes all payments with $O(n \log n + m)$ time complexity (see [384] for detailed algorithm description and analysis).

18.5 Multicast in Selfish Networks

There is a fundamental difference between unicast and multicast problems, which causes different algorithmic tractability of optimization problems: an optimal unicast route between two nodes can be found within a polynomial time, while finding an optimal solution for the multicast problem is known to be NP-hard, since it is equivalent to the minimum Steiner tree problem. Therefore it is necessary to work with sub-optimal multicast structures.

18.5.1 Truthful Multicast Routing in Selfish Wireless Networks

In this section we present the mechanism by Wang, Li and Wang ([386]) which deals with the multicast routing problem in wireless networks with selfish agents.

Least Cost Path Tree in Link Weighted Networks. There are some multicast structures, which perform well in praxis and are algorithmically tractable. In [386] three such structures are discussed: least cost path tree (LCPT), minimum spanning tree in two variations and approximated Steiner tree. Two different network models are considered: edge weighted and node weighted graphs. For each model and multicast structure the authors present a strategyproof mechanism. Here we present the mechanism for edge weighted networks with LCPT as multicast structure.

Problem Statement. We consider a wireless network with nodes $V = (v_1, \ldots, v_n)$ and network links $E = (e_1, \ldots, e_m) \subset V^2$. The resulting graph $G = (V, E)$ is edge weighted.

The economic model is as follows: network links are selfish agents with their types being the costs of transmitting a unit data packet. Each link e_i has its private type c_i and declares the cost to be d_i. Declared costs of all links define the vector $d = (d_1, \ldots, d_m)$. By $d|^k x$ we denote the vector $(d_1, \ldots, d_{k-1}, x, d_{k+1}, \ldots, d_m)$.

If we have a set of nodes, participating in a multicast session, the problem is to find the least cost path tree, connecting all of this nodes and to compute payments to all intermediate nodes. Let $Q = \{q_1, \ldots, q_r\}$ be the set of receivers and $s = q_0$ the source of the data. The output o of the mechanism is then the least cost path tree $LCPT(d)$, connecting the source with the receivers.

Mechanism Description. In the first phase, links are reporting their costs to s using the link-state algorithm. The source node then computes the least cost path to each receiver denoted by $LCP(s, q_i, d)$. All this paths form the least cost path tree $LCPT(d)$. The least cost path tree can be computed with Dijkstra's algorithm in $O(n \log n + m)$ time. The next step is to compute the payments to all intermediate nodes.

VCG Is Not Strategyproof on LCPT. It is important to note, that a VCG payment scheme would lead to untruthfulness in the case of LCPT. Payment to an intermediate link k would be

$$p_k(d) = \omega(LCPT(d|^k\infty)) - \omega(LCPT(d)) + d_k, \qquad (18.8)$$

where ω denotes the total weight of the structure.

Figure 18.1 shows a simple graph with a source node s, two receiver nodes q_1 and q_2 and an intermediate node v_3. Edge costs are M for edges sq_1, sq_2 and sv_3 and ε for edges v_3q_1 and v_3q_2. If all edges report their costs truthfully, LCPT is computed as shown in the middle graph. The utility of the link sv_3 is then 0, since it is not on the LCPT. If sv_3 declares its cost as $M - 2\varepsilon$, the LCPT will be computed as shown on the right side of the figure. Its VCG-payment will be $\omega(LCPT(d|^3\infty)) - \omega(LCPT(d|^3(M - 2\varepsilon)) + M - 2\varepsilon = 2M - (M - 2\varepsilon + 2\varepsilon) + M - 2\varepsilon = 2M - 2\varepsilon$. The utility of sv_3 is then $2M - 2\varepsilon - M = M - 2\varepsilon$, which is greater than zero, if $0 < \varepsilon < M/2$. Thus, there is a possibility for agents to increase their own utility by lying about the costs in the VCG payment scheme.

Strategyproof Payment Scheme. For each receiver q_i the least cost path $LCP(s, q_i, d)$ is computed. The payment to an intermediate link e_k on this path in the unicast case is

$$p_k^i(d) = \omega(LCP(s, q_i, d|^k\infty)) - \omega(LCP(s, q_i, d)) + d_k \qquad (18.9)$$

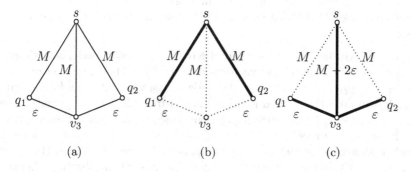

Fig. 18.1. (a) The graph with true costs; (b) the graph with LCPT when costs truthfully reported; (c) the graph with LCPT when sv_3 lies about its costs.

The payment to each intermediate link e in the multicast is then defined as

$$(d) = \max_{i \in} {}^{i}(d), \tag{18.10}$$

Links not on LCPT are paid nothing.

Theorem 18.5.1. *Payment based on (18.9) and (18.10) is truthful.*

Proof. When a link reports its cost truthfully, its utility is non-negative. Since the payment scheme for unicast is truthful, a link e cannot improve its payment ^{i} based on $CP(s, {}_{i}, d)$ and therefore it cannot improve its payment based on (18.10). So this payment scheme is truthful. $\qquad\square$

Computational Complexity. LCPT can be computed in $O(n \log n +)$ time. For each receiver ${}_{i}$, all ^{i} can be computed with the unicast fast VCG payment computation algorithm by Wang and Li from [384] in time $O(n \log n +)$. For all receivers it would take $O(rn \log n + r)$.

18.5.2 General Mechanism Design Rules for Multicast Routing

In the previous section a truthful mechanism has been introduced, which computes the payment scheme for the LCPT multicast structure. The authors of the original paper proposed mechanisms for two other tree based multicast structures: pruning minimum spanning tree and link weighted Steiner tree.

Tree based structures are not the only possibility to approximate the optimal multicast structure, the minimum weight Steiner tree. Structures based on meshes, grids and circles have been proposed in the literature. A natural question arises: is it possible to adapt a given multicast structure to a non-cooperative environment? In other words: is there a truthful payment scheme for a given multicast structure? And, if yes, how can it be found?

In [385] Wang et al. present a necessary and sufficient condition on a structure for the existence of a truthful payment scheme under assumption of non-colluding selfish rational agents in a bi-connected network. Furthermore, they give general rules to design such a scheme and describe a payment sharing model.

Network Model and Economical Model. Network model is as described in Section 18.3. Agents' cost vector is denoted by c. For a given set of receiver nodes $Q = (q_1, \ldots, q_r)$ and a sender node $s = q_0$ we want to find a mechanism which computes the desired multicast structure, calculates payments to relay agents according to a truthful payment scheme and shares these payments fairly among the receivers. The multicast structure may be a tree, a mesh, a grid or some other structure. Agents can be either nodes or links, the graph is then respectively node or edge weighted. Fairness of the sharing scheme will be declared later. The economic model is as described in section 18.2.

Truthful Payment Existence Condition. For a given output function $o(c)$ in a multicast mechanism, the question is, is there a truthful payment scheme? A necessary and sufficient condition is given by the following definition and theorem.

Definition 18.5.2. *An output function $o(c)$ in a multicast routing mechanism satisfies monotone the property (MP) if for every agent a_i and fixed vector c_{-i} following is true: if a_i is selected in the output with cost c_i, then it would also be selected with smaller cost.*

Theorem 18.5.3. *For an output function o in a multicast routing mechanism M exists a payment scheme p such that $M = (o,p)$ is strategyproof iff o satisfies monotone property.*

The proof is given in [385].

The monotone property definition is equivalent to the following: an agent a_i is selected by the mechanism, if there exists a threshold value $\kappa_i(o, c_{-i})$ and the agents' cost is at most this value. This definition can be directly used for the construction of a truthful payment scheme.

We define the payment scheme p as follows: if an agent a_i is selected by the output, it receives $p_i = \kappa_i(o, c_{-i})$; otherwise, its payment is zero. Obviously, such a payment scheme is truthful.

Rules for Finding a Truthful Payment Scheme. Finding the threshold value for each agent solves the problem of finding a truthful payment scheme for a multicast routing mechanism. The question is, how can this value be found? There is no general approach for finding $\kappa_i(o, c_{-i})$, however there are two techniques, which can be applied.

1. *Simple Combination:* We can use this rule if the multicast output o can be represented as a combination of m other outputs $o^1, \ldots, o^m$ of multicast routing mechanisms:

$$o(c) = o^1(c) \vee o^2(c) \ldots \vee o^m(c).$$

If every o^i satisfies the monotone property and $\kappa(o^i, c)$ is the vector of threshold values in o^i, then o also satisfies the monotone property and the threshold value vector for o is

$$\kappa(o, c) = (\max_{1 \leq i \leq m} \kappa_1(o^i, c_{-1}), \ldots, \max_{1 \leq i \leq m} \kappa_n(o^i, c_{-n})).$$

2. *Round-Based Multicast Method:* a method is round-based if the multicast structure is constructed in rounds: in each round some unselected agents are selected, the input problem and the cost profile are modified if necessary.

   Theorem 18.5.5 states a sufficient condition for round-based mechanisms to satisfy the monotone property, which is used in the algorithm 41. The technique for computing threshhold values is given by the algorithm 42.

Input: $G = (V, E, c)$ - graph, Q - receiver set
Output: multicast structure with desired properties
1   $r := 1$
2   $c^{(1)} := c$
3   $Q^{(1)} := Q$
4   **while** *desired property not achieved* **do**
5 | Update the cost vector and the receiver set according to update rule U^r:

$$U^r : o^r \times [c^{(r)}, Q^{(r)}] \rightarrow [c^{(r+1)}, Q^{(r+1)}]$$

6 | where o^r decides, which agents are selected in the round.
7 | $r := r + 1$
8   **end**

Algorithm 41: General structure of round-based multicast method

Definition 18.5.4. *an updating rule U^r is called* crossing-independent *if for any unselected agent:*

- $c_{-i}^{(r+1)}$ *and* $Q^{(r+1)}$ *do not depend on* $c_i^{(r)}$ *and*
- *for fixed* $c_{-i}^{(r)} : d_i^{(r)} \leq c_i^{(r)} \Rightarrow d_i^{(r+1)} \leq c_i^{(r+1)}$

Theorem 18.5.5. *A round-based multicast method o satisfies monotone property if, for every round r, method o^r satisfies monotone property and the update function U^r is crossing-independent.*

Input: Round-based multicast method o, agent a_k
Output: x – payment to a_k

1 $x := \infty$
2 $r := 1$
3 **while** *valid output not found* **do**
4 | Find $\kappa_k(o^r, c^r_{-k})$
5 | $l_r := \kappa_k(o^r, c^r_{-k})$
6 | $l_r := 0$, **if** U **then**
7 | | p
8 | **else**
9 | | d
10 | **end**
11 | ate cost vector and receiver set to obtain $c^{(r+1)}$ and $Q^{(r+1)}$
12 | $r := r + 1$
13 **end**
14 Let $g_i(x)$ denote c^i_k (the cost of agent k in the round i) when the input cost vector (the cost vector before first round) is set to $c|^k x$ (instead of the original c)
15 Find the minimum value for x, which fulfills the following inequalities system: $g_i(x) \geq l_i$, for $1 \leq i \leq t$, where t is the total number of rounds
16 This minimum value is the payment to agent k

Algorithm 42: Payment computation for a round-based method

Fair Payment Sharing Scheme. We consider the payment sharing model among receivers: each receiver $q_i \in Q$ should pay a *sharing* S_i of the total payment to the intermediate agents. Let $P(Q, d) = \sum_k p_k(Q, d)$ denote the total payment to the relay agents and $S_i(Q, d)$ – the sharing of receiver q_i in the scheme S. A scheme S is called *fair* or *reasonable* if it satisfies following conditions:

1. *Nonnegative Sharing:* Each receiver q_i has nonnegative sharing: $S_i(Q, d) \geq 0$.
2. *Cross-Monotone:* For any two receiver sets $R_1 \subseteq R_2$ containing q_i holds: $S_i(R_1, d) \geq S_i(R_2, d)$. In words: growing number of receivers in the network does not increase costs of q_i.
3. *No-Free-Rider:* for each receiver $q_i \in R$ its sharing is not less than $1/|R|$ of its unicast sharing $S_i(q_i, d)$.
4. *Budget Balance:* The payment to all relay agents is shared by all receivers: $P(R, d) = \Sigma_{q_i \in R} S_i(R, d)$.

In [385] authors show that for some multicast topologies and their corresponding truthful payment schemes a fair sharing scheme may not exist.

18.6 Chapter Notes

As we have seen in this chapter, the routing problem in *selfish* wireless networks can be dealt with by means of mechanism design. For the unicast routing problem an optimal solution can be found in polynomial time, while in the multicast case finding optimum is NP-hard. Sub-optimal multicast structures must be considered, which does not allow to use the well known mechanisms of Vickrey-Clarke-Groves family, since this would lead to untruthfulness of the mechanisms' payment scheme. All of the discussed payment schemes are not budget-balanced which means that agents are paid premiums over their actual costs. This is necessary to assure the truthfulness of the schemes.

The applicability of described mechanisms depends on the number of different node owners in the network, battery capacity, processing power, and communication requirements of the nodes. For example, in a typical wireless sensor network with battery-powered nodes and only one owner, it is not reasonable to implement some payment mechanisms in the nodes. It could although be useful to work with variable cost-of-energy parameters and Ad hoc VCG in order to distribute the traffic over the whole network and thus prolongate its life time. This would also give the network good flexibility and scalability.

In a wireless ad hoc network of portable computers, where different persons exchange some data, it makes sense to implement some of the described mechanisms in order to avoid "free riders" and to assure the best possible connectivity of the network. If the data is valuable, it could also make sense to apply some payment scheme to such a network.

In [206] and [385] authors provide a general framework for designing truthful payment schemes for some classes of multicast structures and give criteria for the existence and rules for finding truthful payment scheme for multicast methods. We discussed some rules from this framework in the last section.

In [126] the authors consider the problem of interdomain routing from the point of view of algorithmic mechanism design. In this paper, nodes are treated as selfish agents. The presented mechanism computes routes for all source-destination pairs at a time. A BGP (Border Gateway Protocol) based distributed algorithm for mechanism and payment computation is also presented.

In [385] authors present a distributed algorithm for the payment computation in the multicast case, which is based on the algorithm from [126].

A fast payment computation algorithm for node weighted unicast VCG is presented in [386]. An efficient algorithm for payment computation in edge weighted VCG-based routing mechanisms is also presented in [183]. A comprehensive survey on distributed algorithmic mechanism design is given in [127].

A different approach to deal with selfishness of devices is presented in [353]. Authors define a parameter, called sympathy, which defines the level of selfishness of agents.

An introduction to algorithmic mechanism design is given in [305].

There are still many open questions in the distributed algorithmic mechanism design area: can the computational complexity of the algorithms be reduced? how can the network overhead be minimized? is it possible to reduce the overpayments? We hope, that these and many other questions will be answered in the next few years.

19

Time Synchronization

Marcel Busse and Thilo Streichert

19.1 Introduction

Time synchronization is an essential requirement in actually any distributed system such as sensor networks. Many applications require precise clocks, e.g., for the coordination of wake-up and sleeping times, TDMA schedules, ordering of sensed events, estimation of position information, and so on. The scope of time synchronization is to estimate (i) *packet delay* or *latency*, (ii) the *offset* between clocks, and (iii) the *drift* between clocks.

Packet delay is not easy to obtain but it can be estimated by means of the *round trip time (RTT)* which denotes the time span between issuing a request to another node and getting the response. The RTT includes the time both packets were on their way and additionally the time which passed at the responder side between receiving and answering the request. The two-way packet delay can be derived precisely by subtracting the *idle time* passed at the responder from the RTT. However, it is not easy to tell how it splits up into the delay on the way to the responder and the way back.

The *offset* between two clocks denotes the amount of time one clock is ahead of another or the time they drifted apart as shown in Figure 19.1. If the offset is known, a node can translate events being measured in the time domain of another node into its own time domain, simply by adding or subtracting the clock offset.

If a clock is too fast or too slow by a constant factor, this error is denoted as *clock drift* . The drift can be derived by a linear regression from a larger number of offset measurements, compared against a credible clock. If a clock is more often ahead of real-time than behind it, it runs too fast, otherwise too slow. Note that the global real-time will be denoted with capital letters while the local time will be denoted with small letters in the following.

There are a number of disturbing influences on a message's latency that might lead to asymmetric packet delays due to *non-determinism*. They can be classified as follows [223]:

D. Wagner and R. Wattenhofer (Eds.): Algorithms for Sensor and Ad Hoc Networks, LNCS 4621, pp. 359–380, 2007.

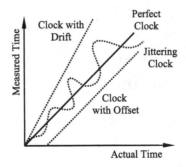

Fig. 19.1. Types of errors that can occur in time measurements. An imaginary perfect clock experiences no deviation between measured time and real-time. A clock with an offset deviates from the actual time only by a constant shift. Exhibiting a drift error means that the clock is either driven too fast or to slow.

- *Sending Time S*: This is the time between the desire to send a packet and the time until it actually reaches the bottom of the network stack, without including access time to the medium.
- *Medium Access Time A*: This is the time a node has to wait until it is allowed to access the medium. Depending on the medium access protocol, the waiting time can be e.g. the time until the node's next time slot is reached or the contention to the medium is won.
- *Propagation Time $P_{A,B}$*: After a message has left a sender A, $P_{A,B}$ denotes the physical propagation time until it reaches the receiver B. Due to the speed of light and comparative short radio ranges, this time is very small.
- *Reception Time R*: Analog to the sending time, this is the time that passes from the message arriving at the transceiver until it reaches the application layer.

There have been different proposals for time synchronization, but mainly they differ in the way they estimate and attempt to correct for these sources of error. RTT-based synchronization estimates packet delay and clock offset by sending a request from one node A to a neighbor B and getting back an answer as depicted in Figure 19.2. The offset and packet delay will be derived from the local time each node measures. However, none of these times has to comply to real-time. Only the drift between the clocks has to be the same since it could not be estimated by the four timestamps. For the packet delay δ and clock offset θ we then get

$$\delta = \frac{(t_4 - t_1) - (t_3 - t_2)}{2}, \tag{19.1}$$

$$\theta = \frac{t_2 - (t_1 + \delta) + t_3 - (t_4 - \delta)}{2} = \frac{(t_2 - t_1) + (t_3 - t_4)}{2}. \tag{19.2}$$

The round trip time $t_4 - t_1$ denotes the time for the whole communication. It is diminished by the idle time $t_3 - t_2$ at the receiver side to obtain the time

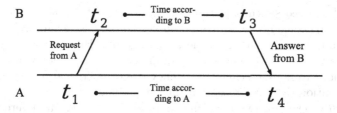

Fig. 19.2. Node A sends a request to node B at time t_1. The packet takes some time until it reaches B at time t_2 according to its clock. After the idle time $t_3 - t_2$, B responds with the request. It will finally arrive at time t_4 according to the A's clock.

both packets actually were on their way. We denoted this time as the two-way packet delay. The single packet delay δ can, unfortunately, only be derived by an estimate as being half as long as the two-way delay. The clock offset is obtained by summing up the difference between B's clock to A's clock twice and dividing it by 2. Here, it is important to see that the packet delay between both nodes is canceled out.

The remainder of this chapter is as follows: In the next section, we present different time synchronization approaches that address sender-receiver as well as receiver-receiver synchronization. We will also have a look at estimated time intervals indicating when an event happened in time. Section 19.3 presents synchronization algorithms in the presence of faulty nodes. These faulty nodes report different time values to other nodes in the network and disturb the synchronization process. Such an inconsistent behavior is called Byzantine error. Section 19.4 shows theoretical bounds in time synchronization and its effect to temporal event ordering. In the last section, we will address the problem of gradient clock synchronization.

19.2 Time Synchronization Approaches

In this section, we will consider three time synchronization schemes, namely *Reference-Broadcast Synchronization (RBS)*, the *Time-Sync Protocol for Sensor Networks (TPSN)*, and a synchronization scheme for temporal relationships that estimates time intervals for events regarding their occurrence.

19.2.1 Reference-Broadcast Synchronization

In Reference-Broadcast Synchronization (RBS) [111], a sender synchronizes a set of receivers with one another by broadcasting reference beacons. This is contrary to traditional models where a sender synchronizes with one receiver by estimating the RTT. RBS uses the beacon's arrival time as a point of reference that can later be used for estimating the clock offsets among recipients.

Unlike RTT-based synchronization, the message contains no timestamp and can be any message broadcast to a group of nodes, e.g., even a RTS or CTS, or any kind of control packet. The advantage of RBS is that all recipients experience the same sending and medium access time, removing the main contributions to non-deterministic latency from the critical path. Thus, only the propagation time and reception time still remain.

The offset error of two recipient nodes A and B and one reference node S can be calculated from $\theta = t_2 - t_3$ with t_2 and t_3 being the arrival time of a reference broadcast sent at time t_1, at which t_i represents the time as it is perceived by node S. Using the classification from the previous section, t_2 and t_3 are

$$t_2 = t_1 + S_S + A_S + P_{S,A} + R_A$$
$$t_3 = t_3 + S_S + A_S + P_{S,B} + R_B.$$

Thus, the clock offset is

$$\theta = t_2 - t_3 = t_1 + S_S + A_S + P_{S,A} + R_A - (t_3 + S_S + A_S + P_{S,B} + R_B)$$
$$= (P_{S,A} - P_{S,B}) + (R_B - R_A). \tag{19.3}$$

Equation 19.3 shows that RBS is only sensitive to the *difference* in propagation and reception time, thus deterministic latencies in processing and reception are irrelevant. All latencies at the sender side cancel off themselves at the receiver. However, the signal propagation times might not be the same since both receivers may have different distances to the synchronizer.

Thus, nodes in the network can synchronize to one another in the following way: After receiving a reference broadcast, each node records its reception time. This can be done at the interrupt time when the packet is received to reduce the reception time R as much as possible. Then, the receivers can exchange their observations and calculate their clock offset to one another.

By sending more than one reference broadcast, the clock offset estimation can be statistically improved. Using m reference broadcasts, a node i can calculate its offset $\theta_{i,j}$ to another node j by computing

$$\theta_{i,j} := \frac{1}{m} \sum_{k=1}^{m} (t_{j,k} - t_{i,k}) \tag{19.4}$$

with $t_{r,b}$ be the arrival time of broadcast b at node r.

However, so far we did not take clock drifts into account. In the presence of different drifts, clocks drift apart immediately after the synchronization process. We can tackle this problem by periodically repeating time synchronization and performing a *least-square linear regression*, instead of averaging several observation points as shown in Equation 19.4. Then, the line's slope and intercept provide an estimate for clock drift, resp. offset, with respect to the remote node's local time domain.

19.2.2 Time-Sync Protocol for Sensor Networks

The Time-Sync Protocol for Sensor Networks (TPSN) [143] is based on RTT synchronization and works in two steps. First, an *initialization phase* is carried out where a hierarchical network structure is established. Using this hierarchy, nodes then perform pairwise synchronization to achieve a global timescale in the *synchronization phase*. In contrast to the receiver-receiver synchronization in RBS, TPSN employs traditional sender-receiver synchronization.

Initialization Phase. The initialization phase is originated by the root node that has the lowest level. It sends a *level discovery* packet including its own level to all neighbors within its radio range. All nodes receiving this announcement adopt their level after increasing it by one. Then, every node starts a random timer. After a node's timer expires, the node will send a level discovery packet itself which is directed to all nodes that did not yet hear an announcement. All other nodes will ignore it. At the end of the flooding process, every node should obtain a level with a decreasing number of hops towards the root node.

Note that it is not guaranteed that all nodes have obtained its level because some messages might be lost, e.g., due to packet collisions. A node with no valid level can send a *level request* packet which is answered by every node that has already a level assigned. The requesting node will then choose the smallest level and increase it by 1.

Synchronization Phase. The root node issues a *time sync* packet which triggers a random timer on all level 1 nodes. On expiration, the node with the shortest timer asks the root for synchronization using a *synchronization pulse*. Synchronization always takes place between a child and a parent node.

After the synchronization pulse arrived at the parent node it will reply with an *acknowledgment*. Now the requesting node knows the round trip time and thus can estimate the single packet delay in the 1-hop communication by Equation 19.1.

Nodes on a higher level that overhear the synchronization pulse start a random timer themselves. It has to be chosen large enough so that the synchronization between their parent and grand parent node can be finished. On expiration of the first timer on the second level the respective node will also send a synchronization pulse to its father node to trigger its own synchronization, and so on.

Synchronization pulses have the side effect that a node can recognize that it did not receive one and that its child nodes may not be synchronized. In this case, a parent node will simply send a time sync packet itself to trigger its children. If a node does not get back an acknowledgment to its synchronization pulse, even after some retransmissions, the node will assume it has lost all its neighbors on the upper level. It then broadcasts a level request packet as in the initialization phase.

As for RBS, we can now investigate which kinds of errors influence the clock offset calculation most and which drop out. Again, consider two nodes, a sender A and a receiver B. With t_i being the time as it is perceived by node A, timestamps t_2 and t_4 are

$$t_2 = t_1 + S_A + A_A + P_{A,B} + R_B$$
$$t_4 = t_3 + S_B + A_B + P_{B,A} + R_A.$$

according to Figure 19.2. Equation 19.1 then leads to

$$\theta = \frac{(S_A - S_B) + (A_A - A_B) + (P_{A,B} - P_{B,A}) + (R_B - R_A)}{2}. \qquad (19.5)$$

In contrast to RBS, we can assume here that the signal propagation time between nodes A and B is approximately the same for both directions, so they compensate each other:

$$\theta = \frac{(S_A - S_B) + (A_A - A_B) + (R_B - R_A)}{2}. \qquad (19.6)$$

In order to minimize the sources of errors, TPSN exploits the possibility of time stamping packets at the MAC layer. Thus sending and access time reduce to the time required for transmitting the whole packet. Since it is assumable that time will be deterministic for both nodes, the difference in sending and access time tends to vanish. Also reception time can be minimized by time stamping at the MAC layer once the first bit was detected. As in RBS, periodical synchronization is necessary to account for clock drifts. Using the same least-square linear regression, clock offset as well as clock drift can be calculated over time.

The synchronization error was evaluated under real experiments. The results in Table 19.9 show the magnitude of the synchronization error averaged over 200 runs. Compared to the error in the clock offset calculation for RBS (see Equation 19.3), TPSN performs approximately twice as good as RBS. This is mostly due to MAC layer time stamping.

The performance of TPSN in a multihop scenario is depicted in Table 19.10. Again, only the magnitude of the synchronization error is shown. With an increasing number of hops, the error slightly increases, too. However, the error does not blow up with hop distance. It almost seems to be constant beyond three hops. The authors assume that this could be due to the probability distribution of the synchronization error [143]. They are currently working on further evaluations and an analytical model for the multihop error.

19.2.3 Estimation of Time Intervals

Kay Römer suggests an approach for estimating time intervals for events in his work on *Time Synchronization in Ad Hoc Networks* [334]. The author argues

	TPSN	RBS
Average error [μs]	16.9	29.13
Worst case error [μs]	44	93
Best case error [μs]	0	0
Percentage of time error is less than or equal to average error	64	53

Table 19.9. Statistics of the synchronization error for TPSN and RBS. TPSN roughly gives a 2x better performance than RBS.

	1-hop	2-hop	3-hop	4-hop
Average error [μs]	17.61	20.91	23.23	21.44
Worst case error [μs]	45.2	51.6	66.8	64
Best case error [μs]	0	0	2.8	0
Percentage of time error is less than or equal to average error	62	57	63	54

Table 19.10. Statistics of the synchronization error for TPSN over multihop. With an increasing number of hops, the synchronization error slightly increases, too.

that keeping clocks synchronized all the time is expensive in terms of number of packet transmissions since it is not always obvious when synchronized clocks will actually be needed. Moreover, networks might be temporary disconnected due to mobility, especially in sparse ad hoc networks.

The idea of Römer's algorithm is to transform timestamps generated by unsynchronized local clocks to the receiver's time domain as nodes exchange messages. However, passing messages always introduces latencies which can never be perfectly estimated, and clocks always exhibit a certain offset and drift, even shortly after a synchronization which is inherently not perfect as well. Thus, instead of considering certain points in time, time intervals are used to give a lower and upper bound of the exact time the event occurred.

In order to account for the inaccuracy of a clock, the clock has to be known in advance. We assume that prior measurements of node k showed an inaccuracy factor of $(1 \pm \rho_k)$, i.e.,

$$1 - \rho_k \leq \frac{\triangle c}{\triangle T} \leq 1 + \rho_k \tag{19.7}$$

for a time difference $\triangle c$ in node k's local time resp. $\triangle T$ in real-time. Thus, a real-time difference $\triangle T$ can be estimated by $[\frac{\triangle c}{1+\rho}, \frac{\triangle c}{1-\rho}]$ with respect to a local time domain. Also vice versa, a time difference in local time $\triangle c$ can be estimated by $[\triangle T(1 - \rho), \triangle T(1 + \rho)]$ with respect to real-time.

Now consider N nodes arranged in-line with node 1 detecting an event whose point in time is passed towards node N. Let r_i be the local time when node i receives the event message, s_i being its local time when it forwards the message, ρ_i its clock drift, and rtt_i and $idle_i$ the round trip time resp. waiting time between node i and its successor $i+1$. An estimate for the message delay between node 1 and 2 would then be $0 \leq \delta \leq rtt_i - idle_i \frac{1-\rho_2}{1+\rho_1}$ in terms of node 2's clock, assuming $idle_i$ is based on the clock of node 1 and rtt_i based on node 2's clock.

Since node 1 sensed the event itself and will not receive an event message, we assume r_1 to be the time when the event occurred. Formally, it will be estimated by $[r_1, r_1]$. A first estimate for node 2 would be

$$\left[r_2 - (s_1 - r_1)\frac{1+\rho_2}{1-\rho_1}, r_2 - (s_1 - r_1)\frac{1-\rho_2}{1+\rho_1} \right] \tag{19.8}$$

transforming the time $(s_1 - r_1)$ the message was stored at node 2 into its local time domain. Considering the upper bound for the message delay, a more accurate estimate would be

$$\left[r_2 - (s_1 - r_1)\frac{1+\rho_2}{1-\rho_1} - \left(rtt_1 - idle_1\frac{1-\rho_2}{1+\rho_1} \right), r_2 - (s_1 - r_1)\frac{1-\rho_2}{1+\rho_1} \right] \tag{19.9}$$

In the same way, node 3 estimates the time span the event occurred as

$$\left[r_3 - (s_1 - r_1)\frac{1+\rho_3}{1-\rho_1} - (s_2 - r_2)\frac{1+\rho_3}{1-\rho_2} \right.$$
$$- \left(rtt_1\frac{1+\rho_3}{1-\rho_2} - idle_1\frac{1-\rho_3}{1+\rho_1} \right) - \left(rtt_2 - idle_2\frac{1-\rho_3}{1+\rho_2} \right),$$
$$\left. r_3 - (s_1 - r_1)\frac{1-\rho_3}{1+\rho_1} - (s_2 - r_2)\frac{1-\rho_3}{1+\rho_2} \right] \tag{19.10}$$

Note that rtt_1 is related to node 2's time domain and must therefore be transformed to the local time of node 3. Continuing that calculation, the interval estimate of node N is then

$$\left[r_N - (1 + \rho_N) \sum_{i=1}^{N-1} \frac{s_i - r_i + rtt_{i-1}}{1 - \rho_i} - rtt_{N-1} + (1 - \rho_N) \sum_{i=1}^{N-1} \frac{idle_i}{1 + \rho_i}, \right.$$
$$\left. r_N - (1 - \rho_N) \sum_{i=1}^{N-1} \frac{s_i - r_i}{1 + \rho_i} \right] \tag{19.11}$$

with $rtt_0 = 0$.

Having time intervals on hand, it is interesting to investigate the temporal order of events expressed by these time spans. Figure 19.3 shows two overlapping events $E_1 \in [T_1, T_2]$ and $E_2 \in [T_3, T_4]$. If $T_2 < T_3$, then E_1 has certainly happened before E_2. If $T_1 > T_4$, then E_1 has happened after E_2. For all remaining cases only a probability can be derived. For example, assuming uniform distributions of E_1 and E_2 within $[T_1, T_2]$ resp. $[T_3, T_4]$, the probability that E_1 happened before E_2 would be

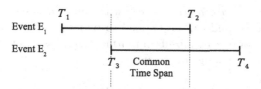

Fig. 19.3. If only fuzzy time intervals can be determined for events, it is not always clear in which order they emerged. In the case of overlapping intervals, only the probability of a specific order can be determined.

$$
\begin{aligned}
P(E_1 < E_2) &= 1 - P(E_1 > E_2) \\
&= 1 - \frac{1}{2} \cdot P(E_1 > T_3) \cdot P(E_2 < T_2) \\
&= 1 - \frac{(T_2 - T_3)(T_2 - T_3)}{2(T_2 - T_1)(T_4 - T_3)}.
\end{aligned}
\tag{19.12}
$$

Furthermore, we can analyze whether two events are at most X time units apart. If the largest lower bound $\max(T_1, T_3)$ is still further apart than X time units from the smallest upper bound $\min(T_2, T_4)$, then the second event happened X time units after the first. On the other hand, if the minimum $\min(T_1, T_3)$ of all lower bounds is not further apart than X time units from the maximum $\max(T_2, T_4)$ of all upper bounds, both events have no chance to be further apart than X time units. Otherwise, only a probability can again be derived.

19.3 Synchronizing Clocks in the Presence of Faults

In the preceding sections, algorithms for synchronizing drifting or jittering clocks or clocks with an offset have been presented. Of course, a drift or a jitter can be seen as a malicious behavior of a clock, which can be treated by exchanging time values between nodes and computing some kind of average, such that all nodes' clocks agree with a certain precision. Unfortunately, this mechanism does not work in the presence of nodes that are faulty and send different time values to different nodes. Such a deceitful incorrect reporting of local time values or inconsistent contradictory behavior is called a *Byzantine error* [248]. In Section 19.3.1 and 19.3.2, we will present two algorithms by Pease, Shostak and Lamport [311, 248, 247] belonging to the class of *Interactive Consistency* algorithms. Another algorithm proposed by Kopetz et al. [222], called *Fault-Tolerant-Average* algorithm, ignores clocks which are outside certain bounds of inconsistency. This algorithm will be described in Section 19.3.3.

The general problem of reaching *consensus* in networks with reliable point-to-point communication has been extensively investigated throughout the last years. Unreliable *failure detectors* [68] with quorum-based algorithms [298] have been proposed and can be applied to many practical problems such as atomic broadcast or decentralized commitment of transactions that be reduced to the problem of reaching consensus.

19.3.1 Interactive Consistency Algorithm

For Interactive Consistency Algorithms, it is presumed to have a fail-safe point-to-point communication with negligible delay. The sender of a message is always identifiable by the receiver of a message, and every node n_i has a private clock value that it wants to synchronize with the other clocks by exchanging this value. Also, it is important that each node in the network is able to directly send messages to every other node. The total number of nodes n_i in such a network is N, such that $1 \leq i \leq N$. Presuming that no more than M nodes are known to be faulty, the following algorithm works for $N \geq 3M + 1$ [34]:

1. Set $d = 1$, such that the node n_1 becomes initially the node n_d which distributes its time value in this round.
2. Call S(M).
3. If $d \neq N$, set $d := d + 1$ and goto step 2, such that the next node n_d can distribute its time value. Otherwise, quit distributing time values.

Algorithm for $S(0)$:

1. Node n_d sends time value to every other node n_i
2. Each node n_i stores the value he receives from the node n_d, or stores a random time value if he receives no value.

Algorithm for $S(M)$, $M > 0$:

1. Node n_d sends its time value to every other node $n_i \neq n_d$
2. For each $i = 1, \ldots, n$ do: Let v_i be the value node n_i received from node n_d. If no value is received from node n_d, node n_i uses a random time value. Then, set $n_d := n_i$ and call the algorithm S(M-1) recursively.
3. For each i, and $j \neq i$, let v_j be the value node n_i received from node n_j in step 2. If no value is received, v_j will be assigned with a random value. Node n_i uses the value $majority(v_1, \ldots, v_{N-1})$. Note that due to the recursion the majority is determined first with the values of $S(0)$ together with the values of $S(1)$. These majority values are considered to be the values v_i of recursion $S(1)$.

The majority function in the last step of the algorithm $S(M)$ returns $majority(v_1, \ldots, v_{N-1})$ for a given set of $N - 1$ time values v_i.

With the described algorithm $S(M)$ we can distribute the clock values of N clocks in a network which is fault tolerant against M Byzantine faults.

The following example presents a scenario where $N = 4$ nodes, of which one is faulty (M=1) synchronize their clocks with each other.

---

[34] In [311, 248, 247] it is proven that this constraint has to be fulfilled.

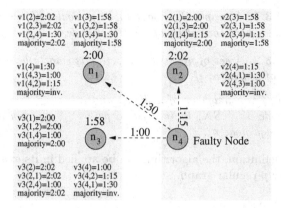

Fig. 19.4. A network with four nodes $n_1, \ldots, n_4$, of which node n_4 is malicious. Then, the Interactive Consistency Algorithm exchanges the time values according to the v_i. This notation $v_i(j, k)$ means that node n_i received the value from n_k and n_k received the value from n_j before.

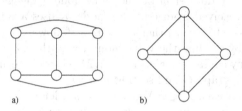

Fig. 19.5. While the graph in a) is 3-regular, the graph in b) is not regular.

Example. In Figure 19.4 node n_1 starts being the node n that distributes its values as described in step 2 of algorithm $S(1)$. Then, each other node sends its value received from node n_1 to all other nodes. This is done by executing algorithm $S(0)$. Now, each node can derive the actual time value of node n_1 by determining the majority of the values received in recursion $S(0)$ and $S(1)$. By looping the steps 1 to 3 of $S(\ )$ each node lets all other nodes know its own time value. Note that the notation $v_i(\ , k)$ means that node n_i received the value v_i from n and n received the value from n_j.

In this example, node n_4 is the faulty node. During the voting step 4 all other nodes recognize that no majority exists. So, a default value *inv.* standing for invalid is assumed for this node. In this small example, it is already recognizable that a huge number of messages has to be sent

In the beginning of this section, we assumed that a node can send messages directly to every other node. Due to this restriction the algorithm would not be applicable. Therefore, a more general type of graph, with nodes representing network nodes and arcs representing communication links, is considered which has to be *-regular*. See Figure 19.5 for an example on -regular graphs.

Definition 19.3.1. *A graph G is said to be p-regular if every node has a regular set of neighbors consisting of p distinct nodes.*

Definition 19.3.2. *A set of nodes $n_{i,1}, \ldots, n_{i,p}$ is said to be a regular set of neighbors SN of a node n_i if*

i.   *each node $n_{i,j}$ is a neighbor of node n_i, and*
ii. *for every node $n_k \notin SN$, there exists a path $P_{n_{i,j}, n_k}$ from $n_{i,j}$ to n_k not passing through node n_i.*

Fulfilling this definition, the algorithm can be applied in its extended version $S(M, p)$ for any p-regular graph.

Algorithm for $S(M, p)$:

1. Choose a regular set of neighbors SN of the node n_d consisting of p nodes $n_i \neq n_d$.
2. Node n_d sends its time value to every other node $n_i \in SN$.
3. For each $n_i \in SN$ do: Let v_i be the value node n_i received from node n_d. If no value is received from node n_d, node n_i uses a random time value. Node n_i sends v_i to every other node n_k as follows:
   i.   If $M = 1$, by sending the value along the path $P_{i,k}$.
   ii. If $M > 1$, by calling the algorithm $S(M-1, p-1)$ recursively, removing n_d from the graph G and setting $n_d := n_i$.
4. For each n_k and each $n_i \in SN$ with $n_i \neq n_k$, let v_i be the value node n_k received from node n_i in step 4. If no value is received from node n_i, n_k uses a random time value. Node n_k uses the value $majority(v_{n_{i,1}, \ldots, n_{i,p}})$, where $SN = n_{i,1}, \ldots, n_{i,p}$.

It has been proven in [248] that the algorithm $S(M, 3M)$ solves the Byzantine fault problem, which is still a strong constraint to the connectivity of a network or graph, respectively.

19.3.2 Interactive Consistency Algorithm with Authentication

The Interactive Consistency Algorithm of the last section assumes that a node may lie about its own time values and about the time values it received from other nodes, which makes the solving of the Byzantine fault problem complex. Restricting this behavior in such a way that a faulty node may lie about its own value but forwards the values it correctly received, simplifies the problem. In order to assure that the faulty node did not modify any received message, *authenticators* are appended to each message. A so-called authenticator A is created only by the originator of a message. It is calculated each time value v_i of node n_j by using, e.g., cryptographic algorithms: $A_{n_j}[v_i]$. For each other node n_k it must be highly improbable that it can produce a message while it must be easy for another node n_k to decode $A_{n_j}[v_i]$.

In summary, a system with the following requirements is assumed:

- Every message that is sent is delivered correctly.
- The receiver of a message knows who sent it.
- The absence of a message can be detected.
- A correctly signed message with authenticators cannot be altered without detection.

Lamport et al. [248] propose an algorithm where each node signs a message that it received and then sends it to other nodes. So, a message consists of a value v_i and a certain number of signs or authenticators, respectively, after it has passed the same number of nodes. This will be denoted with $v : n_i : n_j$, if node n_i sends value v to node n_j signing and sending the value again.

All different received values will be stored in a list called V_i. A function $choice(V_i)$ will be applied to this list and returns one single value v. The definition of this function is arbitrary and could be that $choice(V_i)$ returns the median element of V_i. If the function is called with an empty set, it returns a default value. Note that the list V_i contains each time value only once, thus a majority function cannot be applied. The synchronization algorithm with authenticators $SA(M)$ works as follows:

Algorithm for $SA(M)$, $M > 0$:

1. Set $d := 1$ such that node $n_d = n_1$ will be the node that distributes its clock value in this round.
2. Initialize all nodes' $V_i = \emptyset$.
3. Node n_d signs and sends its time value to every other node $n_i \neq n_d$.
4. For each $i = 1, \ldots, N$ with $i \neq d$ do:
   i. If node n_i receives a message of the form $v : n_d$ and it has not yet received any time value, then
      · it sets $V_i = v$ and
      · it sends the message $v : n_d : n_i$ to every other node.
   ii. If node n_i receives a message of the form $v : n_d : n_j : \cdots : n_k$ and v is not in the set V_i, then
      · it adds v to V_i and
      · if $k < M$, then it sends the message $v : n_d : n_j : \cdots : n_k : n_i$ to every node other then $n_j : \cdots : n_k$
5. For each $i = 1, \ldots, N$ with $i \neq d$ do: If node n_i will receive no more messages, call the function $v = choice(V_i)$.
6. If $d \neq N$, set $d := d + 1$ and goto step 2. such that the next node n_d can distribute its clock value. Otherwise, quit distributing time values.

In step 4 it is said that the algorithm terminates as soon as a node does not receive any more messages. Two possibilities exist to determine this termination criterion. If, within a certain time, no new messages arrive, a time-out will stop the algorithm, or if the sequences $n_d : n_j : \cdots : n_k$ are stored, it can be derived if further messages can be expected or not. In the second strategy, it is important that a node may report that it will not send any more messages.

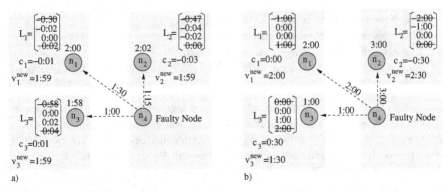

Fig. 19.6. The Fault-Tolerant-Average algorithm records the time differences between the own clock and the received time values in a sorted list L and deletes the M first and last elements. Here, one fault should be tolerated $M = 1$. With the remaining differences in L_i, each node determines its correction value c_i and its new time v_i^{new}.

As shown in the previous section, it is possible to apply this class of algorithm to more general graphs than fully connected once. If node n signs and sends messages only to its neighboring nodes in step 3 and if node n_i only sends the message to every neighboring node not among the n_j in step 4.ii, then the algorithm S $(N-2)$ can be solved.

19.3.3 Fault-Tolerant-Average Algorithm

A system with a total number of N nodes, each with its own clock that tolerates malicious nodes or clocks, respectively, can be synchronized with a so-called Fault-Tolerant-Average (FTA) algorithm. This algorithm requires only one round for message exchanges and, therefore, reduces the number of messages drastically compared to the class of Interactive Consistency Algorithms.

In the first step of the algorithm, each node receives the local time values of all other nodes' clocks. Then, the differences between the own time value and all $N - 1$ received time values are determined and sorted in a list . The actual node itself inserts the difference zero, so that N entries are in the list. Under the assumption that a faulty node sends either too small or too large time values, the smallest and highest differences are deleted from this list. Calculating the average of the $N - 2$ remaining differences leads to the correction value of the node's clock.

Example. Figure 19.6a) shows the same network as presented in the course of the Interactive Consistency Algorithm. Again, four nodes are considered of which one, (n_4), is faulty. Each node n_i records the time values of the other nodes and stores the differences between the own and the time of the other nodes in the sorted List $_i$. Then the highest and lowest values in the list are removed and the average is determined which forms the correction term c_i. Correcting the actual time with c_i leads to the new time v_i^{new}.

Note that the presented example is a best case scenario, where the value of the erroneous clock is always at the boarder of the list and thus removed from the list. If, however, the value of the faulty clock is not ignored as it happens in the following example, the convergence to the same time base between the nodes might be worse.

Example. Figure 19.6b) presents again four nodes of which one (n_4) is faulty. But here, the faulty node sends time values which are within a certain precision interval, such that they are respected for calculating the correction term c_i. As we can see in this figure, the maximal offset between two non faulty clocks is still about one hour after the time correction.

par For analyzing the accuracy of this algorithm, we have to define the term *precision* Π which is the maximal difference between two clocks in an ensemble of non faulty nodes that might occur. Therefore, all differences calculated in the FTA algorithm are within a certain precision window. If a faulty node reports a time value that is close to the upper bound of the precision window to one node and a time value that is close to the lower bound to another node, it can displace another non faulty clock value. This worst case is presented in Figure 19.7.

Here, average values of (i) $4\Pi/5$ and (ii) $3\Pi/5$ are determined which lead to an error of $\Pi/5$. In general, this error E_{byz}, caused by M Byzantine clocks in a network with N nodes is:

$$E_{byz} = \frac{M \cdot \Pi}{(N - 2M)}. \qquad (19.13)$$

Respecting a jitter of the synchronization messages, the offset directly after synchronization is

$$\Phi = E_{byz} + \epsilon = \frac{M \cdot \Pi}{(N - 2M)} + \epsilon. \qquad (19.14)$$

Due to the drift of a clock θ and the time interval R_{int} between two resynchronizations, the precision will be

$$\Pi = \Phi + 2 \cdot \theta + R_{int} = \frac{M \cdot \Pi}{(N - 2M)} + \epsilon + 2 \cdot \theta + R_{int}$$

$$\Rightarrow \Pi = (\epsilon + 2 \cdot \theta + R_{int}) \cdot \frac{N - 2M}{N - 3M}. \qquad (19.15)$$

19.4 Theoretical Bounds for Clock Synchronization and Ordering of Events

While the preceding sections presented algorithms for synchronizing clocks, this section shows a theoretical bound for clock synchronization and its effect

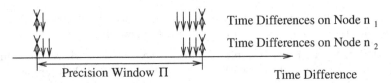

Time Differences on Node n_1

Time Differences on Node n_2

Precision Window Π Time Difference

Fig. 19.7. A worst case scenario is shown, where all differences are either at the upper bound or at the lower bound of a precision interval Π. The time difference with the malicious node and node n_1 is at the upper bound and with node n_2 at the lower bound. Thus, the correction term will be affected drastically.

to *temporal ordering* of events. Interestingly, we will see that only with a certain distance in time, a temporal order of two events can be determined.

19.4.1 A Lower Bound for Clock Synchronization

Two different cases exist where clock synchronization is important: (i) at start-up, when all clocks have initial arbitrary values and (ii) during runtime, when clocks are drifting or new nodes are integrated in a network. For the determination of a lower bound for clock synchronization, a somehow optimal static system is assumed with full connected faultless nodes having no drifting clocks. Then, it can be proven that after the initial synchronization, a difference of $= \kappa(1 - 1/N)$ remains, where N is the number of nodes and κ is the jitter or uncertainty in message delivery time.

As defined in the preceding sections, v_i is the local time of node n_i which is a function of real-time T:

$$v_i(T) = T - \quad_i. \tag{19.16}$$

Figure 19.8 shows the connection between real-time T and the local time v_i, e.g., at local time $v_1 = 0$ of node n_1 the real-time value T is   .

Above,   has been defined as the maximal difference of two local values in real-time. If we assume that between two nodes' clocks n_i, n_j the maximal difference of   can occur, we can state that the shift between n_1 and n has to be less than   :

$$\geq \quad -\ _1$$
$$\Leftrightarrow \quad \leq\ _1 + . \tag{19.17}$$

Now, we will consider the effect of the jitter to the local time values v_i. In Figure 19.9, a system with four nodes is presented. Each node sends and receives messages at a certain local time v_i. Due to a jitter κ, the message delivery time may vary between a lower bound μ and an upper bound $\mu + \kappa$. Assume that all nodes start at the same real-time T ($_i = 0$) and all messages received by a node with a lower node number n_i have the delivery time of μ and all other messages have the delivery time $\mu + \kappa$. Then, we can start node

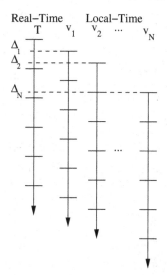

Fig. 19.8. The connection between the local time values v_i and the real-time T. The local time can be seen as a function of real-time: $v_i(T) = T - \Delta_i$.

n_1 κ earlier without any effect on the local send and receive time values of the messages (see Figure 19.9b). Thus, a shifting of a node's time base by κ cannot be recognized by the other nodes, because all events occur at the same local time values v_i. The same holds for Figure 19.9c and 19.9d. If these four pictures present synchronization processes, we can state that all four processes look the same from each node's local point of view. Hence, the correction term of each local clock is the same in all synchronization processes.

Inductively, we can start with Figure 19.9b in which $\quad_{i-1}$ with $i = 2$ is started κ earlier such that $\quad_{i-1}$ in Figure 19.9a equals $\quad_{i-1} + \kappa$ in Figure 19.9b. Since the values $\quad_i$ and $\quad_{i-1}$ in all figures also have to be within the precision , it can be stated:

$$_{i-1} + \kappa - \quad_i \leq$$
$$\Leftrightarrow \quad_{i-1} \leq \quad_i - \kappa + . \tag{19.18}$$

Accumulating all N inequalities of Equations 19.17 and Equations 19.18 leads to

$$\sum_{i=1} \quad_i \leq \sum_{i=1} \quad_i + N \quad - (N-1)\kappa. \tag{19.19}$$

Subtracting the accumulated real-time values $\quad_i$ and dividing by N results in the maximal accuracy a synchronization algorithm is able to reach:

$$= \kappa(1 - 1/N). \tag{19.20}$$

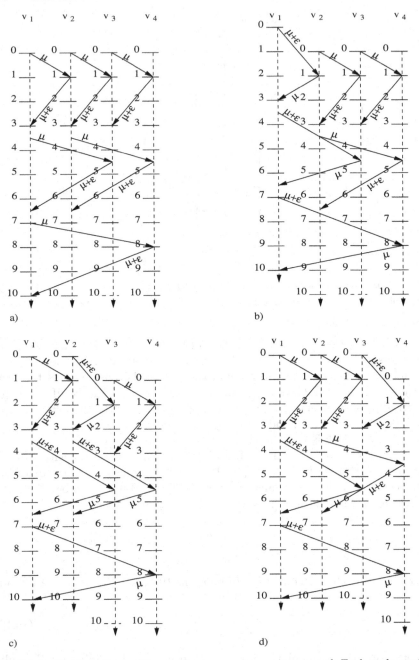

Fig. 19.9. The time lines of four nodes in a system are presented. Each node sends and receives messages at certain instances of time. Due to a jitter ε it is possible to shift the time lines by ε without affecting any time instant (see b, c, d). Assuming, each figure a to c. is a synchronization run, no node would recognize the difference between them.

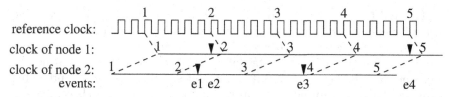

Fig. 19.10. Due to rounding errors and slightly shifted time bases, global times-tamps of events have to differ by at least two. Otherwise a correct order of events cannot be determined.

19.4.2 Precision in Ordering of Events

In Section 19.2, we presented an algorithm which is based on the estimation of time intervals of events. Here, we want to emphasize the restrictions on temporal event ordering again. Due to the impossibility of synchronizing clocks perfectly, a so called *reasonableness condition* [222] ensures that two times-tamps $n_i(e)$ for one event e detected at two different nodes n_i, n_j differ by at most one:

$$| n_i(e) - n_j(e)| \leq 1. \tag{19.21}$$

The difference of one timestamp has an interesting effect on events and their temporal ordering. In Figure 19.10, we see two nodes with their shifted time bases. Node n_2 is a little bit ahead of node n_1, but fulfills the reasonableness condition. It is presumed that the reference clock and the two nodes can trigger events at each time tick or rising edge, respectively. Now, event e_1 occurs at node n_2 at time $n_2(e_1) = 2$ and e_2 occurs at node n_1 at time $n_1(e_2) = 1$. Although the events did not happen simultaneously, it is not possible to derive the correct temporal order. The next two events in Figure 19.10 occur at $n_2(e_3) = 3$ and $n_1(e_4) = 4$ at nodes n_2 and n_1. In this case, the correct temporal order can be derived from the timestamps. Thus, the correct temporal order of two consecutive events cannot be precisely determined if the difference of the referring timestamps is less than two.

19.5 Gradient Clock Synchronization

Up to now, we assumed the synchronization of distributed clocks as being dependent from the topology of a network but being independent from the positioning of the sensing nodes. As an example, suppose that two nodes are distance d apart and are able to detect if and when an object passes. Both nodes exchange their recorded time-values and calculate the difference which contains a certain error due to the clock offsets. In order to determine the speed v of the object, each node computes $v = -$. Assuming that the required accuracy of the speed measurement is 1%, we can accept a higher error in   for a larger Euclidean distance d. Thus, the acceptable clock offset forms a gradient.

Fan and Lynch [121] introduced this so-called *gradient clock synchronization* problem for the first time. The gradient property requires the difference between any two network nodes' clocks to be bounded from above by a non-decreasing function of their distance. Thus, the synchronization to nearby nodes needs to be better than to faraway nodes. Other than in the presented example, the distance is not necessarily the Euclidean distance. It can be also defined as the hop-count or the message delay uncertainty between two nodes.

The considered systems by Fan and Lynch have the following properties:

– reliable communication
– bounded message uncertainty in the bounded message delay
– bounded clock drift

In such systems it is shown in [121] that the clock difference of two nodes is in $\Omega(d + \frac{\log D}{\log \log D})$ where D is the diameter of the network.

Comparable recent work considers networks with a different system model. In [282], the authors assume that the communication frequency in wireless sensor networks is bounded due to restrictions in power consumption. Therefore, the time information is sent together with the application data. The synchronization algorithm thus has no influence on the communication pattern. For such a model a lower bound for the synchronization error can be derived that depends on the number of nodes, the maximal time between communications and the maximal clock drift.

Fan et al. present in [120] a synchronization algorithm that addresses *internal, external* and *gradient synchronization* in an energy-efficient way. While internal synchronization corresponds to synchronizing nodes with each other, external synchronization corresponds to synchronizing nodes with real time. Their algorithm satisfies a relaxed gradient property, i.e., the clock offset of two nodes which are distance d apart is bounded by a linear function of d, almost all the time. Logical clocks are allowed to be constant for some time such that the lower bound achieved in [121] does not apply directly.

Fan's clock synchronization algorithm works as follows: Each node i maintains a *local* clock that represents i's estimate of real time updated by its GPS device and a *global* clock that represents the maximum of other node's local clocks. All nodes perform 1-hop broadcasts once every τ time to maintain internal clock synchronization. Synchronization messages contain the node's local time and the largest GPS value the node has received. If a node receives a new GPS input with a higher time value t, it sets its local and global clock to t and computes the time of the next synchronization round. If a node i receives a synchronization message from another node j, it first examines if j has received a GPS value at least as recent as node i and if j has a larger local clock value than i's global clock. In this case, node i sets its global clock to node j's local clock value and further propagates the message to its neighborhood. It also checks if the time the message was received is within its next synchronization round since then there is no need for node i to perform a synchronization by its own. It thus just calculates the time for the next round.

Using its local and global clock, a node's *logical* clock is then defined as the maximum of both values.

It is shown in [120] that the algorithm satisfies the following requirements, where L_i denotes the logical clock of a node i, $S(t)$ the set of *stable* nodes at real time t, i.e., nodes that recently received an external time signal such as a GPS input, and $d_{i,j}$ the maximal delay between a node i and j:

Requirement 1 (ϵ-Precision). $\forall t \forall i, j \in S(t) : |L_i(t) - L_j(t)| \leq \epsilon$

Requirement 2 (ϵ-Accuracy). $\forall t \forall i \in S(t) : |L_i(t) - t| \leq \epsilon$

Requirement 3 ((T, α, β)-Gradient Precision). $\forall t \in T \forall i, j \in S(t) :$ $|L_i(t) - L_j(t)| \leq \alpha d_{i,j} + \beta$

The first requirement guarantees that the logical clocks of two nodes do not differ more than an upper threshold ϵ. With Requirement 2, the same holds for the offset of logical clocks with respect to realtime. Requirement 3 finally gives a bound for the time difference of two logical clocks, influenced by the distance between both nodes, and thus guarantees the gradient property defined above.

Although Fan and Lynch proved that the clock skew between two nodes is at least $\Omega(d + \frac{\log D}{\log \log D})$, it is still an open problem whether this bound is *tight*, i.e., if a clock synchronization algorithm come close to this bound. In contrast to the upper bound of Fan's algorithm defined by Requirement 1 and 3, Locher and Wattenhofer [262] propose a general gradient clock synchronization algorithm that is bounded by $O(d + \sqrt{D})$ *all the time*. While Fan's lower bound assumes that all nodes have full knowledge concerning the history of sent messages, Locher and Wattenhofer consider a more practical solution. Instead of storing the complete message history, a node only considers the largest clock values received from its neighbors. The algorithm is therefore called *oblivious*. Thus, Locher and Wattenhofer find an gradient clock synchronization algorithm with a clock skew of $o(D)$ between neighboring nodes. Moreover, their algorithm provides a globally asymptotically optimal bound of $\Theta(D)$.

19.6 Chapter Notes

In distributed real-time systems such as sensor networks, body-area-networks, etc., time synchronization is an essential requirement which aims at consistent node states. In this chapter, we have presented several strategies for time synchronization under different circumstances. Beginning with an algorithm called Reference-Broadcast Synchronization (RBS) [111], which is based on broadcasting synchronization messages to a set of receivers, we continued with the Time-Sync Protocol for Sensor Networks (TPSN) [143] which is a two step strategy. While the first step of the protocol establishes a hierarchical structure in the network, the second step performs pairwise synchronizations to achieve a global time scale. Due to the legitimate objection that time synchronization causes high data traffic and clocks need not to be synchronized all the time, we presented an algorithm by Römer [334] that just transforms timestamps from one local time domain of a node to another.

In Section 19.3, we then first introduced a certain kind of error called Byzantine fault that leads to an inconsistent behavior of a node. This kind of error has been treated in two classes of algorithms, the Interactive Consistency Algorithm [311, 248, 247] and the Fault-Tolerant-Average (FTA) [222] algorithm. While the first one tries to find the malicious node, the FTA algorithm determines time values that can be neglected by means of heuristics. The problem of obtaining consistent information at the network's node is also treated by the class of consensus algorithms that are investigated in [68, 298, 129]. Finally, we addressed temporal bounds in time synchronization and its effect on temporal ordering of events [268].

The last section introduced to recently published problem descriptions and their algorithmic treatment. One of the addressed problems deals with gradient clock synchronization where the synchronization of clocks correlates with the geometric topology of a network [121, 120, 282, 262]. A simple algorithm that fulfills the gradient property almost all the time was proposed by Fan et al [120]. While Fan and Lynch provide a lower bound for the clock skew of $\Omega(d + \frac{\log D}{\log\log D})$ [121], an open problem is whether this bound is tight. Locher and Wattenhofer [262] proved that there exits an oblivious clock synchronization algorithm with a worst-case skew of $O(d+\sqrt{D})$. While the algorithm even guarantees a clock skew of $\Theta(D)$, it is currently not known whether a skew of $\Theta(\sqrt{D})$ is globally asymptotically optimal for oblivious algorithm. Furthermore, analyzing about how much this bound can be improved by having more knowledge is still an open problem.

Bibliography

1. Abraham, I., Dolev, D., Malkhi, D.: LLS: a locality aware location service for mobile ad hoc networks. In: Proceedings of the DIALM-POMC Joint Workshop on Foundations of Mobile Computing (DIALM-POMC 2004), pp. 75–84 (2004)
2. Abraham, I., Gavoille, C., Goldberg, A., Malkhi, D.: Routing in networks with low doubling dimension. In: Proceedings of the 26th International Conference on Distributed Computing Systems (ICDCS'06), p. 75. IEEE Computer Society Press, Los Alamitos (2006)
3. Abraham, I., Gavoille, C., Malkhi, D.: Routing with improved communication-space trade-off. In: Guerraoui, R. (ed.) DISC 2004. LNCS, vol. 3274, pp. 305–319. Springer, Heidelberg (2004)
4. Abraham, I., Gavoille, C., Malkhi, D.: Compact routing for graphs excluding a fixed minor. In: Fraigniaud, P. (ed.) DISC 2005. LNCS, vol. 3724, pp. 442–456. Springer, Heidelberg (2005)
5. Abraham, I., Gavoille, C., Malkhi, D.: On space-stretch trade-offs: lower bounds. In: Proceedings of the 18th Annual ACM Symposium on Parallel Algorithms and Architectures (SPAA'06), pp. 207–216. ACM Press, New York (2006)
6. Abraham, I., Gavoille, C., Malkhi, D.: On space-stretch trade-offs: upper bounds. In: Proceedings of the 18th Annual ACM Symposium on Parallel Algorithms and Architectures (SPAA'06), pp. 217–224. ACM Press, New York (2006)
7. Abraham, I., Gavoille, C., Malkhi, D., Nisan, N., Thorup, M.: Compact name-independent routing with minimum stretch. In: Proceedings of the 16th Annual ACM Symposium on Parallel Algorithms and Architectures (SPAA'04), pp. 20–24. ACM Press, New York (2004)
8. Abraham, I., Malkhi, D.: Compact routing on euclidian metrics. In: Proceedings of the 23rd ACM Symposium on Principles of Distributed Computing (PODC'04), pp. 141–149. ACM Press, New York (2004)
9. Abraham, I., Malkhi, D.: Name independent routing for growth bounded networks. In: Proceedings of the 17th Annual ACM Symposium on Parallel Algorithms and Architectures (SPAA'05), pp. 49–55. ACM Press, New York (2005)

10. Abramson, N.: The ALOHA system – Another alternative for computer communications. In: Proceedings of the Fall Joint Compute Conference. AFIPS Conference, vol. 37, pp. 281–285 (1970)

11. Allwright, J.R., Bordawekar, R., Coddington, P., Dincer, K., Martin, C.: A comparison of parallel graph coloring algorithms. Technical report, Northeast Parallel Architectures Center, Syracuse University (1995)

12. Alzoubi, K., Wan, P.-J., Frieder, O.: Message-optimal connected dominating sets in mobile ad hoc networks. In: Proceedings of the 3rd ACM International Symposium on Mobile Ad Hoc Networking and Computing (MOBIHOC'02), Lausanne, Switzerland (2002)

13. Ambühl, C.: An optimal bound for the MST algorithm to compute energy efficient broadcast trees in wireless networks. In: Caires, L., Italiano, G.F., Monteiro, L., Palamidessi, C., Yung, M. (eds.) ICALP 2005. LNCS, vol. 3580, pp. 1139–1150. Springer, Heidelberg (2005)

14. Ambühl, C., Erlebach, T., Mihalák, M., Nunkesser, M.: Constant-factor approximation for minimum-weight (connected) dominating sets in unit disk graphs. In: Díaz, J., Jansen, K., Rolim, J.D.P., Zwick, U. (eds.) APPROX 2006 and RANDOM 2006. LNCS, vol. 4110, Springer, Heidelberg (2006)

15. Anderegg, L., Eidenbenz, S.: Ad hoc-vcg: a truthful and cost-efficient routing protocol for mobile ad hoc networks with selfish agents. In: Proceedings of the 9th Annual International Conference on Mobile Computing and Networking (MOBICOM'03), pp. 245–259. ACM Press, New York (2003)

16. Anderson, R., Chan, H., Perrig, A.: Key infection: Smart trust for smart dust. In: Proceedings of IEEE International Conference on Network Protocols (ICNP 2004), October 2004, IEEE Computer Society Press, Los Alamitos (2004)

17. Arias, M., Cowen, L., Laing, K., Rajaraman, R., Taka, O.: Compact routing with name independence. SIAM Journal on Discrete Mathematics 20(3), 705–726 (2006)

18. Arikan, E.: Some complexity results about packet radio networks. IEEE Transactions on Information Theory 30(4) (1984)

19. Arya, S., Das, G., Mount, D., Salowe, J., Smid, M.: Euclidean spanners: short, thin, and lanky. In: Proceedings of the 27th Annual ACM Symposium on Theory of Computing (STOC'95), pp. 489–498. ACM Press, New York (1995)

20. Aspnes, J., Goldenberg, D., Yang, Y.: On the computational complexity of sensor network localization. In: Nikoletseas, S.E., Rolim, J.D.P. (eds.) ALGOSENSORS 2004. LNCS, vol. 3121, Springer, Heidelberg (2004)

21. ATmega 128: http://www.atmel.com/dyn/resources/prod_documents/doc2467.pdf

22. Attiya, H., Welch, J.: Distributed Computing, 2nd edn. Wiley, Chichester (2004)

23. Ausiello, G., Crescenzi, P., Gambosi, G., Kann, V., Marchetti-Spaccamela, A.: Complexity and Approximation - Combinatorial Optimization Problems and Their Approximability Properties, 2nd edn. Springer, Heidelberg (2002)

24. Awerbuch, B.: Complexity of network synchronization. Journal of the ACM 32(4), 804–823 (1985)

25. Awerbuch, B.: Optimal distributed algorithms for minimum weight spanning tree, counting, leader election and related problems. In: Proceedings of the 19th Annual ACM Symposium on Theory of Computing (STOC'87), pp. 230–240. ACM Press, New York (1987)

26. Awerbuch, B., Bar-Noy, A., Linial, N., Peleg, D.: Compact distributed data structures for adaptive routing. In: Proceedings of the 21st Annual ACM Symposium on Theory of Computing (STOC'89), pp. 479–489. ACM Press, New York (1989)
27. Awerbuch, B., Bar-Noy, A., Linial, N., Peleg, D.: Improved routing strategies with succinct tables. Journal of Algorithms 11(3), 307–341 (1990)
28. Awerbuch, B., Peleg, D.: Sparse partitions. In: Proceedings of the 31th Annual IEEE Symposium on Foundations of Computer Science (FOCS'90), pp. 503–513. IEEE Computer Society Press, Los Alamitos (1990)
29. Axelrod, R.: The Evolution of Cooperation. Basic Books (1984)
30. Bao, L., Garcia-Luna-Aceves, J.J.: Topology management in ad hoc networks. In: Proceedings of the 4th ACM International Symposium on Mobile Ad Hoc Networking and Computing (MOBIHOC'03), Annapolis, MD, USA (2003)
31. Barrière, L., Fraigniaud, P., Narayanan, L.: Robust position-based routing in wireless ad hoc networks with unstable transmission ranges. In: Proceedings of the 5th International Workshop on Discrete Algorithms and Methods for Mobile Computing and Communications (Dial-M'01), Rome, Italy, pp. 19–27 (2001)
32. Bartal, Y., Byers, J., Raz, D.: Global optimization using local information with applications to flow control. In: Proceedings of the 38th Annual IEEE Symposium on Foundations of Computer Science (FOCS'97), October 1997, p. 303. IEEE Computer Society Press, Los Alamitos (1997)
33. Basagni, S., Mastrogiovanni, M., Petrioli, C.: A performance comparison of protocols for clustering and backbone formation in large scale ad hoc networks. In: Proceedings of the 1st IEEE International Conference on Mobile Ad-hoc and Sensor Systems (MASS'04), IEEE Computer Society Press, Los Alamitos (2005)
34. Bellare, M., Goldreich, O., Sudan, M.: Free bits, PCPs and non-approximability — towards tight results. SIAM Journal on Computing 27(3), 804–915 (1998)
35. Bianchi, G.: Performance analysis of the IEEE 802.11 distributed coordination function. IEEE Journal on Selected Areas in Communications 18(3), 535–547 (2000)
36. Björklund, P., Värbrand, P., Yuan, D.: A column generation method for spatial tdma scheduling in ad hoc networks. Ad Hoc Networks 2(4), 405–418 (2004)
37. Blaß, E.-O., Junker, H., Zitterbart, M.: Effiziente Implementierung von Public-Key Algorithmen für Sensornetze. In: Informatik 2005, 2nd Workshop on Sensor Networks (2005)
38. Blaß, E.-O., Zitterbart, M.: An Efficient Key Establishment Scheme for Secure Aggregating Sensor Networks. In: ACM Symposium on Information, Computer and Communications Security (2006)
39. Blough, D., Leoncini, M., Resta, G., Santi, P.: Topology control with better radio models: Implications for energy and multi-hop interference. In: Proceedings of the 8th ACM/IEEE International Symposium on Modeling, Analysis and Simulation of Wireless and Mobile Systems (MSWiM'06), IEEE Computer Society Press, Los Alamitos (2005)
40. Blum, B., He, T., Son, S., Stankovic, J.: IGF: A state-free robust communication protocol for wireless sensor networks. Technical Report CS-2003-11, University of Virginia, USA (2003)

41. Bobba, R., Eschenauer, L., Gligor, V., Arbaugh, W.: Bootstrapping Security Associations for Routing in Mobile Ad-Hoc Networks (2002), http://citeseer.ist.psu.edu/bobba02bootstrapping.html
42. Bock, F.: An algorithm to construct a minimum spanning tree in a directed network. In: Developments in Operations Research, pp. 29–44 (1971)
43. Bose, P., Brodnik, A., Carlsson, S., Demaine, E., Fleischer, R., López-Ortiz, A., Morin, P., Munro, J.: Online routing in convex subdivisions. In: Lee, D.T., Teng, S.-H. (eds.) ISAAC 2000. LNCS, vol. 1969, pp. 47–59. Springer, Heidelberg (2000)
44. Bose, P., Morin, P.: Online routing in triangulations. In: Aggarwal, A.K., Pandu Rangan, C. (eds.) ISAAC 1999. LNCS, vol. 1741, pp. 113–122. Springer, Heidelberg (1999)
45. Bose, P., Morin, P., Stojmenovic, I., Urrutia, J.: Routing with guaranteed delivery in ad hoc wireless networks. In: Proceedings of the 3rd International Workshop on Discrete Algorithms and Methods for Mobile Computing and Communications (Dial-M'99) (1999)
46. Boudec, J.-Y., Vojnović, M.: Perfect simulation and stationarity of a class of mobility models. In: Proceedings of the 24th Annual Joint Conference of the IEEE Computer and Communications Societies (INFOCOM'05), Miami, Florida, March 2005 (2005)
47. Brady, A., Cowen, L.: Compact routing on power-law graphs with additive stretch. In: Proceedings of the 8th Workshop on Algorithm Engineering and Experiments (ALENEX'06), pp. 119–128. SIAM (2006)
48. Brady, A., Cowen, L.: Compact routing with additive stretch using distance labelings. In: Proceedings of the 18th Annual ACM Symposium on Parallel Algorithms and Architectures (SPAA'06), p. 233. ACM Press, New York (2006)
49. Brady, A., Cowen, L.: Exact distance labelings yield additive-stretch compact routing schemes. In: Dolev, S. (ed.) DISC 2006. LNCS, vol. 4167, pp. 339–354. Springer, Heidelberg (2006)
50. Breu, H., Kirkpatrick, D.: Unit disk graph recognition is np-hard. Computational Geometry: Theory and Applications 9(1-2), 3–24 (1998)
51. Broch, J., Maltz, D., Johnson, D., Hu, Y.-C., Jetcheva, J.: A performance comparison of multi-hop wireless ad hoc network routing protocols. In: Proceedings of the 4th Annual International Conference on Mobile Computing and Networking (MOBICOM'98), pp. 85–97 (1998)
52. Bruck, J., Gao, J., Jiang, A.: Localization and routing in sensor networks by local angle information. In: Proceedings of the 6th ACM International Symposium on Mobile Ad Hoc Networking and Computing (MOBIHOC'05), ACM Press, New York (2005)
53. BTnodes rev3 - Product brief: http://www.btnode.ethz.ch/pub/files/btnode_rev3.22_productbrief.pdf
54. Buchegger, S., Boudec, J.-Y.: The Effect of Rumor Spreading in Reputation Systems for Mobile Ad-Hoc Networks. In: Proceedings of the 1th International Symposium on Modeling and Optimizationin Mobile, Ad Hoc and Wireless Networks (WiOpt'03) (2003)
55. Burkhart, M., von Rickenbach, P., Wattenhofer, R., Zollinger, A.: Does topology control reduce interference? In: Proceedings of the 5th ACM International Symposium on Mobile Ad Hoc Networking and Computing (MOBIHOC'04), pp. 9–19. ACM Press, New York (2004)

56. Busch, C., Surapaneni, S., Tirthapura, S.: Analysis of link reversal routing algorithms for mobile ad hoc networks. In: Proceedings of the 15th Annual ACM Symposium on Parallel Algorithms and Architectures (SPAA'03), San Diego, California, USA, (June 2003)

57. Buttyán, L., Hubaux, J.-P.: Nuglets: a virtual currency to stimulate cooperation in self-organized ad hoc networks. Technical Report DSC/2001, Swiss Federal Institute of Technology – Lausanne (2001)

58. Camerini, P., Fratta, L., Maffioli, F.: A Note on Finding Optimum Branchings. Networks 9, 309–312 (1979)

59. Camp, T., Boleng, J., Davies, V.: A survey of mobility models for ad hoc network research. In: Wireless Communications and Mobile Computing (2002)

60. Cao, Q., Abdelzaher, T.: A scalable logical coordinates framework for routing in wireless sensor networks. In: Proc. of IEEE RTSS, IEEE Computer Society Press, Los Alamitos (2004)

61. Capkun, S., Buttyán, L., Hubaux, J.-P.: Self-Organized Public-Key Management for Mobile Ad Hoc Networks. IEEE Transactions on Mobile Computing 2(1), 52–64 (2003)

62. Cardinal, J., Collette, S., Langerman, S.: Region counting graphs. In: Proceedings of the 21st European Workshop on Computational Geometry (EWCG'05), pp. 21–24 (2005)

63. Castro, M., Liskov, B.: Practical Byzantine Fault Tolerance. In: Proceedings of the 3rd Symposium on Operating Systems Design and Implementation (OSDI'99) (1999)

64. Cerpa, A., Estrin, D.: ASCENT: Adaptive Self-Configuring Sensor Networks Topologies. In: Proceedings of the 21st Annual Joint Conference of the IEEE Computer and Communications Societies (INFOCOM'02), IEEE Computer Society Press, Los Alamitos (2002)

65. Cerpa, A., Wong, J., Kuang, L., Potkonjak, M., Estrin, D.: Statistical Model of Lossy Links in Wireless Sensor Networks. In: Proc. ACM/IEEE IPSN (2005)

66. Chan, H., Perrig, A.: PIKE: Peer Intermediaries for Key Establishment in Sensor Networks. In: Proceedings of IEEE Infocom, IEEE Computer Society Press, Los Alamitos (2005)

67. Chan, H., Perrig, A., Song, D.: Random key predistribution schemes for sensor networks. In: IEEE Symposium on Security and Privacy, IEEE Computer Society Press, Los Alamitos (2003)

68. Chandra, T., Toueg, S.: Unreliable failure detectors for reliable distributed systems. Journal of the ACM 43(2), 225–267 (1999)

69. Chen, B., Morris, R.: L+: Scalable Landmark Routing and Address Lookup for Multi-hop Wireless Networks. Technical Report 837, MIT LCS (2002)

70. Cheng, X., Huang, X., Li, D., Wu, W., Du, D.-Z.: Polynomial-time approximation scheme for minimum connected dominating set in ad hoc wireless networks. Networks 42(4), 202–208 (2003)

71. Chiang, C.-C., Wu, H.-K., Liu, W., Gerla, M.: Routing in clustered multihop, mobile wireless networks with fading channel. In: Proceedings of the IEEE Singapore International Conference on Networks (SICON'97), pp. 197–211. IEEE Computer Society Press, Los Alamitos (1997)

72. Chu, Y.-J., Lin, T.-H.: On the shortest arborescence of a directed graph. Science Sinica 14, 1396–1400 (1965)

73. Chudak, F., Erlebach, T., Panconesi, A., Sozio, M.: Primal-dual distributed algorithms for covering and facility location problems. Unpublished Manuscript (2005)

74. Chvátal, V.: A greedy heuristic for the set-covering problems. Operations Research 4(3), 233–235 (1979)

75. Clark, B., Colbourn, C., Johnson, D.: Unit disk graphs. Discrete Mathematics 86, 165–177 (1990)

76. Clausen, T., Jacquet, P. (eds.): Optimized link state routing protocol (OLSR). IETF RFC 3626 (October 2003)

77. Clementi, A.E.F., Crescenzi, P., Penna, P., Rossi, G., Vocca, P.: On the complexity of computing minimum energy consumption broadcast subgraphs. In: Ferreira, A., Reichel, H. (eds.) STACS 2001. LNCS, vol. 2010, pp. 121–131. Springer, Heidelberg (2001)

78. Clementi, A.E.F., Di Ianni, M., Silvestri, R.: The minimum broadcast range assignment problem on linear multi-hop wireless networks. Theoretical Computer Science 299, 751–761 (2003)

79. Clementi, A.E.F., Huiban, G., Penna, P., Rossi, G., Verhoeven, Y.C.: Some recent theoretical advances and open questions on energy consumption in ad-hoc wireless networks (2002)

80. Clementi, A.E.F., Penna, P., Silvestri, R.: The power range assignment problem in radio networks on the plane. In: Reichel, H., Tison, S. (eds.) STACS 2000. LNCS, vol. 1770, pp. 651–660. Springer, Heidelberg (2000)

81. Cole, R., Vishkin, U.: Deterministic coin tossing with applications to optimal parallel list ranking. Information and Control 70(1), 32–53 (1986)

82. WST: Application areas: Condition based monitoring, http://wins.rockwellscientific.com/WST_CBM.html

83. Cormen, T., Leiserson, C., Rivest, R.: Introduction to Algorithms. MIT Press, Cambridge (1990)

84. Cormen, T., Leiserson, C., Rivest, R., Stein, C.: Introduction to Algorithms, 2nd edn. MIT Press, Cambridge (2001)

85. Couto, D.S.J.D., Aguayo, D., Chambers, B.A., Morris, R.: Performance of multihop wireless networks: shortest path is not enough. In: Proc. of ACM HotNets, ACM Press, New York (2003)

86. Cowen, L.J.: Compact routing with minimum stretch. In: Proceedings of the 10th Annual ACM–SIAM Symposium on Discrete Algorithms (SODA'99), pp. 255–260. SIAM (1999)

87. Cowen, L.J., Goddard, W., Jesurum, C.E.: Coloring with defect. In: Proceedings of the 8th Annual ACM–SIAM Symposium on Discrete Algorithms (SODA'97), Philadelphia, PA, USA, pp. 548–557. IEEE Computer Society Press, Los Alamitos (1997)

88. Cristescu, R., Beferull-Lozano, B., Vetterli, M.: On network correlated data gathering. In: Proceedings of the 23nd Annual Joint Conference of the IEEE Computer and Communications Societies (INFOCOM'04), Hong Kong, IEEE Computer Society Press, Los Alamitos (2004)

89. Cristescu, R., Beferull-Lozano, B., Vetterli, M.: Networked slepian-wolf: Theory, algorithms and scaling laws. In: to appear in IEEE Transactions on Information Theory (2005)

90. Cristescu, R., Beferull-Lozano, B., Vetterli, M., Wattenhofer, R.: Network correlated data gathering with explicit communication: Np-completeness and algorithms. In: to appear in IEEE/ACM Transactions on Networking (April 2006)

91. Cristescu, R., Vetterli, M.: On the optimal density for real-time data gathering of spatio-temporal processes in sensor networks. In: Proceedings of the 4th International Symposium on Information Processing in Sensor Networks (IPSN'05), Los Angeles, CA (2005)

92. Culler, D.: Towards the sensor network macroscope. Keynote Talk at the 6th ACM International Symposium on Mobile Ad Hoc Networking and Computing (MobiHoc 05) (2005)

93. Daemen, J., Rijmen, V.: The Design of Rijndael. Springer, Heidelberg (2002)

94. Dai, F., Wu, J.: An extended localized algorithm for connected dominating set formation in ad hoc wireless networks. IEEE Transactions on Parallel and Distributed Systems 15(10), 908–920 (2004)

95. Darwen, P.J., Yao, X.: On Evolving Robust Strategies for Iterated Prisoner's Dilemma. In: Yao, X. (ed.) Progress in Evolutionary Computation. LNCS, vol. 956, pp. 276–292. Springer, Heidelberg (1995)

96. Das, S.M., Pucha, H., Hu, Y.C.: Performance Comparison of Scalable Location Services for Geographic Ad Hoc Routing. In: INFOCOM 2005, IEEE Computer Society Press, Los Alamitos (2005)

97. Datta, S., Stojmenovic, I., Wu, J.: Internal node and shortcut based routing with guaranteed delivery in wireless networks. In: Cluster Computing 5, pp. 169–178. Kluwer Academic Publishers, Dordrecht (2002)

98. De Marco, G., Pelc, A.: Fast distributed graph coloring with $O(\Delta)$ colors. In: Proceedings of the 12th Annual ACM–SIAM Symposium on Discrete Algorithms (SODA'01), Philadelphia, PA, USA, pp. 630–635. Society for Industrial and Applied Mathematics, New York (2001)

99. Deb, B., Nath, B.: On the node-scheduling approach to topology control in ad hoc networks. In: Proceedings of the 6th ACM International Symposium on Mobile Ad Hoc Networking and Computing (MOBIHOC'05), pp. 14–26. ACM Press, New York (2005)

100. Demirbas, M., Arora, A., Gouda, M.: A pursuer-evader game for sensor networks. In: Proc. of SSS (2003)

101. Diestel, R.: Graph Theory, 3rd edn. Springer, Heidelberg (2005)

102. Dolev, D., Yao, A.C.-C.: On the Security of Public-Key Protocols. IEEE Transactions on Information Theory (1983)

103. Doshi, S., Bhandare, S., Brown, T.: An on-demand minimum energy routing protocol for a wireless ad hoc network. SIGMOBILE Mob. Comput. Commun. Rev. 6(3), 50–66 (2002)

104. Doyle, P.G., Snell, J.L.: Random Walks and Electric Networks. The Carus Mathematical Monographs. The Mathematical Association of America (1984)

105. Dubhashi, D., Mei, A., Panconesi, A., Radhakrishnan, J., Srinivasan, A.: Fast distributed algorithms for (weakly) connected dominating sets and linear-size skeletons. In: Proceedings of the 14th Annual ACM–SIAM Symposium on Discrete Algorithms (SODA'03), pp. 717–724 (2003)

106. Dunkels, A., Gronvall, B., Voigt, T.: Contiki - a lightweight and flexible operating system for tiny networked sensors. In: Proceedings of the 29th Annual IEEE International Conference on Local Computer Networks (LCN'04), Washington, DC, USA, pp. 455–462. IEEE Computer Society Press, Los Alamitos (2004)

107. Edmonds, J.: Optimum branchings. Research of the National Bureau of Standards 71B, 233–240 (1967)

108. Eilam, T., Gavoille, C., Peleg, D.: Compact routing schemes with low stretch factor. Journal of Algorithms 46(2), 97–114 (2003)
109. Elkin, M.: A faster distributed protocol for constructing a minimum spanning tree. In: Proceedings of the 15th Annual ACM–SIAM Symposium on Discrete Algorithms (SODA'04), pp. 359–368 (2004)
110. Elkin, M.: Unconditional lower bounds on the time-approximation tradeoffs for the distributed minimum spanning tree problem. In: Proceedings of the 36th Annual ACM Symposium on Theory of Computing (STOC'04), pp. 331–340. ACM Press, New York (2004)
111. Elson, J., Girod, L., Estrin, D.: Fine-grained network time synchronization using reference broadcasts. Proceedings of the 5th Symposium on Operating Systems Design and Implementation (OSDI'02) 36(SI), 147–163 (2002)
112. Ephremides, A., Truong, T.V.: Scheduling broadcasts in multihop radio networks. IEEE Transactions on Communications 38(4), 456–460 (1990)
113. Eppstein, D.: Spanning trees and spanners. In: Sack, J.R., Urrutia, J. (eds.) Handbook of Computational Geometry, chapter 9, pp. 425–461. Elsevier Science Publishers B.V., Amsterdam (2000)
114. Erdős, P., Frankl, P., Füredi, Z.: Families of finite sets in which no set is covered by the union of r others. Israel Journal of Mathematics 51, 79–89 (1985)
115. Eren, T., Goldenberg, D., Whiteley, W., Yang, Y.R., Morse, A.S., Anderson, B., Belhumeur, P.N.: Rigidity, computation, and randomization in network localization. In: Proceedings of the 23nd Annual Joint Conference of the IEEE Computer and Communications Societies (INFOCOM'04), IEEE Computer Society Press, Los Alamitos (2004)
116. Erlebach, T., Jansen, K., Seidel, E.: Polynomial-time approximation schemes for geometric graphs. In: Proceedings of the 12th Annual ACM–SIAM Symposium on Discrete Algorithms (SODA'01), Philadelphia, PA, USA, pp. 671–679. Society for Industrial and Applied Mathematics (2001)
117. Eschenauer, L., Gligor, V.D.: A Key Management Scheme for Distributed Sensor Networks. In: 9th Conference on Computer and Communications Security (2002)
118. Even, S., Goldreich, O., Moran, S., Tong, P.: On the NP-completeness of certain network testing problems. Networks 14, 1–24 (1984)
119. Fall, K., Varadhan, K.: The ns Manual (2002)
120. Fan, R., Chakraborty, I., Lynch, N.: Clock synchronization for wireless networks. In: Higashino, T. (ed.) OPODIS 2004. LNCS, vol. 3544, pp. 400–414. Springer, Heidelberg (2005)
121. Fan, R., Lynch, N.: Gradient clock synchronization. In: Proceedings of the 23rd ACM Symposium on Principles of Distributed Computing (PODC'04), St. John's, Newfoundland, Canada, July 2004, pp. 320–327. ACM Press, New York (2004)
122. Feeney, L.: Energy efficient communication in ad hoc networks. In: Basagni, S., Conti, M., Giordano, S., Stojmenovic, I. (eds.) Mobile Ad Hoc Networking, Wiley, Chichester (2004)
123. Feige, U.: A threshold of $\ln n$ for approximating set cover. Journal of the ACM 45(4), 634–652 (1998)
124. Feige, U.: Approximating the bandwidth via volume respecting embeddings. Journal of Computer and System Sciences 60(3), 510–539 (2000)

125. Feige, U., Halldórsson, M.M., Kortsarz, G., Srinivasan, A.: Approximating the domatic number. SIAM Journal on Computing 32(1), 172–195 (2002)
126. Feigenbaum, J., Papadimitriou, C., Sami, R., Shenker, S.: A bgp-based mechanism for lowest-cost routing. In: Proceedings of the 21st ACM Symposium on Principles of Distributed Computing (PODC'02), pp. 173–182. ACM Press, New York (2002)
127. Feigenbaum, J., Shenker, S.: Distributed algorithmic mechanism design: recent results and future directions. In: Proceedings of the 6th International Workshop on Discrete Algorithms and Methods for Mobile Computing and Communications (Dial-M'02), pp. 1–13 (2002)
128. Finn, G.G.: Routing and addressing problems in large metropolitan-scale internetworks. Technical Report ISI/RR-87-180, USC/ISI (March 1987)
129. Fischer, M.J., Lynch, N., Paterson, M.S.: Impossibility of distributed consensus with one faulty process. Journal of the ACM 32(2), 374–382 (1985)
130. Floyd, S., Jacobson, V., Liu, C.-G., McCanne, S., Zhang, L.: A reliable multicast framework for light-weight sessions and application level framing. In: Proc. of ACM SIGCOMM, ACM Press, New York (1995)
131. Flury, R., Wattenhofer, R.: MLS: An Efficient Location Service for Mobile Ad Hoc Networks. In: 7th ACM International Symposium on Mobile Ad Hoc Networking and Computing (MOBIHOC), Florence, Italy (May 2006)
132. Fonseca, R., Ratnasamy, S., Culler, D., Shenker, S., Stoica, I.: Beacon Vector Routing: Scalable Point-to-Point in Wireless Sensornets. In: Proc. of NSDI (2005)
133. Fraigniaud, P., Gavoille, C.: Routing in trees. In: Orejas, F., Spirakis, P.G., van Leeuwen, J. (eds.) ICALP 2001. LNCS, vol. 2076, pp. 757–772. Springer, Heidelberg (2001)
134. Fraigniaud, P., Gavoille, C.: A space lower bound for routing in trees. In: Alt, H., Ferreira, A. (eds.) STACS 2002. LNCS, vol. 2285, pp. 65–75. Springer, Heidelberg (2002)
135. Frank, C., Römer, K.: Algorithms for generic role assignment in wireless sensor networks. In: Proceedings of the 3rd International Conference on Embedded Networked Sensor Systems (SENSYS'05), San Diego, CA, USA (November 2005)
136. Frank, C.: Distributed facility location algorithms for flexible configuration of wireless sensor networks. In: Proceedings of the 3rd International Conference on Distributed Computing in Sensor Systems (DCOSS'07), Santa Fe, NM, USA (June 2007)
137. Frederickson, G., Janardan, R.: Optimal message routing without complete routing tables. In: Proceedings of the 5th ACM Symposium on Principles of Distributed Computing (PODC'86), pp. 88–97. ACM Press, New York (1986)
138. Frick, M.: A survey of (m,k)-colorings. In: Kennedy, J.W., Quintas, L.V. (eds.) Quo Vadis, Graph Theory? Annals of Discrete Mathematics, vol. 55, pp. 45–58. Elsevier Science Publishers B.V, Amsterdam (1993)
139. Friedman, E., Resnick, P.: The Social Cost of Cheap Pseudonyms. Journal of Economics and Management Strategy 10(2), 173–199 (1998)
140. Füßler, H., Widmer, J., Käsemann, M., Mauve, M., Hartenstein, H.: Contention-based forwarding for mobile ad-hoc networks. Elsevier's Ad Hoc Networks 1(4), 351–369 (2003)
141. Gabow, H.N.: Implementation of Algorithms for maximum Matching on Nonbipatite Graphs. PhD thesis, Stanford University (1974)

142. Gafni, E.M., Bertsekas, D.P.: Distributed algorithms for generating loop-free routes in networks with frequently changing topology. IEEE Transactions on Communications 29 (1981)
143. Ganeriwal, S., Kumar, R., Srivastava, M.B.: Timing-sync protocol for sensor networks. In: Proceedings of the 1st International Conference on Embedded Networked Sensor Systems (SENSYS'03), Los Angeles, CA, USA, November 2003, pp. 138–149. ACM Press, New York (2003)
144. Ganesan, D., Cristescu, R., Beferull-Lozano, B.: Power-efficient sensor placement and transmission structure for data gathering under distortion constraints. In: Proceedings of the 3rd International Symposium on Information Processing in Sensor Networks (IPSN'04), Berkeley, CA (2004)
145. Ganesan, D., Estrin, D., Heidemann, J.: Dimensions: Why do we need a new data handling architecture for sensor networks. In: Proc. of ACM HotNets, ACM Press, New York (2002)
146. Gansner, E.R., Koren, Y., North, S.C.: Graph drawing by stress majorization. In: Pach, J. (ed.) GD 2004. LNCS, vol. 3383, Springer, Heidelberg (2005)
147. Gao, J., Guibas, L., Hershberger, J., Zhang, L., Zhu, A.: Discrete mobile centers. In: Proceedings of the 17th Annual Symposium on on Computational Geometry (SCG'01), pp. 188–196. ACM Press, New York (2001)
148. Gao, J., Guibas, L., Hershberger, J., Zhang, L., Zhu, A.: Geometric spanner for routing in mobile networks. In: Proceedings of the 2nd ACM International Symposium on Mobile Ad Hoc Networking and Computing (MOBIHOC'01), Long Beach, CA, USA (October 2001)
149. Garay, J.A., Kutten, S., Peleg, D.: A sub-linear time distributed algorithm for minimum-weight spanning trees. SIAM Journal on Computing 27, 302–316 (1998)
150. Garey, M.R., Johnson, D.S.: Computers and Intractability. A Guide to the Theory of $\mathcal{NP}$-Completeness. W.H. Freeman and Company (1979)
151. Gasieniec, L., Pelc, A., Peleg, D.: The wakeup problem in synchronous broadcast systems. SIAM Journal on Discrete Mathematics 14(2), 207–222 (2001)
152. Gavoille, C.: A survey on interval routing. Theoretical Computer Science 245(2), 217–253 (2000)
153. Gavoille, C.: Routing in distributed networks: Overview and open problems. ACM SIGACT News—Distributed Computing Column 32(1), 36–52 (2001)
154. Gavoille, C., Gengler, M.: Space-efficiency for routing schemes of stretch factor three. Journal of Parallel and Distributed Computing 61(5), 679–687 (2001)
155. Gavoille, C., Peleg, D.: Compact and localized distributed data structures. Distributed Computing 16(2-3), 111–120 (2003)
156. Gavoille, C., Peleg, D., Pérennès, S., Raz, R.: Distance labeling in graphs. Journal of Algorithms 53(1), 85–112 (2004)
157. Gavoille, C., Pérennès, S.: Memory requirements for routing in distributed networks. In: Proceedings of the 15th ACM Symposium on Principles of Distributed Computing (PODC'96), pp. 125–133. ACM Press, New York (1996)
158. Gehweiler, J., Lammersen, C., Sohler, C.: A distributed $O(1)$-approximation algorithm for the uniform facility location problem. In: Proceedings of the 18th Annual ACM Symposium on Parallel Algorithms and Architectures (SPAA'06), Cambridge, Massachusetts, USA, pp. 237–243. ACM Press, New York (2006)

159. Gentry, C.: Key Recovery and Message Attacks on NTRU-Composite. In: Pfitzmann, B. (ed.) EUROCRYPT 2001. LNCS, vol. 2045, p. 182. Springer, Heidelberg (2001)

160. Gilbert, E.N., Pollak, H.O.: Steiner minimal trees. SIAM Journal on Applied Mathematics 16(1), 1–29 (1968)

161. Girod, L., Elson, J., Cerpa, A., Stathopoulos, T., Ramanathan, N., Estrin, D.: EmStar: a Software Environment for Developing and Deploying Wireless Sensor Networks. In: Proc. of USENIX (2004)

162. Goel, A., Estrin, D.: Simultaneous optimization for concave costs: Single sink aggregation or single source buy-at-bulk. In: Proceedings of the 14th Annual ACM–SIAM Symposium on Discrete Algorithms (SODA'03), Baltimore, Maryland, USA, pp. 499–505 (2003)

163. Goldberg, A.V., Plotkin, S.A., Shannon, G.E.: Parallel symmetry-breaking in sparse graphs. In: Proceedings of the 19th Annual ACM Symposium on Theory of Computing (STOC'87), pp. 315–324. ACM Press, New York (1987)

164. Goldreich, O.: Foundations of Cryptography – Volume I Basic Tools. Cambridge University Press, Cambridge (2001),
http://www.wisdom.weizmann.ac.il/~oded/foc-vol1.html

165. Goldreich, O.: Foundations of Cryptography – Volume II Basic Applications. Cambridge University Press, Cambridge (2004),
http://www.wisdom.weizmann.ac.il/~oded/foc-vol2.html

166. Gotsman, C., Koren, Y.: Distributed graph layout for sensor networks. In: Pach, J. (ed.) GD 2004. LNCS, vol. 3383, Springer, Heidelberg (2005)

167. Green, J., Laffont, J.: Incentives in public desision making. In: Studies in Public Economies, vol. 1, pp. 65–78 (1979)

168. Greenstein, B., Estrin, D., Govindan, R., Ratnasamy, S., Shenker, S.: DIFS: A distributed index for features in sensor networks. In: Proc. of IEEE WSNA, IEEE Computer Society Press, Los Alamitos (2003)

169. Grossglauser, M., Vetterli, M.: Locating nodes with EASE: Mobility diffusion of last encounters in ad hoc networks. In: Proceedings of the 22nd Annual Joint Conference of the IEEE Computer and Communications Societies (INFOCOM'03), IEEE Computer Society Press, Los Alamitos (2003)

170. Groves, T.: Incentives in teams. Econometrica 41(4), 617–631 (1973)

171. Guba, N., Camp, T.: Recent Work on GLS: A Location Service for an Ad Hoc Network. In: Proceedings of the Grace Hopper Celebration (GHC 2002) (2002)

172. Guha, S., Khuller, S.: Approximation algorithms for connected dominating sets. In: Díaz, J. (ed.) ESA 1996. LNCS, vol. 1136, pp. 179–193. Springer, Heidelberg (1996)

173. Guha, S., Khuller, S.: Greedy strikes back: Improved facility location algorithms. Journal of Algorithms 31, 228–248 (1999)

174. Gupta, P., Kumar, P.R.: The capacity of wireless networks. IEEE Transactions on Information Theory IT-46(2), 388–404 (2000)

175. Gura, N., Patel, A., Wander, A.: Comparing Elliptic Curve Cryptography and RSA on 8-bit CPUs (2003),
http://www.research.sun.com/projects/crypto

176. Hajek, B.E., Sasaki, G.H.: Link scheduling in polynomial time. IEEE Transactions on Information Theory 34(5), 910–917 (1988)

177. Hale, W.K.: Frequency assignment: Theory and applications. IEEE Proceedings 68, 1497–1514 (1980)

178. Han, C.-C., Kumar, R., Shea, R., Kohler, E., Srivastava, M.B.: A dynamic operating system for sensor nodes. In: Proceedings of the 3rd International Conference on Mobile Systems, Applications, and Services (MOBISYS'05), pp. 163–176. ACM Press, New York (2005)
179. He, T., Krishnamurthy, S., Stankovic, J.A., Abdelzaher, T.F., Luo, L., Stoleru, R., Yan, T., Gu, L., Hui, J., Krogh, B.: An energy-efficient surveillance system using wireless sensor networks. In: In Proceedings of Second International Conference on Mobile Systems, Applications, and Services (MobiSys) (2004)
180. Hedetniemi, S.T., Jacobs, D.P., Srimani, P.K.: Fault tolerant distributed coloring algorithms that stabilize in linear time. In: 16th IEEE International Parallel and Distributed Processing Symposium (IPDPS '02), Washington, DC, USA, p. 153. IEEE Computer Society Press, Los Alamitos (2002)
181. Heinzelman, W.R., Chandrakasan, A., Balakrishnan, H.: Energy-efficient communication protocol for wireless microsensor networks. In: Proceedings of the 33rd Hawaii International Conference on System Sciences (HICSS'00) (2000)
182. Heissenbüttel, M.: Routing and Broadcasting in Ad-hoc Networks. PhD thesis, University of Bern, Switzerland (2005)
183. Hershberger, J., Suri, S.: Vickrey pricing in network routing: Fast payment computation. In: Proceedings of the 42nd Annual IEEE Symposium on Foundations of Computer Science (FOCS'01), pp. 252–259. IEEE Computer Society Press, Los Alamitos (2001)
184. Hill, J., Szewczyk, R., Woo, A., Hollar, S., Culler, D., Pister, K.: System architecture directions for networked sensors. In: Proc. of ASPLOS, (November 2000)
185. Hochbaum, D.S.: Heuristics for the fixed cost median problem. Mathematical Programming 22(1), 148–162 (1982)
186. Hou, T.C., Li, V.O.K.: Transmission range control in multihop packet radio networks. IEEE Transactions on Communications 34(1), 38–44 (1986)
187. Howgrave-Graham, N., Nguyen, P.Q., Pointcheval, D., Proos, J., Silverman, J.H., Singer, A., Whyte, W.: The Impact of Decryption Failures on the Security of NTRU Encryption. In: Boneh, D. (ed.) CRYPTO 2003. LNCS, vol. 2729, pp. 226–246. Springer, Heidelberg (2003)
188. Huang, H., Richa, A.W., Segal, M.: Approximation algorithms for the mobile piercing set problem with applications to clustering in ad-hoc networks. In: Proceedings of the 6th International Workshop on Discrete Algorithms and Methods for Mobile Computing and Communications (Dial-M'02), Atlanta, GA, USA, pp. 52–61 (2002)
189. Hubaux, J.-P., Buttyán, L., Capkun, S.: The Quest for Security in Mobile Ad Hoc Networks. In: Proceedings of the 2nd ACM International Symposium on Mobile Ad Hoc Networking and Computing (MOBIHOC'01), ACM Press, New York (2001)
190. Humblet, P.A.: A distributed algorithm for minimum weighted directed spanning trees. IEEE Transactions on Communications COM-31(6), 756–762 (1983)
191. IEEE: P802.11, IEEE Draft Standard for Wireless LAN Medium Acess Control (MAC) and Physical Layer (PHY) Specifications, D3.0
192. Marathe, H.B H.I.M.V., Radhakrishnan, V., Ravi, S S., Rosenkrantz, D.J., Stearns, R.E.: NC-approximation schemes for NP- and PSPACE-hard problems for geometric graphs. Journal of Algorithms 26(2), 238–274 (1998)

193. Drysdale III, R.L(S.), McElfresh, S.A., Snoeyink, J.: On exclusion regions for optimal triangulations. Discrete Applied Mathematics 109(1–2), 49–65 (2001)
194. Intanagonwiwat, C., Govindan, R., Estrin, D.: Directed diffusion: a scalable and robust communication paradigm for sensor networks. In: Proc. of ACM MobiCom, ACM Press, New York (2000)
195. Intel Mote2: http://embedded.seattle.intel-research.net/wiki/
196. Jain, K., Mahdian, M., Markakis, E., Saberi, A., Vazirani, V.V.: Greedy facility location algorithms analyzed using dual fitting with factor-revealing LP. Journal of the ACM 50, 795–824 (2003)
197. Jain, K., Padhye, J., Padmanabhan, V.N., Qiu, L.: Impact of interference on multi-hop wireless network performance. In: Proceedings of the 9th Annual International Conference on Mobile Computing and Networking (MO-BICOM'03), pp. 66–80. ACM Press, New York (2003)
198. Jain, K., Vazirani, V.V.: Primal-dual approximation algorithms for metric facility location and k-median problems. In: Proceedings of the 40th Annual IEEE Symposium on Foundations of Computer Science (FOCS'99), October 1999, pp. 2–13. IEEE Computer Society Press, Los Alamitos (1999)
199. Jaromczyk, J.W., Toussaint, G.T.: Relative neighborhood graphs and their relatives. Proceedings IEEE 80(9), 1502–1517 (1992)
200. Jia, L., Rajaraman, R., Suel, T.: An efficient distributed algorithm for constructing small dominating sets. In: Proceedings of the 20th ACM Symposium on Principles of Distributed Computing (PODC'01), pp. 33–42. ACM Press, New York (2001)
201. Joa-Ng, M., Lu, I-T.: A peer-to-peer zone-based two-level link state routing for mobile ad hoc networks. IEEE Journal on Selected Areas in Communications 17(8), 1415–1425 (1999)
202. Johnson, D.B.: Scalable and Robust Internetwork Routing for Mobile Hosts. In: Proc. of IEEE ICDCS, IEEE Computer Society Press, Los Alamitos (1994)
203. Johnson, D.S., Maltz, D.A.: Dynamic source routing in ad hoc wireless networks. In: Imielinski, T., Korth, H.F. (eds.) Mobile Computing, Kluwer Academic Publishers, Dordrecht (1996)
204. Jurdzinski, T., Stachowiak, G.: Probabilistic algorithms for the wakeup problem in single-hop radio networks. In: Bose, P., Morin, P. (eds.) ISAAC 2002. LNCS, vol. 2518, pp. 535–549. Springer, Heidelberg (2002)
205. Kalpakis, K., Dasgupta, K., Namjoshi, P.: Efficient algorithms for maximum lifetime data gathering and aggregation in wireless sensor networks. Computer Networks: The International Journal of Computer and Telecommunications Networking 42(6), 697–716 (2003)
206. Kao, M.-Y., Li, X.-Y., Wang, W.: Towards truthful mechanisms for binary demand games: a general framework. In: Proceedings of the 6th ACM conference on Electronic Commerce (EC'05), pp. 213–222. ACM Press, New York (2005)
207. Karger, D.R., Lehmann, E.L., Leighton, T., Levine, M., Lewin, D., Panigrahy, R.: Consistent hashing and random trees: distributed caching protocols for relieving hot spots on the World Wide Web. In: Proc. of ACM STOC, ACM Press, New York (1997)
208. Kargl, F.: Sicherheit in Mobilen Ad hoc Netzwerken. PhD thesis, University of Ulm (2003)

209. Karn, P.: MACA – A new channel access method for packet radio. In: Proceedings of the 9th ARRL Computer Networking Conference 1990, pp. 134–140 (1990)
210. Karp, B.: Geographic Routing for Wireless Networks. PhD thesis, Division of Engineering and Applied Sciences, Harvard University (2002)
211. Karp, B., Kung, H T.: GPSR: greedy perimeter stateless routing for wireless networks. In: Proceedings of the 6th Annual International Conference on Mobile Computing and Networking (MOBICOM'00) (2000)
212. Keil, J M., Gutwin, C.A.: Classes of graphs which approximate the complete Euclidean graph. Discrete and Computational Geometry 7, 13–28 (1992)
213. Khalili, A., Katz, J., Arbaugh, W.A.: Toward Secure Key Distribution in Truly Ad-Hoc Networks. In: Proceedings of the 2003 Symposium on Applications and the Internet Workshops (SAINT-w'03) (2003)
214. Khuller, S., Raghavachari, B., Young, N.E: Balancing minimum spanning and shortest path trees. Algorithmica 14, 305 (1995)
215. Kim, Y.-J., Govindan, R., Karp, B., Shenker, S.: Geographic routing made practical. In: Proceedings of the Second USENIX/ACM Symposium on Networked System Design and Implementation (NSDI 2005), Boston, MA, USA, May 2005
216. Kim, Y.-J., Govindan, R., Karp, B., Shenker, S.: On the pitfalls of geographic face routing. In: Proceedings of the DIALM-POMC Joint Workshop on Foundations of Mobile Computing (DIALM-POMC 2005), Cologne, Germany, September 2005
217. Kirkpatrick, D.G., Radke, J.D.: A framework for computational morphology. In: Toussaint, G.T. (ed.) Computational Geometry, pp. 217–248. North-Holland, Amsterdam (1985)
218. Kirousis, L.M., Kranakis, E., Krizanc, D., Pelc, A.: Power consumption in packet radio networks. Theoretical Computer Science 243, 289–305 (2000)
219. Kleinrock, L., Kamoun, F.: Hierarchical routing for large networks; performance evaluation and optimization. Computer Networks and ISDN Systems 1, 155–174 (1977)
220. Koblitz, N.: A Course in Number Theory and Cryptography. Springer, Heidelberg (1994)
221. Konjevod, G., Richa, A.W., Xia, D.: Optimal-stretch name-independent compact routing in doubling metrics. In: Proceedings of the 25th ACM Symposium on Principles of Distributed Computing (PODC'06), pp. 198–207. ACM Press, New York (2006)
222. Kopetz, H.: Real-Time Systems - Design Principles for Distributed Embedded Applications. Kluwer Academic Publishers, Dordrecht (1997)
223. Kopetz, H., Schwabl, W.: Global time in distributed real-time systems. Technical Report 15, TU Wien (1989)
224. Korte, B., Vygen, J.: Combinatorial Optimization: Theory and Algorithms, 2nd edn. Springer, Heidelberg (2002)
225. Kranakis, E., Singh, H., Urrutia, J.: Compass routing on geometric networks. In: Proceedings of the 11th Canadian Conference on Computational Geometry, Vancouver, Canada, August 1999, pp. 51–54 (1999)
226. Krishna, P., Chatterjee, M., Vaidya, N.H., Pradhan, D.K.: A cluster-based approach for routing in ad-hoc networks. In: Proceedings of the 2nd Symposium on Mobile and Location-Independent Computing, pp. 1–10 (1995)

227. Krishnamachari, B.: A wireless sensor networks bibliography. http://ceng.usc.edu/~anrg/SensorNetBib.html
228. Krivitski, D., Schuster, A., Wolff, R.: A local facility location algorithm for sensor networks. In: Prasanna, V.K., Iyengar, S., Spirakis, P.G., Welsh, M. (eds.) DCOSS 2005. LNCS, vol. 3560, Springer, Heidelberg (2005)
229. Kuhn, F., Moscibroda, T., Nieberg, T., Wattenhofer, R.: Fast deterministic distributed maximal independent set computation on growth-bounded graphs. In: Fraigniaud, P. (ed.) DISC 2005. LNCS, vol. 3724, pp. 273–287. Springer, Heidelberg (2005)
230. Kuhn, F., Moscibroda, T., Nieberg, T., Wattenhofer, R.: Local approximation schemes for ad hoc and sensor networks. In: Proceedings of the DIALM-POMC Joint Workshop on Foundations of Mobile Computing (DIALM-POMC 2005), Cologne, Germany (September 2005)
231. Kuhn, F., Moscibroda, T., Wattenhofer, R.: Initializing newly deployed ad hoc and sensor networks. In: Proceedings of the 10th Annual International Conference on Mobile Computing and Networking (MOBICOM'04), Philadelphia, USA, pp. 260–274. ACM Press, New York (2004)
232. Kuhn, F., Moscibroda, T., Wattenhofer, R.: Radio network clustering from scratch. In: Albers, S., Radzik, T. (eds.) ESA 2004. LNCS, vol. 3221, pp. 460–472. Springer, Heidelberg (2004)
233. Kuhn, F., Moscibroda, T., Wattenhofer, R.: What cannot be computed locally? In: Proceedings of the 23rd ACM Symposium on Principles of Distributed Computing (PODC'04), St. John's, New Foundland, Canada, July 2004, pp. 300–309. ACM Press, New York (2004)
234. Kuhn, F., Moscibroda, T., Wattenhofer, R.: On the locality of bounded growth. In: Proceedings of the 24th ACM Symposium on Principles of Distributed Computing (PODC'05), July 2005, pp. 60–69. ACM Press, New York (2005)
235. Kuhn, F., Moscibroda, T., Wattenhofer, R.: Unit disk graph approximation. In: Proceedings of the DIALM-POMC Joint Workshop on Foundations of Mobile Computing (DIALM-POMC 2005), Cologne, Germany, September 2005 (2005)
236. Kuhn, F., Moscibroda, T., Wattenhofer, R.: Fault-tolerant clustering in ad hoc and sensor networks. In: Proceedings of the 26th International Conference on Distributed Computing Systems (ICDCS'06), July 2006
237. Kuhn, F., Moscibroda, T., Wattenhofer, R.: The price of being near-sighted. In: Proceedings of the 17th Annual ACM–SIAM Symposium on Discrete Algorithms (SODA'06), ACM Press, New York (2006)
238. Kuhn, F., Wattenhofer, R.: Constant-time distributed dominating set approximation. In: Proceedings of the 22nd ACM Symposium on Principles of Distributed Computing (PODC'03), Boston, MA, USA, pp. 25–32. ACM Press, New York (2003)
239. Kuhn, F., Wattenhofer, R.: On the complexity of distributed graph coloring. In: Proceedings of the 25th ACM Symposium on Principles of Distributed Computing (PODC'06), Denver, Colorado, USA, July 2006, pp. 7–15. ACM Press, New York (2006)
240. Kuhn, F., Wattenhofer, R., Zhang, Y., Zollinger, A.: Geometric ad-hoc routing: Of theory and practice. In: Proceedings of the 22nd ACM Symposium on Principles of Distributed Computing (PODC'03), Boston, MA, USA, ACM Press, New York (2003)

241. Kuhn, F., Wattenhofer, R., Zollinger, A.: Asymptotically optimal geometric mobile ad-hoc routing. In: Proceedings of the 6th International Workshop on Discrete Algorithms and Methods for Mobile Computing and Communications (Dial-M'02), Atlanta, GA, USA (2002)

242. Kuhn, F., Wattenhofer, R., Zollinger, A.: Ad-hoc networks beyond unit disk graphs. In: Proceedings of the DIALM-POMC Joint Workshop on Foundations of Mobile Computing (DIALM-POMC 2003), San Diego, CA, USA (September 2003)

243. Kuhn, F., Wattenhofer, R., Zollinger, A.: Worst-case optimal and average-case efficient geometric ad-hoc routing. In: Proceedings of the 4th ACM International Symposium on Mobile Ad Hoc Networking and Computing (MOBI-HOC'03), Annapolis, MD, USA (2003)

244. Kumar, S., Alaettinoglu, C., Estrin, D.: Scalable object-tracking through unattended techniques (SCOUT). In: Proc. of IEEE ICNP (2000)

245. Kumar, V S A., Marathe, M.V., Parthasarathy, S., Srinivasan, A.: Algorithmic aspects of capacity in wireless networks. SIGMETRICS Perform. Eval. Rev. 33(1), 133–144 (2005)

246. Kutten, S., Peleg, D.: Fast distributed construction of small k-dominating sets and applications. In: Proceedings of the 14th ACM Symposium on Principles of Distributed Computing (PODC'95), pp. 238–249. ACM Press, New York (1995)

247. Lamport, L., Melliar-Smith, P M.: Synchronizing clocks in the presence of faults. Journal of the Association of Computer Machinery 32(1), 52–78 (1985)

248. Lamport, L., Shostak, R., Pease, M.: The byzantine generals problem. IEEE Transactions on Programming Languages and Systems 4(3), 382–401 (1982)

249. Langendoen, K., Halkes, G.: Energy-efficient medium access control. In: Zurawski, R. (ed.) Embedded Systems Handbook, CRC Press (2005)

250. Lenstra, A.K., Verheul, E.R.: The XTR public key system. In: Bellare, M. (ed.) CRYPTO 2000. LNCS, vol. 1880, Springer, Heidelberg (2000)

251. Leong, B., Liskov, B., Morris, R.: Geographic Routing without Planarization. In: Proceedings of the Third USENIX/ACM Symposium on Networked System Design and Implementation (NSDI 2006), San Jose, CA, USA, May 2006 (2006)

252. Leong, B., Mitra, S., Liskov, B.: Path Vector Face Routing: Geographic Routing with Local Face Information. In: Proceedings of the 13th IEEE International Conference on Network Protocols (ICNP'05), Boston, Massachusetts, USA (2005)

253. Li, J., Jannotti, J., Couto, D.S J D., Karger, D.R., Morris, R.: A Scalable Location Service for Geographic Ad-Hoc Routing. In: Proceedings of the 6th Annual International Conference on Mobile Computing and Networking (MOBICOM'00), pp. 120–130 (August 2000)

254. Li, L., Halpern, J.Y., Bahl, P., Wang, Y.-M., Wattenhofer, R.: A cone-based distributed topology-control algorithm for wireless multi-hop networks. IEEE/ACM Transactions on Networking 13(1), 147–159 (2005)

255. Li, X.-Y.: Topology control in wireless ad hoc networks. In: Basagni, S., Conti, M., Giordano, S., Stojmenovic, I. (eds.) Mobile Ad Hoc Networking, chapter 6, Wiley, Chichester (2003)

256. Li, X.-Y., Călinescu, G., Wan, P.-J., Wang, Y.: Localized Delaunay triangulation with application in ad hoc wireless networks. IEEE Transactions on Parallel and Distributed Systems 14(9) (2003)

257. Li, X., Kim, Y.-J., Govindan, R., Hong, W.: Multidimensional range queries in sensor networks. In: Proc. of ACM SenSys, ACM Press, New York (2003)
258. Lindsey, S., Raghavendra, C.S.: Pegasis: Power-efficient gathering in sensor information systems. In: Proceedings of IEEE Aerospace Conference, vol. 3, pp. 3–1130 (2002)
259. Lindsey, S., Raghavendra, C.S., Sivalingam, K.M.: Data gathering in sensor networks using energy delay metric. In: International Workshop on Parallel and Distributed Computing Issues in Wireless Networks and Mobile Computing, San Francisco, CA, April 2001 (2001)
260. Linial, N.: Locality in distributed graph algorithms. SIAM Journal on Computing 21(1), 193–201 (1992)
261. Liu, J., Issarny, V.: Enhanced Reputation Mechanism for Mobile Ad Hoc Networks. In: Proceedings of the 2nd International Conference on Trust Management (iTrust'04), pp. 48–62 (2004)
262. Locher, T., Wattenhofer, R.: Oblivious gradient clock synchronization. In: Dolev, S. (ed.) DISC 2006. LNCS, vol. 4167, Springer, Heidelberg (2006)
263. Lorincz, K., Malan, D.J., Fulford-Jones, T.R.F., Nawoj, A., Clavel, A., Shnayder, V., Mainland, G., Welsh, M.: Sensor networks for emergency response: Challenges and opportunities. In: IEEE Pervasive Computing, IEEE Computer Society Press, Los Alamitos (2004)
264. Lotker, Z., Patt-Samir, B., Peleg, D.: Distributed MST for constant diameter graphs. In: Proceedings of the 20th ACM Symposium on Principles of Distributed Computing (PODC'01), pp. 63–71. ACM Press, New York (2001)
265. Lu, H.-I: Improved compact routing tables for planar networks via orderly spanning trees. In: Ibarra, O.H., Zhang, L. (eds.) COCOON 2002. LNCS, vol. 2387, pp. 57–66. Springer, Heidelberg (2002)
266. Luby, M.: A simple parallel algorithm for the maximal independent set problem. In: Proceedings of the 17th Annual ACM Symposium on Theory of Computing (STOC'85), May 1985, pp. 1–10. ACM Press, New York (1985)
267. Luby, M., Nisan, N.: A parallel approximation algorithm for positive linear programming. In: Proceedings of the 25th Annual ACM Symposium on Theory of Computing (STOC'93), May 1993, pp. 448–457. ACM Press, New York (1993)
268. Lundelius, J., Lynch, N.: An upper and lower bound for clock synchronization. Information and Control 62, 190–204 (1984)
269. Lundquist, J., Dettinger, M., Cayan, D.: Meteorology and hydrology in yosemite national park: a sensor network application. In: Proceedings of Information Processing in Sensor Networks (IPSN) (2003)
270. Luo, H., Luo, J., Liu, Y.: Energy efficient routing with adaptive data fusion in sensor networks. In: Proceedings of the DIALM-POMC Joint Workshop on Foundations of Mobile Computing (DIALM-POMC 2005), pp. 80–88. ACM Press, New York (2005)
271. Madden, S., Franklin, M.J., Hellerstein, J.M., Hong, W.: TAG: a Tiny AGgregation Service for Ad-Hoc Sensor Networks. In: Proc. of OSDI, Dec. 2002
272. Mahdian, M., Ye, Y., Zhang, J.: Improved approximation algorithms for metric facility location problems. In: Jansen, K., Leonardi, S., Vazirani, V.V. (eds.) APPROX 2002. LNCS, vol. 2462, pp. 229–242. Springer, Heidelberg (2002)

273. Mainwaring, A., Polastre, J., Szewczyk, R., Culler, D., Anderson, J.: Wireless sensor networks for habitat monitoring. In: Proceedings of ACM International Workshop on Wireless Sensor Networks and Applications (WSNA'02), ACM Press, New York (2002)

274. Mainwaring, A., Polastre, J., Szewczyk, R., Culler, D., Anderson, J.: Wireless Sensor Networks for Habitat Monitoring. In: Proc. of IEEE WSNA, September 2002

275. Malan, D.J., Welsh, M., Smith, M.D.: A Public-Key Infrastructure for Key Distribution in TinyOS Based on Elliptic Curce Cryptography. In: 1st IEEE International Conference on Sensor and Ad Hoc Communications and Networks, IEEE Computer Society Press, Los Alamitos (2004)

276. Marathe, M.V., Breu, H., Ravi, H.B H.I.S S., Rosenkrantz, D.J.: Simple heuristics for unit disk graphs. Networks 25, 59–68 (1995)

277. Maroti, M., Kusy, B., Simon, G., Ledeczi, A.: The flooding time synchronization protocol. In: Proceedings of ACM Second International Conference on Embedded Networked Sensor Systems (SenSys 04), ACM Press, New York (2004)

278. Marsh, S.: Formalising Trust as a Computational Concept. PhD thesis, Department of Mathematics and Computer Science, University of Stirling (1994)

279. Marti, S., Giuli, T J., Lai, K., Baker, M.: Mitigating Routing Misbehavior in Mobile Ad Hoc Networks. In: Proceedings of the 6th Annual International Conference on Mobile Computing and Networking (MOBICOM'00), pp. 255–265 (2000)

280. Martinez, K., Padhy, P., Riddoch, A., Ong, R., Hart, J.: Glacial environment monitoring using sensor networks. In: Proceedings of Workshop on Real-World Wireless Sensor Networks (REALWSN'05) (2005)

281. Mas-Colell, A., Whinston, M., Green, J.: Microeconomic Theory. Oxford University Press, Oxford (1995)

282. Meier, L., Thiele, L.: Brief announcement: Gradient clock synchronization in sensor networks. In: Proceedings of the 24th ACM Symposium on Principles of Distributed Computing (PODC'05), Las Vegas, Nevada, USA, July 2005, p. 283 (2005)

283. Menezes, A.J., van Oorschot, P.C., Vanstone, S.A.: Handbook of Applied Cryptography. CRC Press (1997)

284. Meyer auf der Heide, F., Schindelhauer, C., Volbert, K., Grünewald, M.: Energy, congestion and dilation in radio networks. In: Proceedings of the 14th Annual ACM Symposium on Parallel Algorithms and Architectures (SPAA'02), pp. 230–237. ACM Press, New York (2002)

285. MICA2: http://www.xbow.com/Products/Product_pdf_files/Wireless_pdf/MICA2_Datasheet.pdf.

286. MICA2DOT: http://www.xbow.com/Products/Product_pdf_files/Wireless_pdf/MICA2DOT_Datasheet.pdf.

287. MICAz: http://www.xbow.com/Products/Product_pdf_files/Wireless_pdf/MICAz_Datasheet.pdf.

288. Miller, S.P., Neuman, B C., Schiller, J I., Saltzer, J.H.: Kerberos Authentication and Authorization System. Project Athena Technical Plan, MIT Project Athena (1998)

289. Mirchandani, P.B., Francis, R.L. (eds.): Discrete Location Theory. Discrete Mathematics and Optimization. Wiley, Chichester (1990)

290. Moscibroda, T., O'Dell, R., Wattenhofer, M., Wattenhofer, R.: Virtual coordinates for ad hoc and sensor networks. In: Proceedings of the DIALM-POMC Joint Workshop on Foundations of Mobile Computing (DIALM-POMC 2004) (2004)

291. Moscibroda, T., von Rickenbach, P., Wattenhofer, R.: Analyzing the energy-latency trade-off during the deployment of sensor networks. In: Proceedings of the 25th Annual Joint Conference of the IEEE Computer and Communications Societies (INFOCOM'06), Barcelona, Spain (April 2006)

292. Moscibroda, T., Wattenhofer, R.: The complexity of connectivity in wireless networks. In: MISSING – WHERE WAS THIS PUBLISHED?

293. Moscibroda, T., Wattenhofer, R.: Coloring unstructured radio networks. In: Proceedings of the 17th Annual ACM Symposium on Parallel Algorithms and Architectures (SPAA'05), pp. 39–48. ACM Press, New York (2005)

294. Moscibroda, T., Wattenhofer, R.: Facility location: Distributed approximation. In: Proceedings of the 24th ACM Symposium on Principles of Distributed Computing (PODC'05), July 2005, pp. 108–117. ACM Press, New York (2005)

295. Moscibroda, T., Wattenhofer, R.: Maximal independent sets in radio networks. In: Proceedings of the 24th ACM Symposium on Principles of Distributed Computing (PODC'05), July 2005, pp. 148–157. ACM Press, New York (2005)

296. Moscibroda, T., Wattenhofer, R.: Maximizing the lifetime of dominating sets. In: Proceedings of the 5th IEEE International Workshop on Algorithms for Wireless, Mobile, Ad Hoc and Sensor Networks (WMAN'05), IEEE Computer Society Press, Los Alamitos (2005)

297. Moscibroda, T., Wattenhofer, R.: Minimizing interference in ad hoc and sensor networks. In: Proceedings of the DIALM-POMC Joint Workshop on Foundations of Mobile Computing (DIALM-POMC 2005), Cologne, Germany, September 2005, pp. 24–33 (2005)

298. Mostfaoui, A., Raynal, M.: Solving consensus using chandra-toueg's unreliable failure detectors: general quorum-based approach. In: Proceedings of the 13th International Symposium on Distributed Computing (DISC'99) (1999)

299. MSP430 MCUs: http://ti.com/msp430

300. Murphey, R.A., Pardalos, P.M., Resende, M.G C.: Frequency assignment problems. In: Du, D.-Z., Pardalos, P.M. (eds.) Handbook of Combinatorial Optimization, vol. 4, pp. 295–377. Kluwer Academic Publishers Group, Dordrecht (1999)

301. Naor, M., Stockmeyer, L.J.: What can be computed locally? SIAM Journal on Computing 24(6), 1259–1277 (1995)

302. Newsome, J., Shi, E., Song, D., Perrig, A.: The Sybil Attack in Sensor Networks: Analysis & Defenses. In: Proceedings of the 3rd International Symposium on Information Processing in Sensor Networks (IPSN'04), pp. 259–268 (2004)

303. Nieberg, T., Hurink, J.: A ptas for the minimum dominating set problem in unit disk graphs. In: Erlebach, T., Persinao, G. (eds.) WAOA 2005. LNCS, vol. 3879, Springer, Heidelberg (2006)

304. Nikaein, N., Bonnet, C., Nikaein, N.: HARP - Hybrid Ad hoc Routing Protocol. In: International Symposium on Telecommunications (IST 2001), Teheran, Iran, September 2001 (2001)

305. Nisan, N., Ronen, A.: Algorithmic mechanism design (extended abstract). In: Proceedings of the 31st Annual ACM Symposium on Theory of Computing (STOC'99), pp. 129–140. ACM Press, New York (1999)

306. Ogier, R.G., Templin, F.L., Lewis, M.G.: Topology dissemination based on reverse-path forwarding (TBRPF). IETF RFC 3684 (February 2004)

307. C.M. O'Rourke: Efficient NTRU Implementations. Master's thesis, Worcester Polytechnic Institute (April 2002), http://www.wpi.edu/Pubs/ETD/Available/etd-0430102-111906/unrestricted/corourke.pdf

308. Osterweil, E., Mysore, M., Rahimi, M., Estrin, D.: The Extensible Sensing System. Technical report, Center for Embedded Networked Sensing (CENS), UCLA (2003)

309. Park, V.D., Corson, M S.: A highly adaptive distributed routing algorithm for mobile wireless networks. In: Proceedings of the 16th Annual Joint Conference of the IEEE Computer and Communications Societies (INFOCOM'97), pp. 1405–1413. IEEE Computer Society Press, Los Alamitos (1997)

310. Parthasarathy, S., Gandhi, R.: Distributed algorithms for coloring and connected domination in wireless ad hoc networks. In: Lodaya, K., Mahajan, M. (eds.) FSTTCS 2004. LNCS, vol. 3328, pp. 447–459. Springer, Heidelberg (2004)

311. Pease, M., Shostak, R., Lamport, L.: Reaching agreement in the presence of faults. Journal of the Association of Computer Machinery 27(2), 228–234 (1980)

312. Peleg, D.: Proximity-preserving labeling schemes and their applications. In: Widmayer, P., Neyer, G., Eidenbenz, S. (eds.) WG 1999. LNCS, vol. 1665, pp. 30–41. Springer, Heidelberg (1999)

313. Peleg, D.: Distributed Computing: A Locality-Sensitive Approach. SIAM Monographs on Discrete Mathematics and Applications (2000)

314. Peleg, D., Rubinovich, V.: A near-tight lower bound on the time complexity of distributed minimum-weight spanning tree construction. SIAM Journal on Computing 30(5), 1427–1442 (2001)

315. Peleg, D., Upfal, E.: A trade-off between space and efficiency for routing tables. Journal of the ACM 36(3), 510–530 (1989)

316. Pemmaraju, S., Pirwani, I.: Energy conservation via domatic partitions. In: Proceedings of the 7th ACM International Symposium on Mobile Ad Hoc Networking and Computing (MOBIHOC'06), pp. 143–154. ACM Press, New York (2006)

317. Perich, F., Undercoffer, J.L, Kagal, L., Joshi, A., Finin, T., Yesha, Y.: In Reputation We Believe: Query Processing in Mobile Ad-Hoc Networks. In: Proceedings of the 1th International Conference on Mobile and Ubiquitous Systems: Networking and Services (August 2004)

318. Perkins, C., Bhagwat, P.: Highly dynamic Destination-Sequenced Distance-Vector routing (DSDV) for mobile computers. In: Proc. of ACM SIGCOMM, ACM Press, New York (1994)

319. Perkins, C., Royer, E.M.: Ad hoc On-Demand Distance Vector Routing. In: Proc. of the 2nd IEEE Workshop on Mobile Computing Systems and Applications (WMCSA), February 1999, IEEE Computer Society Press, Los Alamitos (1999)

320. Perrig, A., Stankovic, J., Wagner, D.: Security in Wireless Sensor Networks. Commun. ACM 47(6), 53–57 (2004)

321. Prakash, R.: Unidirectional links prove costly in wireless ad-hoc networks. In: Proceedings of the 3rd International Workshop on Discrete Algorithms and Methods for Mobile Computing and Communications (Dial-M'99), Seattle, Washington, USA (1999)

322. Priyantha, N.B., Balakrishnan, H., Demaine, E.D., Teller, S.: Anchor-free distributed localization in sensor networks. Technical Report TR-892, MIT, LCS (April 2003)

323. Rajaraman, R.: Topology control and routing in ad hoc networks: a survey. ACM SIGACT News 33(2), 60–73 (2002)

324. Rajendran, V., Obraczka, K., Garcia-Luna-Aceves, J J.: Energy-efficient, collision-free medium access control for wireless sensor networks. In: Proceedings of the 1st International Conference on Embedded Networked Sensor Systems (SENSYS'03), pp. 181–192 (2003)

325. Ramanathan, S.: A unified framework and algorithm for channel assignment in wireless networks. Wireless Networks 5(2), 81–94 (1999)

326. Ramanathan, S., Lloyd, E.L.: Scheduling algorithms for multi-hop radio networks. In: Proceedings of SIGCOMM'92, pp. 211–222. ACM Press, New York (1992)

327. Ramasubramanian, V., Haas, Z.J., Sirer, E.G.: SHARP: A hybrid adaptive routing protocol for mobile ad hoc networks. In: Proceedings of the 4th ACM International Symposium on Mobile Ad Hoc Networking and Computing (MOBIHOC'03), Annapolis, MD, USA (2003)

328. Rao, A., Ratnasamy, S., Papadimitriou, C.H., Shenker, S., Stoica, I.: Geographic routing without location information. In: Proc. of ACM MobiCom, ACM Press, New York (2003)

329. Rappaport, T.: Wireless Communication: Principles and Practice. Prentice-Hall, Englewood Cliffs (2002)

330. Raz, R., Safra, S.: A sub-constant error-probability low-degree test, and a sub-constant error-probability PCP characterization of NP. In: Proceedings of the 29th Annual ACM Symposium on Theory of Computing (STOC'95), pp. 475–484. ACM Press, New York (1997)

331. Reiter, M.K., Stubblebine, S.G.: Authentication Metric Analysis and Design. ACM Transactions on Information and System Security 2(2), 138–158 (1999)

332. Roberts, F.S.: T-colorings of graphs: Recent results and open problems. Discrete Mathematics 93(2-3), 229–245 (1991)

333. Roberts, L.G.: ALOHA packet system with and without slots and capture. ACM SIGCOMM Computer Communication Review 5(2), 28–42 (1975)

334. Römer, K.: Time synchronization in ad hoc networks. In: Proceedings of the 2nd ACM International Symposium on Mobile Ad Hoc Networking and Computing (MOBIHOC'01), Long Beach, CA, USA, October 2001, pp. 173–182. ACM Press, New York (2001)

335. Römer, K., Mattern, F.: The design space of wireless sensor networks. IEEE Wireless Communications 11(6), 54–61 (2004)

336. Rothammel, K., Krischke, A.: Rothammels Antennenbuch. DARC-Verlag (2002)

337. Royer, E.M., Toh, C.-K.: A review of current routing protocols for ad-hoc mobile wireless networks. IEEE Personal Communications 6 (1999)

338. Santi, P.: Topology control in wireless ad hoc and sensor networks. ACM Computing Surveys 37(2), 164–194 (2005)

339. Santi, P.: Topology Control in Wireless Ad Hoc and Sensor Networks. Wiley, Chichester (2005)
340. Santoro, N., Khatib, R.: Labelling and implicit routing in networks. The Computer Journal 28(1), 5–8 (1985)
341. Savarese, C., Rabaey, J.M., Langendoen, K.: Robust positioning algorithms for distributed ad-hoc wireless sensor networks. In: USENIX Technical Annual Conference'02 (2002)
342. ScatterWeb: http://www.scatterweb.de/
343. Schmid, S., Wattenhofer, R.: Algorithmic Models for Sensor Networks. In: 14th International Workshop on Parallel and Distributed Real-Time Systems (WPDRTS), April 2006
344. Schneier, B.: Applied Cryptography, 2nd edn. John Wiley & Sons Inc., Chichester (1995)
345. Shang, Y., Ruml, W.: Improved mds-based localization. In: Proceedings of the 23nd Annual Joint Conference of the IEEE Computer and Communications Societies (INFOCOM'04), IEEE Computer Society Press, Los Alamitos (2004)
346. Shenker, S., Ratnasamy, S., Karp, B., Govindan, R., Estrin, D.: Data-centric storage in sensornets. In: SIGCOMM Comput. Commun. Rev. (2003)
347. Shmoys, D.B., Tardos, E., Aardal, K.: Approximation algorithms for facility location problems (extended abstract). In: Proceedings of the 29th Annual ACM Symposium on Theory of Computing (STOC'95), pp. 265–274. ACM Press, New York (1997)
348. Simmons, G.: An Introduction to Shared Secret and/or Shared Control Schemes and their Application. Contemporary Cryptology. IEEE Computer Society Press, Los Alamitos (1992)
349. Simon, G., Ledeczi, A., Maroti, M.: Sensor network-based countersniper system. In: In Proceedings of ACM Second International Conference on Embedded Networked Sensor Systems (SenSys 04), ACM Press, New York (2004)
350. Slepian, D., Wolf, J.K.: Noiseless coding of correlated information sources. In: IEEE Transactions on Information Theory, pp. 471–480. IEEE Computer Society Press, Los Alamitos (1973)
351. Smart kindergarten: http://nesl.ee.ucla.edu/projects/smartkg/
352. Solis, I., Obraczka, K.: The impact of timing in data aggregation for sensor networks. In: Proceedings of the IEEE International Conference on Communications (ICC'04), IEEE Computer Society Press, Los Alamitos (2004)
353. Srinivasan, V., Nuggehalli, P., Chiasserini, C.F., Rao, R.R.: Energy efficiency of ad hoc networks with selfish users. In: Proceedings of the 4th European Wireless Conference (EW'02), February 2002
354. Szegedy, M., Vishwanathan, S.: Locality based graph coloring. In: Proceedings of the 25th Annual ACM Symposium on Theory of Computing (STOC'93), pp. 201–207. ACM Press, New York (1993)
355. Takagi, H., Kleinrock, L.: Optimal transmission ranges for randomly distributed packet radio terminals. IEEE Transaction on Communications 32(3), 246–257 (1984)
356. Tan, H.Ö., Körpeoğlu, İ.: Power efficient data gathering and aggregation in wireless sensor networks. SIGMOD Rec. 32(4), 66–71 (2003)
357. Tarjan, R.E.: Finding optimum branchings. Networks 7, 25–35 (1977)
358. Tarjan, R.E.: Data Structures and Network Algorithms. Society for Industrial and Applied Mathematics (1983)

359. TelosB: http://www.xbow.com/Products/Product_pdf_files/Wireless_
     pdf/TelosB_Datasheet.pdf
360. Thorup, M.: Quick and good facility location. In: Proceedings of the 14th
     Annual ACM–SIAM Symposium on Discrete Algorithms (SODA'03), SIAM
     (2003)
361. Thorup, M.: Compact oracles for reachability and approximate distances in
     planar digraphs. Journal of the ACM 51(6), 993–1024 (2004)
362. Thorup, M., Zwick, U.: Compact routing schemes. In: Proceedings of the
     13th Annual ACM Symposium on Parallel Algorithms and Architectures
     (SPAA'01), pp. 1–10. ACM Press, New York (2001)
363. Thorup, M., Zwick, U.: Approximate distance oracles. Journal of the
     ACM 52(1), 1–24 (2005)
364. TinyOS community forum: http://www.tinyos.net/.
365. Tobagi, F.A., Kleinrock, L.: Packet switching in radio channels: Part II – The
     hidden terminal problem in carrier sense multiple-access and the busy-tone
     solution. IEEE Transactions on Communications 23, 1417–1433 (1975)
366. Tsuchiya, P.F.: The landmark hierarchy: a new hierarchy for routing in very
     large networks. In: Proc. of ACM SIGCOMM,, ACM Press, New York (1988)
367. Tutte, W.T.: How to draw a graph. Proceedings of the London Mathematical
     Society 13, 743–768 (1963)
368. SETI@Home: http://setiathome.ssl.berkeley.edu/
369. Urrutia, J.: Routing with guaranteed delivery in geometric and wireless net-
     works. In: Stojmenovic, I. (ed.) Handbook of Wireless Networks and Mobile
     Computing, chapter 18, pp. 393–406. Wiley, Chichester (2002)
370. van Emde Boas, P.: Preserving order in a forest in less than logarithmic time
     and linear space. Information Processing Letters 6(3), 80–82 (1977)
371. van Hoesel, L G W., Havinga, P J M.: A lightweight medium access protocol
     (LMAC) for wireless sensor networks. In: 1st Int. Workshop on Networked
     Sensing Systems (INSS) (2004)
372. van Leeuwen, J., Tan, R.B.: Routing with compact routing tables. Technical
     Report RUU-CS-83-16, Dept. of Computer Science, Utrecht University (1983)
373. van Leeuwen, J., Tan, R.B.: Interval routing. The Computer Journal 30(4),
     298–307 (1987)
374. van Leeuwen, J., Tan, R.B.: Compact routing methods: A survey. In: Pro-
     ceedings of the 1st International Colloquium on Structural Information and
     Communication Complexity (SIROCCO'94), pp. 99–110. Carleton University
     Press (1994)
375. Vazirani, V.V.: Approximation Algorithms, 2nd edn. Springer, Heidelberg
     (2003)
376. Veltkamp, R.C.: The γ–neighborhood graph. Computational Geometry: The-
     ory and Applications 1(4), 227–246 (1992)
377. Vizing, V.G.: On an estimate of the chromatic class of a p-graph. Diskret.
     Analiz. 3, 25–30 (1964)
378. von Rickenbach, P., Schmid, S., Wattenhofer, R., Zollinger, A.: A robust in-
     terference model for wireless ad-hoc networks. In: 19th IEEE International
     Parallel and Distributed Processing Symposium (IPDPS '05), p. 239. IEEE
     Computer Society Press, Los Alamitos (2005)
379. von Rickenbach, P., Wattenhofer, R.: Gathering correlated data in sensor
     networks. In: Proceedings of the DIALM-POMC Joint Workshop on Founda-
     tions of Mobile Computing (DIALM-POMC 2004), Philadelphia, Pennsylva-
     nia, USA (October 2004)

380. Vygen, J.: Approximation algorithms for facility location problems. Technical Report 05950-OR, Research Institute for Discrete Mathematics, University of Bonn (2005)

381. Walters, J.P., Liang, Z., Shi, W., Chaudhary, V.: Wireless sensor network security: A survey (2000), http://www.cs.wayne.edu/-XXX-tilde-weisong/papers/walters05-wsn-security-survey.pdf.

382. Wan, P.-J., Alzoubi, K., Frieder, O.: Distributed construction of connected dominating set in wireless ad hoc networks. In: Proceedings of the 21st Annual Joint Conference of the IEEE Computer and Communications Societies (INFOCOM'02), IEEE Computer Society Press, Los Alamitos (2002)

383. Wan, P.-J., Călinescu, G., Li, X.-Y., Frieder, O.: Minimum-energy broadcast routing in static ad hoc wireless networks. In: Proceedings of the 20th Annual Joint Conference of the IEEE Computer and Communications Societies, pp. 1162–1171. IEEE Computer Society Press, Los Alamitos (2001)

384. Wang, W., Li, X.-Y.: Truthful low-cost unicast in selfish wireless networks. In: Proceedings of the 4th IEEE International Workshop on Algorithms for Wireless, Mobile, Ad Hoc and Sensor Networks (WMAN'04), IEEE Computer Society Press, Los Alamitos (2004)

385. Wang, W., Li, X.-Y., Sun, Z., Wang, Y.: Design multicast protocols for non-cooperative networks. In: Proceedings of the 24th Annual Joint Conference of the IEEE Computer and Communications Societies (INFOCOM'05), Miami, Florida (March 2005)

386. Wang, W., Li, X.-Y., Wang, Y.: Truthful multicast routing in selfish wireless networks. In: Proceedings of the 10th Annual International Conference on Mobile Computing and Networking (MOBICOM'04), pp. 245–259. ACM Press, New York (2004)

387. Wang, Y., Vassileva, J.: Bayesian Network-Based Trust Model. In: Proceedings of the IEEE/WIC International Conference on Web Intelligence (WI'03), pp. 372–378 (2003)

388. Wang, Y., Wang, W., Li, X.-Y.: Distributed low-cost backbone formation for wireless ad hoc networks. In: Proceedings of the 6th ACM International Symposium on Mobile Ad Hoc Networking and Computing (MOBIHOC'05), pp. 2–13. ACM Press, New York (2005)

389. Wattenhofer, M., Wattenhofer, R., Widmayer, P.: Geometric routing without geometry. In: Pelc, A., Raynal, M. (eds.) SIROCCO 2005. LNCS, vol. 3499, Springer, Heidelberg (2005)

390. Wattenhofer, R.: Sensor networks: Distributed algorithms reloaded - or revolutions? In: Flocchini, P., Gąsieniec, L. (eds.) SIROCCO 2006. LNCS, vol. 4056, Springer, Heidelberg (2006)

391. Wattenhofer, R., Li, L., Bahl, P., Wang, Y.-M.: Distributed topology control for wireless multihop ad-hoc networks. In: Proceedings of the 20th Annual Joint Conference of the IEEE Computer and Communications Societies (INFOCOM'01), pp. 1388–1397. IEEE Computer Society Press, Los Alamitos (2001)

392. Wattenhofer, R., Zollinger, A.: XTC: A practical topology control algorithm for ad-hoc networks. In: 18th IEEE International Parallel and Distributed Processing Symposium (IPDPS '04), IEEE Computer Society Press, Los Alamitos (2004)

393. Watts, D.J.: Small Worlds. Princeton University Press, Princeton (1999)

394. Whang, D.H., Xu, N., Rangwala, S., Chintalapudi, K., Govindan, R., Wallace, J.W.: Development of an embedded networked sensing system for structural health monitoring. In: In Proceedings of International Workshop on Smart Materials and Structures Technology (2004)
395. Woo, A., Tong, T., Culler, D.: Taming the underlying challenges of reliable multihop routing in sensor networks. In: Proc. of ACM SenSys, ACM Press, New York (2003)
396. Woodall, D.: Improper colourings of graphs. In: Nelson, R., Wilson, R.J. (eds.) Graph Colourings, Longman Scientific and Technical (1990)
397. Wu, J.: Dominating-set-based routing in ad hoc wireless networks. In: Stojmenovic, I. (ed.) Handbook of Wireless Networks and Mobile Computing, chapter 20, pp. 425–450. Wiley, Chichester (2002)
398. Wu, J., Li, H.: On calculating connected dominating set for efficient routing in ad hoc wireless networks. In: Proceedings of the 3rd International Workshop on Discrete Algorithms and Methods for Mobile Computing and Communications (Dial-M'99), pp. 7–14 (1999)
399. Intel PXA27x0 processor family: http://download.intel.com/design/pca/applicationsprocessors/datashts/28000304.pdf.
400. Yao, A.C.-C.: On constructing minimum spanning trees in k–dimensional spaces and related problems. SIAM Journal on Computing 11(4), 721–736 (1982)
401. Yu, Y., Krishnamachari, B., Prasanna, V.K.: Energy-latency tradeoffs for data gathering in wireless sensor networks. In: Proceedings of the 23nd Annual Joint Conference of the IEEE Computer and Communications Societies (INFOCOM'04), March 2004, IEEE Computer Society Press, Los Alamitos
402. Zacharia, G., Moukas, A., Maes, P.: Collaborative Reputation Mechanisms in Electronic Marketplaces. In: Proceedings of the 32nd Hawaii International Conference on System Sciences (HICSS'99) (1999)
403. Zebranet: http://www.princeton.edu/~mrm/zebranet.html
404. Zhao, J., Govindan, R.: Understanding packet delivery performance in dense wireless sensor networks. In: Proc. of ACM SenSys, ACM Press, New York (2003)
405. Zhong, S., Chen, J., Yang, Y.R.: Sprite: A simple, cheat-proof, credit-based system for mobile ad-hoc networks. In: Proceedings of the 22nd Annual Joint Conference of the IEEE Computer and Communications Societies (INFOCOM'03), pp. 1987–1997. IEEE Computer Society Press, Los Alamitos (2003)
406. Zhou, L., Haas, Z.J.: Securing Ad Hoc Networks. IEEE Network 13(6), 24–30 (1999)
407. Zimmermann, P.R.: The Official PGP User's Guide. MIT Press, Cambridge (1995)
408. Zollinger, A.: Networking Unleashed: Geographic Routing and Topology Control in Ad Hoc and Sensor Networks. PhD thesis, ETH Zurich, Switzerland, Diss. ETH 16025 (2005)
409. Zorzi, M., Rao, R.R.: Geographic random forwarding (GeRaF) for ad hoc and sensor networks: Multihop performance. IEEE Transaction on Mobile Computing 2(4), 337–348 (2003)

This bibliography is available in BIBTEX format from
http://i11www.iti.uni-karlsruhe.de/gisa/gisa.bib.

Author Index

Subject Index

Lecture Notes in Computer Science

Sublibrary 1: Theoretical Computer Science and General Issues

For information about Vols. 1– 4445
please contact your bookseller or Springer

Vol. 4621: D. Wagner, R. Wattenhofer (Eds.), Algorithms for Sensor and Ad Hoc Networks. XIII, 415 pages. 2007.

Vol. 4619: F. Dehne, J.-R. Sack, N. Zeh (Eds.), Algorithms and Data Structures. XVI, 662 pages. 2007.

Vol. 4618: S.G. Akl, C.S. Calude, M.J. Dinneen, G. Rozenberg, H.T. Wareham (Eds.), Unconventional Computation. X, 243 pages. 2007.

Vol. 4616: A. Dress, Y. Xu, B. Zhu (Eds.), Combinatorial Optimization and Applications. XI, 390 pages. 2007.

Vol. 4613: F.P. Preparata, Q. Fang (Eds.), Frontiers in Algorithmics. XI, 348 pages. 2007.

Vol. 4600: H. Comon-Lundh, C. Kirchner, H. Kirchner (Eds.), Rewriting, Computation and Proof. XVI, 273 pages. 2007.

Vol. 4599: S. Vassiliadis, M. Berekovic, T.D. Hämäläinen (Eds.), Embedded Computer Systems: Architectures, Modeling, and Simulation. XVIII, 466 pages. 2007.

Vol. 4598: G. Lin (Ed.), Computing and Combinatorics. XII, 570 pages. 2007.

Vol. 4596: L. Arge, C. Cachin, T. Jurdziński, A. Tarlecki (Eds.), Automata, Languages and Programming. XVII, 953 pages. 2007.

Vol. 4595: D. Bošnački, S. Edelkamp (Eds.), Model Checking Software. X, 285 pages. 2007.

Vol. 4590: W. Damm, H. Hermanns (Eds.), Computer Aided Verification. XV, 562 pages. 2007.

Vol. 4588: T. Harju, J. Karhumäki, A. Lepistö (Eds.), Developments in Language Theory. XI, 423 pages. 2007.

Vol. 4583: S.R. Della Rocca (Ed.), Typed Lambda Calculi and Applications. X, 397 pages. 2007.

Vol. 4580: B. Ma, K. Zhang (Eds.), Combinatorial Pattern Matching. XII, 366 pages. 2007.

Vol. 4576: D. Leivant, R. de Queiroz (Eds.), Logic, Language, Information and Computation. X, 363 pages. 2007.

Vol. 4547: C. Carlet, B. Sunar (Eds.), Arithmetic of Finite Fields. XI, 355 pages. 2007.

Vol. 4546: J. Kleijn, A. Yakovlev (Eds.), Petri Nets and Other Models of Concurrency – ICATPN 2007. XI, 515 pages. 2007.

Vol. 4545: H. Anai, K. Horimoto, T. Kutsia (Eds.), Algebraic Biology. XIII, 379 pages. 2007.

Vol. 4533: F. Baader (Ed.), Term Rewriting and Applications. XII, 419 pages. 2007.

Vol. 4528: J. Mira, J.R. Álvarez (Eds.), Nature Inspired Problem-Solving Methods in Knowledge Engineering, Part II. XXII, 650 pages. 2007.

Vol. 4527: J. Mira, J.R. Álvarez (Eds.), Bio-inspired Modeling of Cognitive Tasks, Part I. XXII, 630 pages. 2007.

Vol. 4525: C. Demetrescu (Ed.), Experimental Algorithms. XIII, 448 pages. 2007.

Vol. 4514: S.N. Artemov, A. Nerode (Eds.), Logical Foundations of Computer Science. XI, 513 pages. 2007.

Vol. 4513: M. Fischetti, D.P. Williamson (Eds.), Integer Programming and Combinatorial Optimization. IX, 500 pages. 2007.

Vol. 4510: P. Van Hentenryck, L.A. Wolsey (Eds.), Integration of AI and OR Techniques in Constraint Programming for Combinatorial Optimization Problems. X, 391 pages. 2007.

Vol. 4507: F. Sandoval, A.G. Prieto, J. Cabestany, M. Graña (Eds.), Computational and Ambient Intelligence. XXVI, 1167 pages. 2007.

Vol. 4501: J. Marques-Silva, K.A. Sakallah (Eds.), Theory and Applications of Satisfiability Testing – SAT 2007. XI, 384 pages. 2007.

Vol. 4497: S.B. Cooper, B. Löwe, A. Sorbi (Eds.), Computation and Logic in the Real World. XVIII, 826 pages. 2007.

Vol. 4494: H. Jin, O.F. Rana, Y. Pan, V.K. Prasanna (Eds.), Algorithms and Architectures for Parallel Processing. XIV, 508 pages. 2007.

Vol. 4493: D. Liu, S. Fei, Z. Hou, H. Zhang, C. Sun (Eds.), Advances in Neural Networks – ISNN 2007, Part III. XXVI, 1215 pages. 2007.

Vol. 4492: D. Liu, S. Fei, Z. Hou, H. Zhang, C. Sun (Eds.), Advances in Neural Networks – ISNN 2007, Part II. XXVII, 1321 pages. 2007.

Vol. 4491: D. Liu, S. Fei, Z.-G. Hou, H. Zhang, C. Sun (Eds.), Advances in Neural Networks – ISNN 2007, Part I. LIV, 1365 pages. 2007.

Vol. 4490: Y. Shi, G.D. van Albada, J.J. Dongarra, P.M.A. Sloot (Eds.), Computational Science – ICCS 2007, Part IV. XXXVII, 1211 pages. 2007.

Vol. 4489: Y. Shi, G.D. van Albada, J.J. Dongarra, P.M.A. Sloot (Eds.), Computational Science – ICCS 2007, Part III. XXXVII, 1257 pages. 2007.

Vol. 4488: Y. Shi, G.D. van Albada, J.J. Dongarra, P.M.A. Sloot (Eds.), Computational Science – ICCS 2007, Part II. XXXV, 1251 pages. 2007.

Vol. 4487: Y. Shi, G.D. van Albada, J.J. Dongarra, P.M.A. Sloot (Eds.), Computational Science – ICCS 2007, Part I. LXXXI, 1275 pages. 2007.

Vol. 4484: J.-Y. Cai, S.B. Cooper, H. Zhu (Eds.), Theory and Applications of Models of Computation. XIII, 772 pages. 2007.

Vol. 4475: P. Crescenzi, G. Prencipe, G. Pucci (Eds.), Fun with Algorithms. X, 273 pages. 2007.

Vol. 4474: G. Prencipe, S. Zaks (Eds.), Structural Information and Communication Complexity. XI, 342 pages. 2007.

Vol. 4459: C. Cérin, K.-C. Li (Eds.), Advances in Grid and Pervasive Computing. XVI, 759 pages. 2007.

Vol. 4449: Z. Horváth, V. Zsók, A. Butterfield (Eds.), Implementation and Application of Functional Languages. X, 271 pages. 2007.

Vol. 4448: M. Giacobini (Ed.), Applications of Evolutionary Computing. XXIII, 755 pages. 2007.

Vol. 4447: E. Marchiori, J.H. Moore, J.C. Rajapakse (Eds.), Evolutionary Computation, Machine Learning and Data Mining in Bioinformatics. XI, 302 pages. 2007.

Vol. 4446: C. Cotta, J.I. van Hemert (Eds.), Evolutionary Computation in Combinatorial Optimization. XII, 241 pages. 2007.